MEGA-EVENT CITIES:
URBAN LEGACIES OF GLOBAL SPORTS EVENTS

Mega-event Cities:
Urban Legacies of Global Sports Events

VALERIE VIEHOFF
University of Bonn, Germany

&

GAVIN POYNTER
University of East London, UK

Routledge
Taylor & Francis Group

LONDON AND NEW YORK

First published 2015 by Ashgate Publishing

Published 2016 by Routledge
2 Park Square, Milton Park, Abingdon, Oxon OX14 4RN
711 Third Avenue, New York, NY 10017, USA

First issued in paperback 2018

Routledge is an imprint of the Taylor & Francis Group, an informa business

British Library Cataloguing in Publication Data
A catalogue record for this book is available from the British Library.

The Library of Congress has cataloged the printed edition as follows:
Mega-event cities : urban legacies of global sports events / edited by Valerie Viehoff and Gavin Poynter.
    pages cm. -- (Design and the built environment)
 Includes bibliographical references and index.
    ISBN 978-1-4724-4017-4 (hardback) -- ISBN 978-1-4724-4018-1 (ebook) -- ISBN 978-1-4724-4019-8 (epub) 1. Hosting of sporting events--Social aspects. 2. Sports tournaments--Social aspects. 3. Urbanization. 4. City planning I. Viehoff, Valerie. II. Poynter, Gavin, 1949-
    GV712.M44 2015
    796.06'94--dc23
                                                                                                2015010332

ISBN 13: 978-1-138-54678-3 (pbk)
ISBN 13: 978-1-4724-4017-4 (hbk)

Contents

List of Figures

List of Tables

List of Plates

List of Contributors

Cristiano M. Belem holds a PhD in Exercise Science and Sports and currently works as a researcher at the State University of Rio de Janeiro (UERJ) and for DIESPORTE at the Federal University of Bahia, Brazil.

Einar Braathen is Senior Researcher at the Norwegian Institute of Urban and Regional Research (NIBR), Oslo, Norway.

Allan Brimicombe is Professor and Head of the Centre for Geo-Information Studies at the University of East London and a chartered geographer.

Anne-Marie Broudehoux is Professor at the School of Design, Université du Québec à Montréal in Canada.

Constantinos Cartalis is an expert in sustainability research and Associate Professor at the University of Athens, Greece.

Hüseyin Cengiz is Professor and Head of the Department of Urban and Regional Planning, Yildiz Technical University, Istanbul, Turkey.

Berta Cerezuela is Head of Projects at the Olympic Studies Centre, Autonomous University of Barcelona, Spain.

Phil Cohen is Emeritus Professor in Cultural Studies at the University of East London, UK and Visiting Professor in Urban Cultures at the University of Umea, Sweden.

Lamartine Pereira DaCosta holds a PhD in Philosophy and is currently Professor of Sport History and Sport Management at the University Gama Filho in Rio de Janeiro, Brazil.

Egidio Dansero is Professor in the Department of Cultures, Politics and Society at the University of Turin, Italy.

Juliet Davis is Senior Lecturer at the Welsh School of Architecture, University of Cardiff, UK.

Alvaro de Miranda is Senior Research Fellow at the London East Research Institute, University of East London, UK.

Ailton Fernando Santana de Oliveira is Professor in the field of Sports Policies at the Federal University of Sergipe, Brazil.

Michael Dobbins, FAIA, FAICP served as Commissioner of Planning and Development for the City of Atlanta, 1996–2002, and subsequently joined the Georgia Tech College of Architecture as a Professor of Practice, where he continues to teach and practice urban design and planning.

Ozlem Edizel recently completed her PhD at Brunel University and is also a visiting lecturer at the University of Westminster, UK.

Leon S. Eplan, FAICP, was Atlanta's Commissioner of Planning and Development leading up to its 1996 Summer Olympics and is now President of Urban Mobility Consult, Ltd. A former Professor at Georgia Tech, he also served twice as President of the American Institute of Planning.

Semiha Sultan Eryilmaz is Assistant Professor in the Department of Urban and Regional Planning at Necmettin Erbakan University, Turkey.

Vyasha Harilal is a PhD candidate in Geography at the University of KwaZulu-Natal, South Africa.

Neil Herrington is Principal Lecturer at the Cass School of Education and Communities, University of East London, UK.

Chris Kennett is a full-time professor at the Business Engineering School, La Salle (Universitat Ramon Llull), in Barcelona, Spain.

Iain MacRury is Professor and Head of Research and Knowledge Exchange at the Bournemouth Media School, Bournemouth University, UK.

Brij Maharaj is Professor of Geography at the University of KwaZulu-Natal, South Africa.

Gilmar Mascarenhas de Jesus is Professor in the Department of Geography at the State University of Rio de Janeiro (UERJ), Brazil.

Alfredo Mela is Professor at the Interuniversity Department of Regional and Urban Studies and Planning (DIST), a joint department of the Polytechnic and the University of Turin, Italy.

Francesc Muñoz is Professor and Director of the Observatori de la Urbanització at Universitat Autònoma de Barcelona, Spain.

David Powell is Visiting Professor at the London East Research Institute, University of East London and founder of DPA, an art and urban regeneration consultancy.

Gavin Poynter is Professor Emeritus in the School of Social Sciences and Chair of the London East Research Institute at the University of East London, UK.

Matthew Richmond is completing a PhD at the Brazil Institute, King's College London, UK.

H. Randal Roark, FAICP, is Professor Emeritus in the School of Architecture and Planning at Georgia Tech. During the Olympic period he was Director of Planning and Design for the Corporation for Olympic Development in Atlanta (CODA).

Cristiana Rossignolo is Associate Professor of Geography at the Interuniversity Department of Regional and Urban Studies and Planning, a joint department of the Polytechnic and the University of Turin, Italy.

Michael Rustin is Professor of Sociology at the University of East London and a Visiting Professor at the Tavistock Clinic.

Fernanda Sánchez is completing her PhD in Architecture and Urbanism (PPGAU) at the Universidade Federal Fluminense (UFF), Rio de Janeiro, Brazil.

Celina Sørbøe is a researcher working on several projects in Brazil related to mega-events, including Chance2Sustain and Cities against Poverty: Brazilian Experiences.

Celi Nelza Zulke Taffarel is Professor in the Department of Education at the Universidade Federal da Bahia, Brazil.

Valerie Viehoff is a Research Fellow at the University of Bonn. She previously worked at the London East Research Institute, University of East London, UK.

Ralph Ward is Visiting Professor at the London East Research Institute, School of Social Sciences, University of East London, UK.

Acknowledgements

Staff based at the University of East London, including the London East Research Institute, the Centre for Geo-Information Studies and the University's Olympic Office have engaged in research on the impacts of the Olympic Games on the area and its communities since 2005. Over the last decade, our research has brought us into contact with colleagues, now friends, from many parts of the globe, especially, Professor Lamartine Da Costa (Rio de Janeiro State University), Berta Cerezuela (Olympic Studies Centre, Autonomous University of Barcelona) and Francesc Muñoz (Urban Planning Observatory, AUB). It was these and other colleagues from cities across the world that provided the inspiration and support for undertaking this publication.

The primary focus of our research has been on the transformation of the east side of a global city; examining, from different disciplinary perspectives, the complex processes of social, cultural and economic change that has and continues to unfold. Necessarily, our studies have drawn us into trying to understand how other cities have responded to being hosts of sports mega-events and our enquiries have led us to the production of a text that, we hope, provides some rich insights into the 'mega-event city'.

This book would not have been written without the contributors giving their time, in the midst of busy schedules, to share their research with us and write their chapters to tight deadlines, and without the support of Valerie Rose at Ashgate Publishing with whom we now have a long association that extends to four edited volumes on themes related to London and the city's east side. We would also like to thank Paul Brickell, Graham Kinshott and Richard Derecki who have engaged with and supported our work in and on East London for several years. The editors wish to convey special thanks to John Lock, Alvaro de Miranda, David Powell, Richard Sumray and Ralph Ward for their continuing engagement with our research on cities and, in particular, East London and our warmest thanks to our UEL colleagues, Penny Bernstock (LERI), and Andrew Calcutt (LERI) for their insights and contributions to our publications in this field of urban studies. We would also like to thank Sue Isaac and Karen Wilton for undertaking so many administrative tasks to enable us to create this book. Finally, very special thanks to our families for all their support throughout the time it has taken to bring the idea for this volume to fruition.

Valerie Viehoff and Gavin Poynter

List of Abbreviations

ACOG	Atlanta Committee for the Olympic Games
ACRJ	Rio de Janeiro Trading Association
ANC	African National Congress
BOA	British Olympic Association
BRT	Bus Rapid Transit
CAP	Central Atlanta Progress
CDA	Critical Discourse Analysis
CEPACS	Additional Construction Potential Certificates
CODA	Corporation for Olympic Development in Atlanta
DCLG	Department of Communities and Local Government
DCMS	Department of Culture, Media and Sport
DIA	Durban International Airport
DIESPORTE	Diagnostico Nacional do Esporte (National Sports Diagnostics)
EIA	Environmental Impact Assessment
ESRC	Economic and Social Research Council
FIFA	Fédération Internationale de Football Association
FIRJAN	Federation of Industries of the State of Rio de Janeiro
GLA	Greater London Authority
GWCC	Georgia World Congress Center
HOCR	Home Office Counting Rules
IOC	International Olympic Committee
ISO	International Standards Organisation
KISA	King Shaka International Airport
KZN	KwaZulu-Natal
LCS	Legacy Communities Scheme
LIS	London International Sport
LLDC	London Legacy Development Corporation
LMF	Legacy Masterplan Framework
LOCOG	London Organising Committee of the Olympic and Paralympic Games
MARTA	Metropolitan Atlanta Regional Transit Authority
NAO	National Audit Office
NCRS	National Crime Recording Standards
NOC	National Olympic Committee
NOL	Notifiable Offences List
OCOG	Organising Committee of the Olympic Games
ODA	Olympic Delivery Authority
OGI	Olympic Games Impact Study
OGKM	Olympic Games Knowledge Management
OLSPG	Olympic Legacy Supplementary Planning Guidance
OM	Olympic Movement
OPLC	Olympic Park Legacy Company
PAC	*Programa de Aceleração do Crescimento* (Growth Acceleration Programme)
PSFP	Public Sector Funding Package
SRF	Strategic Regeneration Framework

TOK   Transfer of Knowledge
TOP   The Olympic Partners Programme
UEL   University of East London
UPP   *Unidades de Polícia Pacificadora* (Police Pacification Units)
USOC   United States Olympic Committee

Introduction: Cities and Sports Mega-events

Gavin Poynter and Valerie Viehoff

The idea for this book arose in late summer 2013. It was an important juncture – one year after London 2012 and in the month, September, in which the International Olympic Committee (IOC) was to announce the host city for the 2020 Olympic and Paralympic Games. It was possible to reflect upon post-2012 developments in east London whilst also considering the tumultuous events that were taking place in the cities of the next host nation, Brazil, and those that had occurred in the early summer of 2013 in a city (Istanbul) and a nation (Turkey) that sought to host the Summer Olympics in 2020. London appeared to provide insights into successful ways of explicitly linking the mega-event to a long term programme of urban social and economic renewal while, paradoxically, public responses to a broadly similar linkage in Brazil and Turkey met with widespread resistance.

It seemed that the extension of the IOC's Olympic values to embrace the longer term sporting, socio-economic and cultural benefits to be derived in the post-event phase by host cities had been successfully institutionalised in the conceptualisation of legacy developed by London. London 2012 appeared to provide a compelling rebuff to those who criticised the modern version of the Games as a costly, highly commercialised 'five ringed circus' (Shaw 2008). But within months of the closing ceremony in London, events in the streets of cities in Brazil and Turkey demonstrated the capacity of ordinary citizens to contest 'legacy' – to contest the narrative that emphasised the long term beneficial effects of hosting the games; a narrative that had been so readily adopted, in the wake of London's example, by the political and sporting elites in cities such as Rio de Janeiro and Istanbul. Legacy, and the positive socio-economic benefits it implied in its association with a mega-event, no longer provided the solution that served to legitimise the hosting of the world's leading sporting festival; it became, at least for those citizens who protested in Brazil and Turkey, part of the problem.

Contributors to this book address some of the complex urban issues and paradoxes exemplified by the cases of London 2012, Rio de Janeiro (2016) and Istanbul's candidacy for the 2020 Summer Olympic and Paralympic Games. The book also draws upon case studies from other cities and nations that have hosted or seek to host major sporting events to examine, in particular, the implications for urban development and for those who live within them. This introduction briefly identifies the main features of the mega-event and its contemporary relationship to the sports industry before considering in a little more detail the variety of social responses to such events that have emerged in the period since London 2012. The final section introduces the key themes and issues that shape the structure of the book.

Mega-events and Cities

The mega-event is defined by its scale, its brief duration and its lasting effects (Roche 2000). For Roche, the mega-event, as festival, is historically associated with modernity, the 20th century's embrace of universalism and scientific progress and the human achievements – cultural, social, economic – that the event celebrates. For others, the recent association of the festival, sporting or otherwise, has been linked to the transformation from the modern to the post-modern, the transition from the dominance of industry and production to services and consumption. The rapid de-industrialisation that accompanied the crises in western economies in the 1970s and 1980s left many cities with large areas of industrial wasteland, sites that required urgent renewal if a city was to restore itself. That cities, especially those in western societies, sought to use mega-events, and especially 'expos' and sports events, to catalyse or accelerate such change, is an important recent trend. This necessity to embrace the post-industrial, and the parallel emergence of the rapid globalising tendencies of late 20th century capitalism, gave rise to an order in which nations and, particularly, cities, have increasingly sought to compete to secure economic advantage (Roche 2000; Smith 2012). Whilst cities in the developed 'West' of the globe sought to renew and regenerate, cities in many other continents experienced rapid expansion of the urban population and dynamic economic development. Seoul in South Korea, for example, doubled its population from just over five million in 1970 to over ten million by 2000, with the Seoul Olympics (1988) being a significant catalyst for political transition and infrastructural development (Yoon 2009). Over the past two decades this process of urbanisation and city growth has given rise to some important trends.

First, the territorial expansion of the city has generated a concentration of technological and physical infrastructure which, according to Graham and Marvin (Graham and Marvin 2001), has served the interests of political and business elites rather than the whole urban population. Such development, Harvey notes, enables the city to constantly tear down the old and construct the new in an endless attempt to keep up with the demands of the process of capital accumulation (Harvey 2001). Second, this process has tended to reinforce and exacerbate patterns of social inequality within the city as, according to Saskia Sassen the city's political and business elites employ a new, subordinate class to service their needs (Sassen 1991). Third, city and national governments have sought to assure the competitive or entrepreneurial nature of the city through wrestling with new forms of local governance and infrastructure investment involving partnerships between the public and private sectors, some with greater success than others and, finally, the world's leading cities have established relationships that increasingly reflect the new global networks of enterprises and value chains that have arisen (Kim and Short 2008:106). The mega-event, especially the sports mega-event, has been used by political and business elites to engage with these trends, with the festival being deployed to legitimise to their wider publics an extensive programme of urban change.

The hosting of sports mega-events is no longer largely confined to the so-called advanced or developed economies. Over recent decades, rapidly developing cities in the East, Middle-East and South have joined the race to host such events. The sports 'industry' has become a big, global business. In 2011, the industry (defined to include infrastructure, media rights, licensed products, sporting goods and the events) was valued globally by Kearney at US$480–620 billion, an estimate that included major events whilst a PWC report that focused on 'on-going' sports events – those that continue annually – estimated that the industry will expand by about five per cent per annum over the period 2010–2014, despite the global impact of economic recession (Kearney 2011:1; PWC 2011).

Accompanying the emergence of a global sports industry, many city-wide and national leaders have recognised the importance of city regions as sources of economic and employment growth in which activity may be concentrated. The hosting of a sports mega-event provides the opportunity to promote rapid urban development that favours new rounds of accumulation, typically enabled by major improvements in infrastructure. Finally, urban infrastructure improvements and new stadia have been increasingly likely to be delivered by consortia of international corporations who combine a range of design, technical and construction expertise that enables them to meet the contractual obligations entered into by the host city. The complexity of the contracts and their completion within tight timeframes has important implications for the governance of the mega-event and the capacity of affected citizens to engage with the transformation of their communities (Raco 2012).

The sports mega-event has become, therefore, the site of two significant and conflicting trends. It has provided the opportunity for political, sporting and business elites to pursue large scale projects of urban development and renewal. Such projects typically seek to tear down the 'old' and construct the 'new' – a process of creative destruction that is legitimised, it is hoped, by the popular appeal of the event and the elite's projection of its positive, long term legacies. Secondly, sports mega-events have provided the context and sometimes the focus for expressions of social discontent with the prevailing political order that has sanctioned and supported them.

Social Responses

On behalf of the International Olympic Committee (IOC), at the close of London 2012, Jacques Rogge, then the IOC President, could declare:

> 'these were happy and glorious games … The legacy of the Games of the XXX Olympiad will become clear in many ways. Concrete improvements in infrastructure will benefit the host nation for years to come. The human legacy will reach every region of the world'. (Rogge 2012)

For Rogge, London 2102 offered insights into how a host city and nation could attain public support for significant levels of public investment to be used to put on a successful Games and for the sporting festival to provide the blueprint for achieving the regeneration of a long neglected brownfield area of the city. The narrative of legacy, incorporated into government, city-wide and local plans, helped to cohere the complex framework of governance required to ensure effective preparation for the event and avoid the spectre of 'white elephants' in their wake.

Within months, however, this narrative was contested by social forces in the next host city of the Summer Olympics, a city that was also host to the 2104 FIFA Football World Cup and in the cities of one of the leading contenders to host the 2020 Summer Olympics:

'Brazil has lost a great opportunity with the World Cup. Fifa (sic) asked for stadiums and Brazil has only delivered stadiums … We should have used the opportunity to deliver good services too'.
(Mayor Paes, Rio de Janeiro, 28th June 2013, cited in: Whewell T. 'Protest-hit Brazil "missed chance" to improve services', BBC News, http://www.bbc.co.uk/news/world-latin-america-23093630)

'On Thursday June 20th, over 300,000 Rio residents marched down Avenida Presidente Vargas, draped in Brazil flags and chanting for improved education, health, transport, and an end to corruption, evictions and overspending on mega-events. The act constituted a powerful collective demand for a better Brazil'.
(RioonWatch Police Violence During Historic Protest Raises Serious Questions, http://rioonwatch.org/?p=9797 June 27 2013)

'If the unrest sparked by the attempted demolition of Gezi Park continues … [hosting] the 2020 Olympics would be just a dream. Who would lose? Us, as the people, and Istanbul'.
(Anatolia News Agency [2013] '2020 Olympics will be just a dream if protests continue: Istanbul mayor', June 24 Anatolia News Agency)

And, in response to the unrest in Brazil and Turkey, the Russian Sports minister was pressed into public statements about the public and private sector investment required to host the Sochi Winter Olympics in 2014:

'More than a million Brazilians have taken to the streets in recent weeks, clashing with police as they vent their anger at the amount of public money being spent on the Confederations Cup and the 2014 World Cup rather than on improving aging public services. But the situation in Russia, which will host these events in 2017 and 2018, differs significantly because most of the $20 billion budget is being spent on upgrading infrastructure in host cities … We live in a democratic society and in the framework of democratic processes everything is possible … We just need to anticipate and understand these things … .But the development program is very well balanced. About 25 per cent is being spent on the sporting side, and the rest is going on a program to modernize 11 regions of the country, transport, security, healthcare, airport upgrades, hotel construction, city infrastructure improvements – all of this is part of the program, Russians should be further placated by the fact that just over $10 billion of the total budget is to come from state coffers – the rest from private sources'.
(Russian Sports Minister Vitaly Mutko [RIA Novosti] quoted in Around the Rings, RIA-Novosti Roundup – 'No Brazil-Style Protests for 2018 World Cup', July 1st, 2013; http://www.aroundtherings.com/articles/view.aspx?id=43727)

For Rogge, London had provided a 'model' that might be emulated by future host cities – venues completed on time, effective governance, detailed attention given to the planning and implementation of the post-games sporting and urban policies and programmes. London's 'legacy Games' appeared to provide a compelling narrative that allied the outlay of significant public investment for a global sporting festival to delivering positive benefits for the host city. For the IOC, London's success illustrated the capacity of the Games to reconcile its commercial 'brand' operation with its avowed, and historically unique, commitment to the universalistic values of Olympism first expounded by the founder of the 'modern' Games over a century earlier. But within a few months, the achievements attributed to 'London 2012' by the host nation and city government and the IOC were overshadowed as attention turned to future Olympic host and bid cities. As 2012 receded to memory, 2013 proved a turbulent year for the IOC and future host and bidding nations.

The strategy of the host city to link a major sporting event to achieving urban transformation – a theme evident in previous host cities and nations, but one that largely structured and characterised London's successful Olympic bid – came under severe pressure. In 2013, cities across Turkey and Brazil witnessed extensive demonstrations and social unrest. In the former, Istanbul's emergence as a leading contender to host the 2020 Summer Olympic and Paralympic Games and in the latter, Brazil's hosting of the 2014 FIFA World Cup and the 2016 Summer Olympics, provided a context and catalyst for the widespread expression of public opposition to the political elites and the policies they pursued in attracting the mega-events.

There is a long history of sports mega-events attracting social protest. The scale and extent of the social unrest in Turkey in the summer of 2013 and in Brazil over the pre-event phases of the FIFA World Cup (2014) and the summer 2016 Olympics were, however, unprecedented. Preparations for Istanbul's bid to host the 2020 Olympic and Paralympic Games and for Brazil's hosting of the 2014 FIFA World Cup and 2016 Olympics were not the sole cause of the extensive urban unrest that erupted in the towns and cities of these two nations but they did provide an important focus for protesters to express their anger at their respective governments. The discontent, sparked initially by transport cost increases in São Paulo and the proposed re-development of one of the few green spaces left in the centre of Istanbul, quickly spread to encompass a much wider set of grievances about urban development plans, the commercial and sporting rather than social priorities of the Turkish and Brazilian governments and their willingness to comply with the contractual obligations placed upon bidding and host nations by FIFA and the IOC, rather than address the more pressing social needs and priorities of their own citizens.

Equally, in the wake of the financial crisis that commenced in 2007–8, other potential and host cities have experienced challenges. In February 2012, the sovereign debt problems facing Italy caused the then Prime Minister, Mario Monti, to withdraw Rome's candidature to host the 2020 Olympics. In October 2012, the Dutch government scrapped plans for Amsterdam to bid for the 2028 Olympic Games, citing the financial risks. By October 2014 the only remaining candidates still interested in hosting the 2022 Winter Olympics were Beijing (PR China) and Almaty (Kazakhstan), after the withdrawal of Stockholm (Sweden), Oslo (Norway) and Krakow (Poland). The Swiss canton, Graubünden, had successfully opposed a joint bid for Davos and St Moritz and a potential bid from Germany was stopped by public referendum. More recently, the huge costs associated with Sochi hosting the 2014 Winter Olympics have met with domestic criticism, including allegations of corruption arising from the contractual arrangements agreed to fund the large scale infrastructure projects that are associated with hosting the Games. The events in Turkey and Brazil and the issues raised in cities such as Sochi, Amsterdam and Rome have not escaped the attention of senior IOC officials. Though publicly they play down the potential risks and implications of social unrest associated with mega-events and of dwindling numbers of candidate cities interested in hosting the Games, there has been some recognition of the need for reform. For example, the new IOC President since September 2013, Thomas Bach, suggested that the bidding process could be made more flexible to ensure that cities continue to be attracted to hosting the summer and winter games and consultations in 14 working groups are underway to develop a new 'Olympic Agenda 2020', a strategic roadmap for the future of the Olympic Movement, to be approved in December 2014.

But the Olympics have not been the only mega-event to attract social protests and cause organisers to reflect upon their organisation of the event. The Bahrain Formula One Grand Prix race was cancelled in 2011 in the wake of social unrest and, in April 2013, its return to Bahrain prompted street protests that sought to highlight the country's lack of democracy. The 2010 Commonwealth Games in New Delhi prompted protests concerning cost over-runs, the failure of the private sector to help finance the event and the displacement of the poor from areas of the city in which the games took place. On a different scale, opposition groups within Iran have used the national team's involvement in domestic FIFA World Cup qualifying matches to demonstrate within the stadiums their opposition to the government regime. In Iran's case the large numbers of spectators provide relative safety in which to conduct public protests whilst the popularity of football within the country ensures widespread media coverage of the games and the protests they facilitate. It would seem, from these few recent examples, that sports mega-events may provide either context or cause of social unrest in host cities and that sometimes the two combine to generate extensive displays of opposition to the existing political order.

In the 20th century protests at major international sporting events focused largely upon the conflicting ideologies arising from the cold war and the international dimensions of regional conflicts. In the 21st century, despite the apparent success of London 2012 in harnessing the Games to a programme of urban renewal, the strategy was, arguably, relatively short lived as a means to legitimate public investment in the sporting festival. The social unrest that has arisen, particularly since the closing ceremonies in London, has been driven by the proposed and actual transformation of the urban fabric that the host and aspirant cities incorporated into the bidding and preparation for these global events.

For opponents, these events have highlighted the self-serving priorities of international sports federations, such as the IOC and FIFA, and of the domestic political and business elites for whom sport has become an important vehicle for achieving their own aggrandizement and that of their cities or nations. However, for many cities across the globe, bidding for and hosting a global sporting event, despite the social unrest in Brazil, Turkey and other nations, continues to be an attractive strategy for achieving international recognition and urban transformation. This contested future for the sports mega-event is a major theme that runs through the chapters that follow. The relationship between the sports mega-event and the city is explored, in particular, through an analysis of the different thematic dimensions of the legacies or long term impacts of these events on the cities and nations that have hosted or seek to host them in the future.

The Structure of the Book

The book divides into five parts in which contributors explore legacy from its meaning and measurement to its implications for the processes of urban transformation that the mega-event may catalyse. Section 1 commences with Francesc Muñoz's overview of the role of mega-events in the 21st century city and his compelling account of the trend toward the 'urbanalisation' of the city. This is followed by chapters from Iain MacRury, Allan Brimicombe and Alvaro de Miranda, who draw upon their extensive research on London 2012, to explore the meaning of legacy, how it may be measured and the significance of the discourse of legacy when allied by host cities to, in particular, a mega-event such as the Olympics.

In Part II, Juliet Davis and Ralph Ward draw upon the experience of London 2012 to discuss the state's approaches to planning that seek to address the spatial dimensions of the social divide between east London and the rest of the city. Davis discusses the ways in which urban spaces may be construed as socially deprived as a means of legitimating master plans that disrupt existing patterns of social life and fail to deliver the benefits to the poorer communities that they purportedly set out to achieve. Ralph Ward uses his own professional experience to examine a broadly similar theme, the legacy promises of London 2012 and how their transformational claims have proven difficult to translate and implement through planning processes. Philip Cohen discusses the imagineering of the Olympic Park and, in particular, the East Village (the former Athletes' Village) as a process of urban fabrication with meanings invested via the official promotional literature – meanings devoid of the richness of the kind of story making through which spaces become places. Egidio Dansero, Alfredo Mela and Cristiana Rossignolo's study of the impact of the Turin Winter Olympics focuses on the production of territory in terms of its symbolic, physical and organisational dimensions, especially as these unfold in the post-event or legacy phase. They present new insights into the complexities of the processes of territorialisation and re-territorialisation in Turin. Finally, Anne-Marie Broudehoux and Fernanda Sánchez argue that the production of the Olympic City is accompanied by spatial reconfigurations that fundamentally transform a city and its government, creating, in the case of Rio de Janeiro, a reconfiguration of the urban fabric that generates socio-spatial exclusions on a considerable scale

Part III focuses upon the intangible, the capacity for host cities to develop, share and manage the knowledge that arises from hosting a mega-event. As Berta Cerezuela and Chris Kennett persuasively argue, such opportunities have often been missed particularly by public authorities in host cities; a view broadly supported by Oliveira et al. in their analysis of Brazil's preparations for the 2016 Olympic Games. Neil Herrington's research into the perceptions of legacy held within the educational setting in east London suggests a research method (Q sort) that may provide more nuanced insights into the conceptions of legacy held by local communities; insights that, if captured and acted upon, may assist cities to address more effectively the cultural and educational opportunities that arise from hosting a mega-event.

The broad theme of sustainability informs the chapters in Part IV. Each chapter focuses on case studies of past sports mega-events, with the authors providing detailed analyses of the post-event phases of urban development in Munich (1972 Olympics), Atlanta (1996 Olympics), Athens (2004 Olympics) and Durban (2010 FIFA World Cup). The historical span of the studies provides insights into the emergence of the concept of sustainability as a policy issue over recent decades, particularly via the comparative study of Munich and London 2012 conducted by Valerie Viehoff, whilst Costas Cartalis (Athens) and Michael Dobbins, Leon Eplan, and Randal Roark (Atlanta) provide detailed analyses of the post-event impacts of hosting the Summer Olympics in their respective cities. These chapters provide some interesting correctives to the view that these host cities secured few benefits from hosting the Olympic Games, especially in the case of Atlanta as Dobbins et al. recount in their discussion of the innovative public interventions that have continued to renew and re-develop key areas of the city in the years following 1996. However, as Brij Maharaj and Vyasha Harilal outline in the case of Durban (2010), achieving the social dimensions of the sustainability 'agenda' proved to be largely elusive in so far as the infrastructure developments that took place tended to reinforce rather than reduce existing social divisions in the city – an outcome that is also acknowledged in the case of Atlanta.

Part V takes the social dimensions of legacy and the impacts of mega-events on communities as a central theme. The experience of holding the Formula One Grand Prix in Istanbul and the urban impacts arising from the hosting of the 2014 FIFA World Cup and 2016 Olympic Games in Rio de Janeiro provide the contexts for three chapters that provide compelling insights into the effects on local communities. As Semila Erylimaz and Hüseyin Cengiz explain, Formula One races took place between the years 2005 and 2011 at the new Istanbul Park Circuit, situated in the Tuzla area of Istanbul. The creation of the track and its accompanying infrastructure provided the context for the construction of new areas of residential development for upper income residents and generated significant increases in land values in the area. Opponents to the developments failed in their legal challenges despite the strength of their arguments based on constitutional, urban planning and environmental grounds.

The chapters on Rio de Janeiro also focus on the urban transformations occurring in the city. Matthew Richmond identifies the underlying trends arising from his analysis of two, as he calls them, 'unspectacular favelas' (Tuiuti and Asa Branca) located in different parts of the city. The chapter examines the objectives of state policies and their implications for favela communities and draws attention to the 'democratic deficit' that has arisen through their selective implementation. Einar Braathen, Gilmar Mascarenhas and Celina Myram Sørbøe also provide evidence of such a 'deficit' in their study of Rio's preparations for the 2014 FIFA World Cup and the 2016 Olympic Games. The authors consider conventional and critical (state of exception) perspectives of the urban transformations taking place in the city and, from their detailed analysis of governance – policy promises and their implementation – they tend toward the critical view. The authors, however, also conclude that the social mobilisation and political engagement of citizens who seek to challenge the trajectory of urban development being pursued by the state may, indeed, provide an important, lasting and contested legacy for the city.

As perhaps this introduction to the themes and chapters of the book suggests, the study of the sports mega-event, its host cities and societies and the range of legacies it bequeaths may be approached through the lens of different intellectual disciplines and theoretical perspectives. The chapters contained in this text reflect the authors' critical engagement with their subjects of study, making, we feel, an important contribution to understanding the sports mega-event as, what Michael Rustin has called in his Epilogue, a new global mode of production.

References

Graham, S. and S. Marvin (2001). *Splintering Urbanism*. Abingdon: Routledge.
Harvey, D. (2001). *Spaces of Capital*. Edinburgh: Edinburgh University Press.
Kearney, A.T. (2011). The Sports Market: Major trends and challenges in an industry full of passion. http://www.atkearney.co.uk/documents/10192/6f46b880-f8d1–4909–9960-cc605bb1ff34; accessed June 10, 2014.
Kim, Y-H. and J.R. Short (2008). *Cities and Economies*. London: Routledge.

PWC (2011). PWC Outlook for the Global Sports Market to 2015. http://www.pwc.com/gx/en/hospitality-leisure/changing-the-game-outlook-for-the-global-sports-market-to-2015.jhtml

Raco, M. (2012). 'The privatisation of urban development and the London 2012 Olympics'. *City*, 16(4), pp. 452–60.

Raco, M. (2014). 'Delivering Flagship projects in an era of regulatory capitalism: State-led privatization and the London Olympics 2012'. *International Journal of Urban and Regional Research*, 38(1) January, pp. 176–197.

Roche, M. (2000). *Mega-events and Modernity*. London: Routledge.

Rogge, J. (2012). 'The Closing Ceremony of the XXX Olympiad' London 12th August, 2012. http://www.olympic.org/Documents/Games_London_2012/London_2012_Closing_Ceremony_English.pdf

Sassen, S. (1991). *The Global City*. Princeton: Princeton University Press.

Shaw, C. A. (2008*). Five Ringed Circus: Myths and Realities of the Olympic Games*. British Columbia: New Society Publishers.

Yoon, H. (2009). 'The Legacy of the 1988 Seoul Olympic Games', in Poynter, G. and I. MacRury (eds.) (2009) *Olympic Cities: 2012 and the Remaking of London*. Farnham: Ashgate, pp. 87–96.

PART I
Urbanism and Legacy in the 21st Century

Urbanalisation and City Mega-events:
From 'Copy&Paste' Urbanism to Urban Creativity

Francesc Muñoz

Contemporary Urban Mega-events

The urban transformation of cities in the Western world over the course of the 20th century cannot be separated from the organisation of urban mega-events. International exhibitions, world fairs and the Olympic Games are three paradigmatic examples of this form of guiding the growth and transformation of cities.[1] This chapter explores the management process of a model of urban intervention that evolved during the 20th century, forming a specific type of policy that will be considered in the first section. Given the limited space available, I will only be considering the Olympic Games. These serve as a clear example of an urban mega-event which brings forth a truly unique type of intervention in the city, to such an extent that we may even speak of 'Olympic urbanism' (Muñoz, 1997, 2005, 2006).

The second section looks at a significant change in the last third of the 20th century in relation to the role of mega-events in processes of urbanisation in cities. Two dynamics are especially responsible for this change in how cities have reformulated the role of mega-events in relation to urban policies. Firstly, the definitive rise in the dynamics of economic, political, cultural, and urban globalisation and, secondly, the progressive emergence of what I have called *urbanalisation* (Muñoz, 2008, 2009), referring to the global expansion of *copy&paste* urbanism, achieves its development all over the world through urban mega-events.

The main results of these changes in terms of major urban events contain four primary elements:

- Firstly, the banalisation of the events, which cease to present an inaugural, colonising, unique, and one-off nature, to become just another element in a festivalised urban normality.
- Secondly, from a model of mega-events closely associated with large modern industrial metropolises, there is a shift towards the multiplication of all types of urban events taking place in different types of cities, fed by the sectoral expansion of which the events with the greatest impact are the protagonists. The normality with which medium-sized cities end up becoming cultural capitals or world headquarters for events linked to all types of economic sectors and issues of social interest – from fashion to sports, from politics to art – is convincing evidence of this.
- Thirdly, there is a change in the role that the urban image played in urban planning and policies in the 20th century. If previously the image served to economically publicise, socially communicate, or culturally build the narrative of urban change, its role would develop until it becomes the first and main requirement to enable processes of change and transformation in the city.
- Lastly, the importance of the image causes the large-scale event to assume a primal function as an element that changes not so much the space of the city as, above all else, its image. That is, even if a physical transformation is produced in the urban space, something which is obvious in the majority of experiences of the organisation of urban mega-events, the main benefit for the city is not, as was

---

1 This is not the place for a detailed explanation of the nature of these canonical urban events and their global hierarchy. A deeper discussion could allow us to establish conclusions on the differences between the three mega-events mentioned here and also to discuss the peculiarity of events that start as local but that, thanks to their repetition over time and to their capacity to connect with issues of a global interest, have acquired importance and predominance in recent decades. This would be the case of the different International Building Exhibitions, the German IBAs, whose last edition was closed in 2013 in HafenCity, Hamburg.

the case in the 20th century, its mere physical transformation. Instead, it provides the urban space with a new image, contributing to the overexposure of the city in the world market of urban snapshots on which the sectors that sustain the globalised economy feed: from tourism to urban entertainment-leisure, and the wide array of products linked to what authors such as Andrew Darley have called the new, and no less global, 'visual digital culture' (Darley, 2000).

Finally, in the conclusion, three hypotheses are outlined concerning the future of the urban mega-event in the city of the 21st century as well as some alternatives proposed to guide it in terms of urban innovation and creativity, thus escaping the *copy&paste* urbanism typical of processes of *urbanalisation*.

The City and Large-scale Events: The Example of Olympic Urbanism

This is not the place to delve into a detailed discussion of the Olympic Games as an urban mega-event. There have already been discussions of how the organisation of the Olympics in the 20th century goes far beyond the limits of being simply a sporting event, For example, how the management of the Olympics came to determine a model of urban intervention that allowed for the formulation of a specific type of urbanism: 'Olympic urbanism'. Within the array of actions that shape this Olympic urbanism, the construction of the Olympic Village clearly stands out. This is a new urban artefact that allows for the clear establishment of the relationship between urbanism and the communication of a specific urban image. It is, in fact, in the Olympic Village that architecture is more clearly used for the image that the city organising the event wants to project internationally.

The building typologies, the formal languages, and the design itself of the spaces form part of an urban landscape specifically conceived to highlight both the current values of modernity and those that are specific to the place. At the same time, the architecture of the Olympic Villages manifests the ambition to reproduce – in a controlled way, and in a reduced and enclosed space – the urban models and architectural proposals that encounter too much rigidity and too many difficulties within the real space of the city to be put into practice.

Thus, in the course of the 20th century, Olympic Villages ceased to be ephemeral constructions; like the military barracks or camps that characterised the first modern Olympics. They also ceased to be defined as temporary accommodation in the city. Since the decade of the 1930s and, above all, after the Second World War, the Olympic Villages would become new constructions, first thought of as part of the process of expanding the city over the territory and, later, as transformations of the existing built urban fabric, characterised by durability and the ambition to build new urban areas.

In this way, the process of urban construction linked to the celebration of the Olympic Games gained in complexity throughout the 20th century. Initially, we were dealing with a series of well-defined elements strictly related to the practice of the different competitions, as is the case of the stadium and the specialised facilities for the different Olympic sports. However, urban protocols and strategies would appear over time, allowing for talk of the consolidation of a specific type of urbanism related to the Games, in which the role of the Olympic Village as a primary urban element would become more and more relevant.

Defining the Olympic Village Model: A 20th Century Story[2]

The Olympic Games of before and after the first world war – London (1908), Stockholm (1912), Antwerp (1920), Paris (1924), and Amsterdam (1928) – responded to the problem of accommodation with what we could define as an emergency residential menu, formed by different types of temporary residences, sometimes including even the same boats that had carried the athletes from their countries of origin.

---

2 For a detailed study of the history, typology, and urban functionality of the Olympic Villages during the 20th century, see Muñoz, 1997. For an illustration of the architectural projects and urban planning programmes of the most relevant and most recent Olympic Villages with plenty of images, see Muñoz, 2005.

While in Paris in 1924 a first and modest Olympic Village was tested – a group of wooden barracks close to the Stadium of Colombes with some additions like postal and telegraph services – the first real Olympic Village was built in Los Angeles for the 1932 Olympics. Together with the Olympic Village of Berlin in 1936, it represents what I have called 'inaugural Olympic Villages' (Muñoz, 1997, 2005, 2006), establishing the basic model of accommodation that would be reproduced in subsequent Olympic Games: a multifunctional structure that, as well as providing accommodation and basic care for the athletes, included facilities such as entertainment areas, spaces for rest and physical care, and leisure areas.

After the Second World War, it was not until the 1960s that the Villages in Rome (1960) or Mexico (1968) would introduce new complex elements to the morphology and functions of the basic model of the Olympic Village inaugurated in Los Angeles and Berlin.[3] Certainly, the Olympics in Rome meant a change in relation to previous events from the point of view of Olympic urbanism. For the first time, there was a regional conception of the urban and behind the decisions of the location of the Olympic facilities and constructions there was also a project for the territorial expansion of the city. Thus, the Olympic Village inserted itself into a programme of residential zoning that went far beyond the immediate solution to the temporary accommodation of the athletes; something that, from then on, would be a common denominator of subsequent Olympic Villages.

With Rome (1960), the architectural form of the Olympic Village became a key element in the projection of the city's image. In fact, the Olympic Games themselves, as an urban mega-event, would change considerably from the second half of the 20th century on, experiencing a process of internationalisation and commercialisation of their contents that would acquire a worldwide dimension over time. Thus, the globalisation of the consumption of sporting events went hand in hand with the planning of real sports districts understood as yet another formula for extending or transforming the city. Olympic urbanism would become, consequently, part of ambitious urban development programmes and the architecture involved would guarantee not only the functionality of the projects but, above all, the spectacle of the urban image as added value. Evidence of this evolution is found in the Olympic Villages of the 1970s – Munich (1972) and, above all, Montreal (1976) – which, unlike those of the previous decade, were not considered from the point of view of urban growth but of re-zoning and renovation in the already existing city.

The Olympic Villages of the last two decades of the 20th century consisted of a mixed set of actions: the experience of programming planned in Moscow (1980) and Barcelona (1992) contrast with the more ephemeral approaches of the university villages in Los Angeles (1984) or Atlanta (1996). In the case of Moscow, the Olympic Village was considered as part of the capital's 1971–1990 Development Plan and within the 10th Five-Year Plan for Economic and Social Development. This included programmes that divided the city into eight planning zones, the development of which was accelerated by Moscow being named host of the Olympics. The interventions in Seoul and Barcelona, although different, share this integration of Olympic urbanism into plans for the recuperation of large urban areas like Jamsil – a flood zone on the bank of the Han River – and the waterfront of Poblenou, the old industrial heart of the 19th century city in the case of Barcelona. On the other hand, in Los Angeles (1984) we find the opposite, a minimum impact Olympic urbanism with only four new constructions among the twenty one sports facilities. Thus, no Olympic Village was built and the athletes were housed in three university campuses. This is a minimalist format that was reproduced in Atlanta (1996), where the facilities of the Georgia Institute of Technology were used as the Village, complemented by the construction of the 'Village Festival Centre', a large shopping-mall, and the 'Olympic towers' – two apartment blocks that completed the residential offer.

Lastly, the Olympic Villages of the beginning of the 21st century – Sydney (2000), Athens (2004), Beijing (2008), and London (2012) – clearly show examples of some of the main urbanism approaches found today. These examples reveal the impact of globalisation on the architecture and image of urban projects, with a clear predominance in the role of international firms that have the capacity to create a recognised brand, the tendency towards spectacle in the shaping of the architectural design and definition of spaces, the concern with environmental sustainability, and the landscape integration of the buildings.

---

3   Thus, the Olympic Villages in London (1948), Helsinki (1952), or Melbourne (1956) are characterised by a return to the use of ephemeral constructions or of military camps, or by programmes that took advantage of part of the social housing included in the cities' urban plans.

Large-scale Urban Events and Urbanalisation

As shown in the case of Olympic urbanism, large-scale events have formalised a series of quite stable urban intervention protocols during the 20th century in terms of the organisation and territorial and physical results of the event itself. At the same time, during the last third of the 20th century, cities, encouraged by the intensity of the dynamics of economic and urban globalisation, have redefined the role of large events in relation to urban policies. Four major changes stand out:

- The banalisation of large-scale urban events.
- The sectoral expansion of urban events and their territorial multiplication in the cities that organise them.
- The change of the place occupied by the urban image during the 20th century in terms of the process of the production of cities.
- The reduction and simplification of the large urban event, now merely serving to update the city's image.

Before looking at each of these four transformations, it is worth briefly presenting an explanation of the context of urban landscape production in recent decades on the basis of the development of architecture and urbanism projects that shape what I have called *urbanalisation*.

Urbanalisation: The Urban Form of the Post-industrial City?

Landscape has traditionally been understood as the morphological translation of the physical features and the social and cultural relationships that define a place and shape the so-called *genius loci*. However, cities are currently facing the emergence of landscapes which are clearly independent from place in the sense that they can be replicated in any other city. In this context, landscapes no longer translate the features of the place as could be expected, nor do they contain cultural or symbolic attributes related to social identification and cohesion.

This process of disconnection between place and landscape can be summarised with the idea of *urbanalisation* that has characterised the recent evolution of cities. The main results of *urbanalisation* can be easily observed in some specific urban scenarios such as historical inner cities or urban waterfronts. Regarding the transformation of historical neighbourhoods, gentrification processes have gone hand in hand with the progressive orientation of urban space towards leisure and consumption, resulting in a very dramatic transformation of the local urban landscape. Different urban regeneration programmes and renewal projects have been developed in a very similar way in different cities bringing about a kind of *copy&paste* urban form, which the global visitor to the local historical area has in mind and hopes to find when perceiving this specific part of the urban landscape.[4]

In terms of urban renewal in waterfronts and riverside areas, common highly standardised architectural and urban design programmes have been implemented worldwide revealing a very restricted menu of options when we look at the urban structures resulting from those projects: the aquarium and the new marina, the shopping area, the leisure sector, the IMAX cinema, the local museum or cultural centre, and the high-rise residential areas facing the water. These elements shape a brand-new urban seafront which also appears as though 'copied and pasted' from one waterfront to another when comparing different key examples since the 1980s, from the well-known Baltimore experience to the long list of different projects in Europe and, more recently, in Asia. This general evolution is the result of a structural trend in the recent urban history of contemporary cities: the progressive conversion of urban historical centres and waterfronts into places for consumption, entertainment, and other activities linked to global tourism. That is to say, the city's traditional spaces – the architectural setting, the topological elements like streets and squares, which have historically

---

4　This is something that was anticipated by authors like John Urry many years ago when discussing the specific nature of the visitor's gaze with regard to the cityscape. Urry's *tourist gaze* concept still explains very well today the way in which previously constructed expectations of the urban landscape strongly shape the visitor's perception of the urban experience.

characterised the compact city as vibrant public spaces – are transformed following very similar patterns of intervention that present a highly standardised type of urban experience. A very interesting paradox arises here.

During the last half century, leisure and consumption spaces have been intensively recreating and imitating urban atmospheres and formal features of cityscapes: the street, the square, the boulevard, the park, etc. Nowadays, it seems that cities, in order to be successful as places to be visited and consumed, need to imitate that urban form, already based on imitations of the city spaces themselves, found in shopping malls, festival markets, or theme parks. This is a process that contributes to and reinforces the standardisation of urban landscapes. In this sense, the majority of urban renovation experiences have created common results such as the economic and functional specialisation of formerly complex urban areas, the morphological segregation of the urban form, and the thematisation of the urban landscape. These three elements characterise *urbanalisation*.

Finally, the recognisable urban form of the compact city, those areas where topological elements like streets or squares contribute to the urban fabric, are also converted into specialised containers. Despite the morphology of the city being maintained, the urban functions have been simplified in a thematic way. Even though the residential function remains, historical areas affected by regeneration programmes have acquired a new function: they have been renewed as spaces not to be inhabited but to be visited.

A recent example illustrating this process of *urbanalisation* can be observed in the urban renewal affecting Jewish ghettos in Eastern Europe. These old neighbourhoods have been renovated following a very similar pattern in different cities and offering a final scene where the historical urban form is merely the visual support for a highly specialised use of the space aimed towards leisure, entertainment, and consumption (Murzyn-Kupisz, 2009). The results of these renovation experiences reveal a city which has been simplified in terms of its attributes and contents. Similar results can be observed in many experiences of urban renovation in historical centres and waterfronts in Europe.[5] Paradoxically, both had been the spaces most culturally identified with the attributes characterising the urban form of the traditional city. Nowadays, they show the progressive loss of urban diversity and complexity due to the recent evolution of the urban form.

As mentioned many times and in fact proposed by important names in European architecture like Aldo Rossi, the urban regeneration projects developed in European cities since the 1980s were based on the morphology of the historic city, proposing its renovation in accordance with new forms of life at the end of the 20th century. On the other hand, the urbanism linked to *urbanalisation* represents the opposite: the definitive abandonment of the historicist model and an embrace of the logo-architecture typical of globally triumphant tourist resorts, visible and 'clonable' from the coast of Florida to Moscow, from Macao to any of the *Nike towns* built in the Bahamas or the coast of Vancouver.

Four Changes in the Relationship Between the Urban Mega-event and the City

As stated earlier, we are interested in highlighting four important changes that have affected the nature of the large-scale urban event and its place in urban policies, considering the global context of the processes of *urbanalisation* commented on above.

Now we briefly look at each of these issues.

The Banalisation of the Urban Mega-event

The process of the festivalisation of urban space, explained by some authors like Marco Venturi (1994) or Darrel Crilley (1993) 20 years ago, has ended up assimilating urban policies into the organisation and management of different types of events. This explains many interesting issues like, for example, the importance of urban marketing and branding initiatives and the need to almost constantly update the city's image. The large event has lost a significant part of the quasi epic elements it presented during the last century, when it was attributed

---

5 The urban iconography created by cinema, for example, has always shown key contents of urban life present in these two specific landscapes: density, intensity, relationships, hazard, chance, or conflict. A film with a very meaningful title, *On the Waterfront*, by Elia Kazan (1954), is a very good example. From the very beginning to the end, the association between the city and the port is clearly present and the previously mentioned city attributes characterise the action.

with an inaugural – in the sense of the beginning of a new urban era – or a colonising meaning, in the sense of incorporating new territories into the city. It could be said that the large-scale event is thus integrated, in a quasi-natural way, into what is nothing other than a banalised urban normality by means of festivals and hotspots, in which what is extraordinary becomes ordinary almost at the same time as it is formalised in the urban space[6].

The idea of *urbanalisation* frames this process according to which the events, organised on the basis of the same recipes, are replicated in some cities when their celebration has not yet ended in others. Without a doubt, these processes clearly show the different faces of the globalisation of economies and cultures that holds a privileged niche in cities. It is something that authors like John Hannigan or Naomi Klein have shown when explaining the processes of brandification, both of modes of consuming urban spaces and of transforming the city.

The Sectoral Expansion of the Urban Event and its Territorial Multiplication

But this banalisation of the major urban event is, in fact, fed by the second change that I highlighted earlier: the sectoral and territorial expansion of these events. The event model associated with urban projects and the cultural affirmation of large modern industrial metropolises has been substituted in less than 30 years by the spread of different types of events related to the most diverse economic, political, social, and cultural interests, which take place, moreover, in a wide array of cities of different types and with different positions in global hierarchies.

For example, the different types of 'capital status' – of culture, of art, of sport, or of a specific economic framework – that exist today, and the new possibilities for organising urban events indicate extreme stratification and specialisation. We have cities like Bruges, as the European Capital of Culture, Valencia, with global sporting events like the celebration of the America's Cup, or the long chain of cultural events in cities like Edinburgh, with its performing arts festival, or Barcelona, with the Festival of Advanced Music, SONAR. These examples are among those that confirm that practically every economic sector today has its own event, from mobile telephone fairs to those of the construction or tourism sectors. Every topic of interest in urban life thus has an event serving as a point of reference, from biennial art and fashion shows to those for cooking and architecture. Of course, these small-scale types of events cannot be compared to the canonical urban mega-events cited at the beginning. However, they do contribute, with their almost daily and persistent occupation of urban space and time, to taking the symbolic and cultural dimension away from major modern events that seem today to be dissolved in the festivalised normality we alluded to earlier. The theorists of the 'liquid society', like Marshall Berman or Zygmunt Bauman, are finally being proved right.

The Alteration of the Nature of the City's Image in the Urban Process

The third big change is that today it has become essential for urban policies to establish a certain type of urban image or, to be more precise, an image understood as a brand for the global façade projected by the city. This third change can be summarised as the definitive alteration of the place occupied by image in the process of urban production (Muñoz, 2008). Its complexity and importance demand a sufficiently detailed explanation.

From the middle of the 19th century and during a large part of the 20th century, image was a secondary element in the process of the construction of cities, in the sense that there was a clear differentiation between the circuit of production, urbanisation and occupation of urban land, and the narrative device that used images to explain to the population how the city was transforming. This is a mechanism that became progressively more efficient as photography, first, and moving images, later, reduced the time previously needed by literature

---

6   John Ploger (2001) explored these ideas, taking into consideration the example of the urban projects proposed in different cities when celebrating the new millennium in the year 2000. The results of this 'millennium urbanism' clearly show the relationship between the urban policies and an urban narrative based upon the organisation of city events.

or painting to capture the process of urban change inevitably linked to the idea of modernity and closely associated with the realisation of large events.[7]

Thus, when it came to the physical transformation of the city that the organisation of the mega-event brought with it, first the city was transformed. Afterwards, the image either documented the process of destruction-construction or narrated it in terms of collective history, using exercises in nostalgia or criticism of the new modern urban space resulting from the change in the urban landscape. In the majority of cases, however, the image often validated the urban change as irrefutable evidence of material progress, especially if that urban transformation derived from the celebration of large events.

Today, the role of the image in relation to that process of urban production has been completely inverted. The image is no longer that documentary or narrative tool, external to urban transformation, but has become the main requirement in order for the transformation to be produced. Consequently, its temporal location also changes so that the image is not constructed and socialised after the urban change but the complete opposite: it is the construction of an urban image which enables, in fact, the physical and real transformation of the city.

We are dealing with a clearly complex phenomenon whose detailed explanation cannot be tackled in this text,[8] but in whose origins the advertising use of the image, which was consolidated in the 20th century, plays a relevant role. Over time, the association of the use of the image with the promotion and sale of a consumer good would also become common in the case of urban images. The symbolic sale of the city would thus be the main priority of a series of innovative proposals that, especially since the second half of the 20th century, would consolidate a model of mass communication that promotes and illuminates a series of urban images marginalising others and relegating them to the shadows.

From the promotional videos already developed in the first decades of the century in relation to international fairs and universal exhibitions to the birth of urban marketing, first, and city branding, later, we can trace the limits of the branding of the urban image that explains the alteration of the role and positioning of the image in relation to the city, as alluded to earlier.

In effect, *brandification* operates in the same way whether we are dealing with a trainer, a yoghurt, or a city. Having a brand image or, to be clearer, a brand made into an image, is a necessary condition for any type of product; it is that image that provides the optimum conditions for symbolic identification and individual appropriation needed by global consumption. This is a consumption habit that does not represent, in reality, the consumption of objects, but that of the experiences and emotions associated with the images that advertise them.

Thus, in the same way that the brand image of a pair of trainers or a yoghurt ensure their commercial success, every city searches for its own brand image to sustain the process of commercialisation of urban attributes and the sale of place associated with the main processes of current forms of urban transformation. This includes the gentrification of urban centres and urban sprawl in residential peripheries, as well as the renovation of waterfronts and the creation of commercial value for old, first generation industrial districts. All of this explains why, as with the trainers or the yoghurt, the city needs to build an image with sufficient brand potential, an urban brand with the capacity to become an image, because the possibility of the physical transformation of the urban space depends on it.

In the same way that the image associated with a successful brand substantially improves the perception of the physical attributes of the trainer or the yoghurt, regardless of their real quality, the brandified urban image allows for the physical transformation of the city to be validated automatically. Thus, there is no evaluation of the meanings, impacts, and consequences of the urban change; it only needs to correspond to some extent with the brand image previously created and promoted. In this sense, when we confirm that the urban image is now nothing other than an image-brand, we also see that the international market of urban images has developed

---

7   This is something we can clearly appreciate if we compare the different narrative-explicative powers of two contemporary works that tackle the topic of urban change in Berlin, one of the European metropolises that experienced the most and the fastest changes in the first decades of the 20th century: the documentary film *Berlin, Symphony of a Great City* (1927), by Walter Ruttmann, in comparison to the work *In Berlin* by Franz Hessel (1929).

8   For a more detailed explanation of the reversal of the role of image in relation to processes of city production, see Muñoz (2008, 2009).

to such an extent that practically everything can be used to create an attractive brand: from the local cuisine to the architecture, from the attributes of the environment to the characteristics of the inhabitants themselves.

We now know of many examples of specific real estate projects and even new neighbourhoods that appear in different cities promoted and associated with a specific brand, sometimes even accompanied by a commercial logo that can be used subsequently in the commercialisation of other products. This happened in New York in areas like TRIBECA – Triangle Below Canal – or in the north of the Little Italy neighbourhood, christened by developers as NOLITA (Northern Little Italy), an urban project that has even generated a clothing brand for young people with the same name (i.e., with the same brand) and which can be found in any shopping area in any European city. Other examples include: Berlin, after the fall of the Wall, where it is currently difficult to differentiate whether *Mitte* is the name of a neighbourhood or of a sponsor; London's *Brick Lane*, a famous street in London's East End,; and Bilbao, where perhaps we will find future urban projects with names like *Mosel*, the decoration and accessories shop specialising in Italian and Nordic design now installed in the city. Urban branding thus represents another step on the path started decades ago by marketing, a step which consists of summarising urban images in labels, in brands, and making the city landscape subsequently adjust itself to them.

The Role of the Mega-Event as a Mobiliser and Catalyst of the Urban Image

The last of the four changes that are proposed here to understand the current relationship between large-scale events and urban policies has to do with the simplification of the urban event in terms of merely serving to update the city's image.

That is, precisely because of everything explained in the previous point, the role of the large-scale urban event has gone from constituting a highly significant moment for the transformation and history of the city to presenting a much more instrumental function, as a support element for the necessary updating of the brand image of the city, regardless of the type of physical transformation that may be produced as a consequence of its celebration. The case of the last Olympic Games in London in 2012 is very significant in this sense as, apart from all of the important physical transformations that the Olympic project meant for London's East End the main asset of the transformation programme referred to the promotion of a change in the urban image of the areas that border the River Thames in that part of the city. This association between event and urban image, and not so much the physical reality of the city, is what enables and gives rise to a whole series of proposals that would have been unthinkable previously, when the event required stronger links with the material urban substratum.[9]

Thus, from 2008 to 2012, Valencia hosted a part of the Formula One World Championship thanks to a new urban circuit created in the city. The final image so clearly recalled the situation in Monte Carlo that, in fact, for a few weeks the image of that city was reproduced as well as, to a certain extent, its appearance as an urban brand. What is interesting about this case is that the city already had an existing circuit in use in the suburbs, but the pull of the brand image, and what attracted the global capital represented by the current Formula One business lobby, was the possibility of having the circuit in the city; in the most central urban area. On the other hand, the last editions of the Winter Olympic Games have stood out due to their celebration in cities that do not have the natural and landscape attributes strictly necessary for the practice of those sports. In fact, both in Torino (2006) and in Vancouver (2010), the resorts and facilities were in other places but the cities assumed the brand image as Winter Olympics cities despite the fact that their physical and material reality literally could not host the event. This is appreciated with even greater clarity in the case of Barcelona, which, despite not exactly having the necessary features, presented itself as a candidate in the initial phase of the Winter Olympic Games for 2022, proposing to carry out the corresponding competitions in the ski resorts of the Pyrenees. These examples demonstrate the real possibilities of relocation and de-anchoring presented

---

9   The association between big events and the urban image has at times come to determine even small details of spatial physical shape based on the needs and requirements of visual consumption. Thus, in the Olympic Village in Los Angeles (1984) sophisticated simulation exercises were used to recreate the urban landscape using decorative elements in public spaces, from using a palette of 'Mediterranean' colours in the design of visual elements and the urban furniture to the use of signage conceived not only to be seen by visitors but also on the television screen. See Muñoz (1997, 2005).

today by major urban events, which are turned into something portable and transportable from one city to another and considered almost in 'take-away' terms.

Events in the City-event: Three Hypotheses on the Future of the Urban Mega-event

Throughout the previous pages, I have explained a fact that I consider especially relevant for understanding the current relationship between large-scale events and the city. The last 150 years have given rise to the consolidation of an urban mega-event model, loyally represented by the universal and international exhibitions or the Olympic Games. This same period has witnessed that model's crisis as it has lost some of its foundational characteristics, redefined its functionality and attributes, and shaped a new kind of scenario in its relationship with urban policies in general, and with the processes of the transformation of the city, in particular.

Traditional large events have lost most of their unique or inaugural nature in order to become urban moments. While they are clearly mobilising – in terms of economic energies, collective imagination, or political consensus – they have tended more and more to dissolve into a temporary urban nature characterised by the quasi continuous presence of events of all types and formats. This banalisation of the event has to do with its sectoral expansion and territorial multiplication but, above all, it is a result of the new role acquired by image in relation to the process of urban production. That, and nothing else, is what explains the simplification of the major event, progressively reduced until merely serving as a means to update the city's image.

In this context of change, it is worth questioning the future of large urban events and their relationship with the city. Lacking the epic aura associated with the construction of the city, on the one hand, and associated, on the other, with the process of production and updating of the city's brand image, can we understand them today as nothing more than a mere instrumental support tool for globalised urban branding? In that regard, and as a final discussion, I propose three hypotheses that may be validated by a simple look at the urban scenarios of the current period.

Firstly, the reduction of the contents of large events, which prioritise their commitment to the mobilisation of the local image and imagination over all other functions, will continue and will become even more evident. This is largely due to the fact that the process of the emergence and dissemination of events has by no means come to an end. It would almost appear that the multiplication of urban events now forms part of a new instruction manual on how to plan and inhabit the city, in such a way that we could imagine a future urban space almost continuously characterised by the attention and the interest demanded by one event or another. In fact, this would be a logical and expected step if we look at what has occurred with other equally defining and characteristic dynamics of the modern city, which now in the new *post-metropolitan* phase, in the words of Edward Soja, have lost their old exceptional nature and have acquired a quasi spatial-temporal permanence. This happens, in effect, with global tourism or the access, no less global, to information. It can be foreseen, therefore, that the same may happen with large urban events.

Secondly, although urban mega-events need a strong local consensus and their success is still measured in terms of the local response during and after their organisation, it is likely that the new century is inaugurating a new event model, characterised by a shift towards the progressively more important presence of a global public in cities. This is a confirmed fact which can be explained on the basis of two of the processes that most strongly feed the dynamics of globalisation: firstly, the strength of global tourism flows, which allows us to speak of 'major events tourism' – a specific type of tourism that, on occasion, is even positioned as the cornerstone for consolidating a concrete event in the urban calendar; secondly, transnational migrations, which are modifying the urban and social structure of many cities, at times even reformulating cultural categories which until recently were fenced in. This 'cosmopolitanism' ensures a type of population defined by a very clear awareness of the global world we live in. If this is the case, a fundamental change could be produced in the shape and nature of the large urban event, since, as an important image mobiliser, we could question the urban imagination that may be projected by major events in the future. This imagination could be closer to that anchored in the place, belonging to the inhabitants of the city, or to that of the *territoriantes* (Muñoz, 2002; 2008) – those populations that, occasionally but in an intense and almost continuous way, maintain the share quota of urban events with their loyalty. In this sense, the closing ceremony of the last Olympic Games in London in 2012 would confirm these perspectives if we look at the clearly global character – the emphasis placed on things like commercial pop music or digital technology – of much of the spectacle.

Thirdly, the growing ubiquity of urban events invites us to consider a final hypothesis that looks at the relationship between these events and urban policies. Thus, on the one hand, it seems obvious that there are clear difficulties for proposing new types of large urban events. The trilogy of consolidated events formed by international exhibitions, world fairs, and the Olympic Games, has been maintained throughout the last 150 years and still enjoys good health, which can be seen in the unceasing requests from many very different cities to host these events. Leaving aside these main mega-events, the truth is that only some sporting events – like the FIFA World Cup and cultural events – like the European capitals of culture or some major film festivals, for example – reach a truly global audience.

On the other hand, the perspective is exactly the opposite in the case of smaller events associated with specific elements of economic, political, social, and cultural activity. My hypothesis in this case is that practically any city in the world at any time could host an urban event, given that the only important element will be its capacity to mobilise and update the city's image and the urban imagination.

Urban Mega-events and the New Agenda of Cities

The three hypotheses coincide in highlighting the role of the urban mega-event as an instrument for managing the urban image in its different aspects. This is a function which brings the mega-event too close to the coordinates of *urbanalisation* as I have explained it here. Against this clearly reductionist and simplifying role and function of the major event, there are at least two challenges for the current city that could be visualised first and materialised later on the basis of the organisation of urban mega-events. Both challenges have to do with the necessity of updating the idea of urban regeneration.

The first is understanding that the transformation of an urban area is not an end in itself but a means to generate new dynamics in the city understood as a whole. Barcelona constitutes a very good example of how an Olympic mega-event not only changes the image of the city but also transforms a peripheral area both in its physiognomy and in its nature, thus equally changing its relationship with the rest of the city. Therefore, associating the organisation of urban mega-events with the transformation not only of isolated urban elements but to that of the city as a whole, including urban, social, and cultural issues, constitutes an important challenge for cities that are willing to organise mega-events in the immediate future.

The second is to accept that the idea of canonical urban regeneration developed since the 1980s in the urban policies of European cities needs urgent updating to introduce a series of questions that were not put on the urban agenda of cities at all 30 years ago. Perhaps the hottest issue is that of the new problems associated with the combination 'sustainability-technology' and, more specifically, the new requirements of urban resilience that the now recognised risk of climate change represents for more than a few cities. Thus, the merely environmental questions normally labelled with the adjective 'sustainable' are giving way to clearly innovative topics related to the resilience capacity of cities and to the urgent need to consider urbanism in terms of energy efficiency, low levels of carbon emissions, and criteria for adaptation to or mitigation of climate change. These are challenges that are now becoming popular on the basis of concepts like 'low carbon' or 'climate proof' urbanism.

The 2012 Olympic Games in London explored these questions connecting sustainability and resilience through the design of Olympic facilities that would reduce their dimensions and capacity after the games, adapting themselves thus in a resilient way to the post-Olympic moment. Also, the criteria of environmental sustainability were clearly present up to the point that their consideration explained in part the relatively low presence of iconic and spectacular architecture, so characteristic of the celebration of these types of global urban mega-events.

It is true that London is already a city with an elevated quota of unique and remarkable architecture, but it is also true that the global display that the Olympic Games represents always presents the temptation to resort to architecture-spectacle as one of the ingredients, if not the main one, of the urban transformation project. Far from this, the urban planning programme in the East End clearly showed an environmental aesthetic that, beyond technical issues like the building materials chosen or the energy efficiency and bioclimatic criteria implemented in the construction, characterised the formal appearance of the majority of the buildings and facilities related to the Olympics.

But the challenge represented by climate change for cities is one of much greater magnitude and, in this sense, perhaps Japan, with its Olympic Games in 2020 in Tokyo, could take advantage of the organisation of the Olympic mega-event to promote these issues. The fact that some of the risks of climate change will be more evident in that territory than in others, grants coherence to the idea that the Olympic Games in Tokyo in 2020 could be the first 'climate-proof' mega-event in history. That would represent a forceful qualitative step towards considering the Olympic mega-event not as the management of the city's urban image, but as something linked to current questions around the future that cities must contemplate in the years to come.

Without a doubt, the processes of *urbanalisation* are real and the current dependence of urban policies on the global requirements that the urban image represents is no less real. All of this creates patterns of inertia in terms of the ways in which we understand and guide urban mega-events, some of which have been explained here. At the same time, understanding that large-scale events can also be a catalyst for tackling new challenges in cities would allow us to guide future mega-events in terms of urban innovation and, above all, creativity. This way, advantage could be taken of the legacy of over a century of organising these types of urban events to provide content for a new urban agenda for cities in the 21st century.

References

Crilley, D. (1993). 'Architecture as advertising: Constructing the image of redevelopment', in Kearns, G. and C. Philo (eds.) (1993). *Selling Places. The city as cultural capital, past and present*. Oxford: Pergamon Press, pp. 231–252.

Darley, A. (2000). *Visual Digital Culture. Surface play and spectacle in new media genres*. London: Routledge.

Hannigan, J. (1998). *Fantasy City. Pleasure and profit in the postmodern metropolis*. London: Routledge.

Hessel, F. (1929). *In Berlin. Day and Night in 1929*. Readux books (1st German ed. *Spazieren in Berlin*. Munich: Rogner & Bernhard, 1929).

Klein, N. (2000). *No Logo. Taking Aim at the Brand Bullies*. Toronto: Knopf Canada.

Muñoz, F. (1997). 'Historic evolution and urban planning typology of Olympic Villages', in M. Moragas SPÀ, M. Llinés, B. Kidd (eds.) (1997). *Olympic Villages: A hundred years of urban planning and shared experiences*. Lausanne: International Olympic Committee/Museé Olimpique, pp. 27–51.

Muñoz, F. (2002). 'The multiplied city: Metropolis of *territoriants*', in Musco, F. (ed.) (2002). *City, Architecture, Landscape*. Venezia: Istituto Universitario di Architettura di Venezia, pp. 75–109.

Muñoz, F. (2005). 'Olympic Villages Urbanism, 1908–2012', in *Quaderns d'Arquitectura i Urbanisme*, num. 245. Escola Técnica Superior d'Arquitectura de Barcelona, ETSAB, Barcelona, pp. 110–131.

Muñoz, F. (2006). 'Olympic urbanism and Olympic villages: Planning strategies in Olympic host cities (London 1908 to London 2012)', in Horne, J., W. Manzenreiter, (eds.) (2006). *Sports Mega-Events: Social scientific analyses of a global phenomenon*. Oxford and Carlton: Blackwell Publishing, pp. 175–187.

Muñoz, F. (2008). *urBANALización: paisajes comunes, lugares globales*. Barcelona: Gustau Gili.

Muñoz, F. (2009). 'Urbanalisation: Common landscapes, global places'. *The Open Urban Studies Journal*, num. 2, pp.75–85.

Murzyn-Kupisz, M. (2009). 'Reclaiming memory or mass consumption? Dilemmas in rediscovering Jewish heritage of Krakow's Kazimierz', in Murzyn-Kupisz, M and J. Purchla (eds.) (2009). *Reclaiming Memory. Urban regeneration in the historic Jewish quarters in Central Europe*. Krakow: MCK, pp. 363–396.

Ploger, J. (2001). 'Millenium urbanism-discursive urbanism'. *European Urban and Regional Studies*, 8, pp. 63–72.

Rossi, A. (1984). *The Architecture of the City*. London UK and Cambridge MAS: MIT Press (1st Italian ed. *L'architettura della città*. Padova: Marsilio, 1966).

Soja, E. (2001). *Post-metropolis. Critical Studies of Cities and Regions*. Oxford: Blackwell.

Urry, J. (1995). *Consuming Places*. London: Routledge.

Ventury, M. (ed.) (1994). *Grandi eventi. La festivalizzazione della politica urbana*. Venice: Il Cardo.

Chapter 1
Olympic Legacy

Iain MacRury

Introduction

The IOC has lately produced a dedicated account of 'legacy' (IOC 2013). This is, in part, a response to the problem identified by both critics and advocates of 'legacy'. Highly frequent as the term has become in the Olympic city lexicon, it is hard to define, hard to pin down and hard to evaluate. Some excerpts from the IOC's *Olympic Legacy* document provide a useful flavour of their approach. The document begins with an imposing introduction:

> The Games are more than just an important sporting event. Aside from the dreams and achievements of young athletes, the Games provide a setting for champions to sow the seeds for future generations. They also enshrine the social responsibility of ensuring that the host cities bequeath a positive legacy. The IOC is firmly committed to guaranteeing that this legacy is as positive as it can possibly be. (Jacques Rogge, IOC President [IOC2013: 2])

The IOC's account highlights types of legacy in thematic areas, as well as distinguishing between 'tangible' and 'intangible' legacies:

- Sporting
- Social
- Environmental
- Urban
- Economic.

The IOC does not include, as it might have, specific headings for, for instance, political or cultural legacies, although these areas are sometimes identified as separate areas for action and review. Nor, as might be expected, does the IOC document explicitly mention the Paralympics in its template view of legacy, in part because, although many host cities place various Paralympic-related legacy ambitions in their bids and plans (e.g. Paralympic sports development and disability awareness), at the global level the Paralympics lies outside the IOC's working definition of its top level responsibilities. The Paralympic legacy is a matter for the IPC (International Paralympic Committee). The account includes a definition distinguishing 'tangible and intangible' legacies. Tangible Olympic legacies 'can include new sporting or transport infrastructure or urban regeneration and beautification which enhance a city's appeal and improve the living standards of local residents'(IOC 2014: 9).

On the other hand:

> Intangible legacies, while not as visible, are no less important. For instance an increased sense of national pride, new and enhanced workforce skills, a "feel good" spirit among the host country's population or the rediscovery of national culture and heritage and an increased environmental awareness and consciousness. (IOC 2014: 9)

This is a relatively developed discourse which reflects – in suitably high-level and generic-overview style – an intensive decade of thinking, reflection and activity around the concept of 'legacy'. It is only since

2003[1] that the following lines appeared in the IOC's Olympic Charter; an undertaking that the role of the IOC includes an expectation that it:

> takes measures to promote a positive legacy from the Olympic Games to the host city and the host country, including a reasonable control of the size and cost of the Olympic Games, and encourages the Organizing Committees of the Olympic Games (OCOGs), public authorities in the host country and the persons or organizations belonging to the Olympic Movement to act accordingly. (IOC 2003: 12)

Amongst the measures taken is a knowledge-based one. The Olympic Games Impact Study was instituted to capture data about areas of impact, and, also, legacy, across three broad areas:

- Economic
- Socio-cultural
- Environmental.

As the IOC describes:

> By this means the IOC will build up a powerful and accurate knowledge base of the tangible effects and legacy of the Games in turn this will enable the IOC to fulfil two of its principal objectives as enshrined in the Olympic Charter. (2009: 41)

The OGI study offers an important potential influence on how legacy is understood. The IOC provides a useful piece of guidance that is relevant for other aspects of the legacy discussion.

> The Olympic Games Impact study is not a projection of the potential impact of the games, as could be conducted i.e. by bid cities. The Olympic Games Impact study is an object of study measuring facts as they occur on the basis of predefined indicators, and analysing them. Impact should not be estimated in advance (there could be an impact/there could be an impact based on suppositions or experience) as only observation over time will show if an impact occurred, or not. (IOC 2009: 40)

This commitment to the long-term and to post-Games review is, in intention at least, a potentially significant element in the OGI since it offers one important set of longitudinal measures 'before, during and after' a Games across many 'legacy' areas. This stands as a potential counter to the more prolific legacy discourse – a discourse which, from 2003 to 2012 and, in the present, certainly in London, has been frequently mobilising 'legacy' as a projective and promissory note. The measurement and longitudinal analysis of legacy is an important component in its proper constitution. Host cities should take this responsibility seriously and as a long-term commitment.

The origins of Olympic 'legacy' lie in the IOC's attempts to reaffirm the value and values of Olympism in a period when its activities were attracting some negative scrutiny. Specifically, 'legacy' seeks to operationalize the notion of 'long-term investment' by host cities and governments – investment inspired by the Olympic Games and Olympic values. The centrality of 'legacy' in the bid process and candidate city's advocacy as they seek to become selected is a direct response to the IOC's decade-long project asserting 'legacy' as a core aim. Asserting 'legacy' was part of longer term strategic effort seeking to better ensure the evidence of, and belief in, the durability of the Olympic Games' global and city-based achievements and contributions.

'Legacy', as concept and practice, emerged as part of an attempt to respond to critique in the 1990s, following a series of scandals linked to ethical matters, commercialisation, 'gigantism', political vicissitudes and a history of boycotts (notably Moscow 1980), and in an era where an economic discourse, broadly neo-liberal (on a global scale) was driving vigorous financialisation and a cost-accounting approach into all government and state activity. There was anxiety about white elephant developments and environmentally damaging, low-utility new buildings – left (post-Games) to haunt host cities and standing as a negative

---

1   Sjoquist, D. (1997) is an early published instance of legacy being discussed – in relation to Atlanta.

contribution with regards to the global reputation of the IOC and of Olympism. The Olympics (at national and international levels) had to work on instituting both rhetoric and practices designed to legitimate the evolving scope and the financial scale of the Olympic mega-event.

But 'legacy' has even outstripped these large ends. The IOC's legacy concept captured the municipal imagination – attracting governments to jump on the rhetorical power of 'legacy' as they, themselves, sought to re-frame a set of challenges linked to the need to bind public and private investment, to appear 'lean' and to deliver value for money. This included the wish, on the part of some cities, to assert their value above and beyond the nation state (Sassen 1991). 'Legacy' is the language of mature, grown up, government. It was, no doubt, appealing to mayors and city leaders to be publicly engaged in enterprise for 'the long-term' (see Andranovich, G., M. Burbank et al. 2001).

The further and potent meaning, the currency and value of 'legacy', then, can also be found in the character of cities' desire for it. It is not just a concoction of the IOC. Olympic 'legacy' marks out, in the city and for the city, a conception and a conversation. The city is motivated or mobilised, if not exactly driven, by a hunger for and a grasping after different means, a change of character in the administration of investment, accumulation and sociality around the (anticipated) Games. The Olympics becomes an emblematic hope for the city future, occasioning promises and warnings; the Games and the 'legacy' become a territory in which to assert political programmes – programmes that might be inhibited within stagnating instituted structures and entrain spatial strategies. The legacy is a future to be colonised – and a site of contest over the definition of the city future.

The seeming enormity of the Olympics, and legacy infrastructure projects in particular, and consequent prominence in debate and in news media, is in part to do with the Games' planning providing a stage for some of these dramatic anxieties to play out. The global and historic prominence of the Olympics and the size of its media audiences provide another and different scale. Legacy discourse mixes limelight and scrutiny.

Metabolising Legacy: Containing the-Games-in-the-City

The 'hard', tangible 'proto-elements' of legacy (the village, the stadium, new infrastructure development and so on) alongside 'human factors' or intangible legacies (the energy, the emergent information environments, the networks, the promises and the projected initiatives around the mega-event); they arrive, all together resembling a package that is at once attractive and contentious. In prospect and in development the Olympic mega-event (as it were) enters the city well before it is fully planned and made. It is ingested 'whole', a complex object in the popular imaginary and in the city-plan. In practical activities and in projections, immediately the Olympic prospect stimulates and disturbs local and city-wide rhythms, places and structures. This complex object can be called 'the-Games-in-the-city'.

This in contrast to a counterfactual scenario: the everyday city, the ordinary un-Olympicised city, with horizons foreshortened by pragmatics and inertias, out of the limelight, away from historicity, global-network glamour and grand planning, the city seemingly idling, towards the short-term. For instance, such a city may have adopted a fairly informal, disorganised and distributed approach (creative or otherwise) to its present and future growth. It may, on the contrary and even at the same time, as a consequence of embedded structures (institutional and spatial) and in its establishments, have reached a point of actual or anticipated stagnation – and with a hard-to-articulate desire for a step change leading to small scale and frustrated localities sporadically chasing various means towards renewed growth. Arguably some, indeed much, of this was true to an extent of London in 2005, and, in a different way, of Rio and its regions.

On the other hand, consider the now-successful hosts/host-in-waiting, the bid document in hand, and preparing for hosting the Games: this poses (in the languages of necessity and inspiration) just the needed address to incipient disorganisation and stagnation – offering purpose, action and focus. 'The-Games-in-the-city' offers solutions to stagnation under the rubrics of the emergent host-Olympic-city-ideal and in related city branding imperatives, just as projects are taken up concretely and immediately in the form of urgent infrastructure development.

In its action, then, the Games and the connected planning (its emergent hard and soft infrastructures) quickly together constitute and become constituted in this relatively more highly organised 'object'; 'the-Games-in-the-city'. This object is complex and multiple but conspicuously identifiable. In turn it allows for the identification of a notional trajectory – 'a direction of travel' if not a consensus. Major spaces in the city are reformed, and quickly.

The city seems to realise then (or to imagine) that it is, and that it has been (for some time) hungry for such an object, to abate real and anticipated disorganisation/fragmentation, to overcome stagnation/frustration. This is evident in the passionate pursuit of the Games (bidding), and in the evident euphoria (popular and political) customarily entailed to the award of the hosting rights (in Singapore/London 2005 and Copenhagen/Rio 2009 respectively).

Nevertheless, after the Games it can be hard for a city to 'swallow' this complex, highly organised (and large) object, and harder still to adequately metabolise the Olympic project – to so quickly ingest and then convert and distribute its energy through the 'body' of the city (and not to mention the nation). The Olympic legacy (in all its artefacts) promises much, but it can sometimes not be 'taken in', not fully, ever. The fast-paced, force feeding (before and during the Games) of a diet of optimism and action may be difficult – unreal or 'incredible'. Negative legacies might be anticipated, or might actually accrue. The legacy of disappointment, failure, missed opportunity or 'waste' might cast a shadow in the wake of the limelight.

Metabolising 'Legacy': Practical Indices of the City in Transition

In this context 'legacy' potentially marks a new way for the city to grow and invites new ways to think about a city's thriving in relation to its futures. This is felt in public discussion and political discourse, but also in private sentiments regarding the apprehension of 'the- Games-in-the-city' – hopes for new jobs, fears of changes and of being side-lined. There is a realisation that the Olympic prize comes in a complex form, that it ties opportunity to responsibility – but provides little clue about how to manage the contradictory entailments of 'legacy'.

This work, this process of metabolising the Games-in-the-city is a helpful metaphor, but it can also be traced in more practical indices, e.g. of emergent connectivity, mobility, morbidity, health, employment and social cohesion or fragmentation echoing some classic approaches in urban sociology and, also, in a manner akin to a number of legacy assessment and evaluation projects, such as the OGI study, the DCMS meta-evaluation (DCMS 2013) or other studies carried out by city-wide or borough-level bodies. The change to the 'metabolism' of the city can be registered in the measure of changing rates and quantities in the throughput of investment and people: foreign direct investment, tourists, elite migrant workers (in service industries) and other indices tracking the emergence of and (further) opening out and exposure of the city to the flows (and ebbs) of global finance. The impact of the-Games-in-the-city on networks and on the way development happens is a further index.

Circulating 'Legacy': A 'Foreign' Term in Established Idioms of Accumulation and Disbursement

The particular language and heightened currency of 'legacy' in this context is widely noted (Gratton and Preuss 2008; Mangan 2008). However, 'legacy' is an unexpected word to have become so widespread in the contemporary contexts of urban development – especially in an era in which we have concurrently learnt (across contexts) to talk routinely about 'leveraging' and 'securitisation' and in which financialisation is made vivid across institutions of all kinds. It is unexpected that this era has produced an idiom and an attendant conceptual frame stimulated for a long period by boom and, more lately, by bust and austerity. Since 2003 in particular and under the auspices of the IOC's concerted action, this 'old' word, 'legacy' has become 'new'.

'Legacy' indexes different, competing, developmental narratives. Its etymology points us to the movement of (familial) objects through the generations. 'Legacy' connotes an atavistic sense of place, property and 'habitation' (Casey 1993), yet Olympic legacy is also future-oriented, with a meaning most often integrated with 'sustainability' in London's Olympic planning, while also tied to 'creative-destructive' (Berman 1983) visions of transformative change.

'Legacy' gains some of its power in the discourse of urban change because it hints at a set of arrangements other than those of exchange and the market – complementing or subverting familiar and powerful discourses of city-space-planning, speculation, development and 'return on investment'. 'Legacy' speaks of a narrative, of a development that is in part irrational and somewhat unpredictable. Legacy is not chosen; nor, then, is it exactly planned (e.g. in relation to ordinary municipal needs and wants – politically set down or otherwise socially registered). However, Olympic 'legacy' can be planned for – and should be. It is important to note, however, that planning legacy and planning for legacy are different things.

Legacy is appealing as an idea because many articulations of 'legacy' allude to an emergent settlement in the city that differs from the hyper-financial registers surrounding non-Olympic urban development ventures. However, though split off in our emotional apprehension of games, a festival or prize (as bonus or gift), the financial discourse does not disappear; far from it: the realisation of costs, the accounting-discourse returns to the Olympic discourse energised by feelings of guilt (at the indulgence of the city and of our part in this), or by anxieties about exclusion and exploitation.

Legacy: Concrete and Liquid Legacies

Expensive as it is, valuable as it might be, 'legacy' is not, simply, readily bought (or sold), not readily exchanged: it is conferred, passed on. In essence 'legacy' offers to resist and frustrates 'liquidity' both in the (pre-crunch) financial sense and as captured in Bauman's suggestive sociological metaphor (Bauman 2000). In its most traditional modes, legacy promises to build historic structures, bind communities to key values and assert long standing commitments to social and environmental ends. This is why it is hard for the financialised city to metabolise the-Games-in-the-city too readily. The contradictions lead to legacy as promise (pre-Games) and as compromise (post-Games).

To try and illustrate a rather abstract point we might look at the London 2012 stadium. The recent argument about the stadium in London neatly captures this dilemma – of liquidity and solidity in the after-Games city – the stadium in legacy mode:

- The West Ham-led bid, which would leave the stadium intact and in use, was more committed to the hard legacy and to preserving the iconicity of the Olympic venue. The stadium was to stand, indexing the Games-past. It required a public financial subsidy – partly from local government and the borough of Newham – to preserve the solid legacy proposition and to affirm a local-territorial community link; this notwithstanding the specific dissent of some of the West Ham football club fans, anxious about the capacity of the generic stadium to support a specific activity: football.
- The rival bid, spearheaded by Tottenham, a north London club, was seeking to (as it were) liquidate the legacy – to decompile and redistribute the legacy-object financially and geographically – replacing, displacing and dispersing the physical stadium (and its embedded opportunity) to Crystal Palace and rededicating the venue to football (only). The proposition as it stood posed financial fluidity against a more solid and emplaced version of legacy.
- With the support of LOCOG and a good deal of popular sentiment expressed in favour of the legacy bid, OPLC[2] and the Government supported the Newham and West Ham proposal. The logic of this debate has been and will continue to be reproduced in the series of further decisions to be made about other venues, places and projects entailed to the Games; notably in the settlement balancing commercial or community acquisition of the Village as housing stock.

Legacy shares much in the definition of inheritance – linking it back to the notion of heritage via the etymology of 'inheritance' and, as Olympic historian John MacAloon has pointed out, to the IOC's French translation of legacy – as 'heritage' (MacAloon 2009). As sociologist Pierre Bourdieu (1984; 2000) reminds us, as individuals and as social groups it is usual to both resist and seek out inheritances – struggling to inherit, but characteristically too, fighting against 'being inherited' by narratives connected to projected inheritance. 'Legacy' is a source and locus for (dramatic) anxiety and competition. In this context its genius is as much about inspiration as about stimulus.

2　Olympic Park Legacy Company; in 2012 this organisation became the London Legacy Development Corporation.

In this sense, 'legacy' is more about 'gift' than it is about 'commodity' (MacRury 2008; MacRury and Poynter 2009). 'Legacy' offers a supplemental idiom and energises promises at once disturbing and restorative in the practical narration of the city-future. Olympic 'legacy' demands the super-imposition of a long view within focus of the short-term, the vision framed by the historicity, experience, sociality and the glamour of the global Olympic movement. For instance, one reading of the legacy promises would see them as referring to a process, in time, of a fuller sense of (young) people inheriting the re-made spaces and places of their city – based in a securing and durable enhancement of the systems and opportunities attaching to and framing the experience of urban habitation. Critique of legacy plans and rhetoric often lies in either a disbelief in the reality of an inheritance and fear of the form and force of an inheritance that displaces or transforms an embedded community-identity/identification and which threatens to occlude (MacRury 2012: 155–8).

Promise and Compromise: Legacy, Gift and Commodity

'Legacy' debate and the complex of organisational and political practicalities of preparing for an Olympic legacy in the city together invite work akin to translation. Either the attempt is to neutralise the economic alterity of 'legacy-as-gift' (MacRury 2008; MacRury and Poynter 2009), re-absorbing it into a market-economic centred conception of the city, its governance and polity and providing summative accounts of (financial) or other tightly defined achievements; or, on the other hand 'legacy' (preserving the alterity it connotes) stands as an orientation point inviting into the city a focus for a politics and a governance attentive not only to 'market realities' emerging from exchange and investment in city-as-commodity space, but to realities entailed to other forms of human and social accumulation: reciprocity and redistribution for instance – emplaced and inhabiting city-as-community and/or polity (Polanyi 1968:148–9). Here legacy references social justice and community cohesion.

On the one hand 'legacy' is co-opted to support socially minded commitments (reciprocity and redistribution) such as to place-making, sustainable communities, environmental protection and equality of opportunities (e.g. in relation to disability) and connected achievements (Valera and Guardia 2002; García 2003; Shipway 2007; Smith 2007; Chappelet 2008; Girginov and Hills 2009; MacKenzie et al. 2008). On the other hand, 'legacy' becomes desirable to the city, as an amplification of and resource towards the market-conception in the urban realm (the dominance of exchange as an integrative mechanism): 'legacy' here is absorbed as a component in the urban-marketing planning concept (Gold and Gold 2008; Zhang and Zhao 2009) recorded in increased land values and tourist receipts, in FDI and in the city's global brand equity (however measured).

In a seminal account of 'legacy' Cashman (1998) points out:

> Given that the local community invests so much in the Games, it is important that the wider benefits of legacy should be canvassed and articulated. Too often, costs and benefits narrowly focus on economics. Legacy involves casting the gaze wider, to poetry and art, architecture, the environment, information, and many other non-tangible factors. (Cashman 1998:112)

His account reminds us of the multiple modes of action and productivity that go into producing the Games and which form the platform for any lasting legacy. It is important to govern the-Olympics-in-the-city with an eye on an acknowledgement that the Games are an extraordinary urban presence, one not entirely translatable into the ordinary flow of life or the forms of accounting utilized to plan and apprehend the city; nor, without effort, into extant institutional forms. As Rustin suggests, borrowing from Deleuze and Guattari, the social ecology of the mega-event is rhizomatic, its events being the efflorescences of energies which circulate between whiles through networks which function largely underground and out of sight (Rustin 2008)

Building the Legacy: Components and a Narrative

There are many accounts detailing the ways Olympics-related projects have changed the urban landscapes of past hosts. Building an athletes' village, constructing large stadia and other sporting venues in a short time scale typically makes for a high profile, sizeable and costly set of developments (Chappelet 1997; Preuss 2004; Liao and Pitts 2006; Muñoz, Horne et al. 2006; Baim 2009). Olympic projects open up productive reflection about the city and its direction as well as provoking controversy, scepticism and protest. As 2012 approached each of these were in evidence, with 'cost' and 'legacy' providing major terms of a series of seesaw debates and analyses.

Larger spatial and amenity-infrastructure developments usually underpin the Olympic-event developments. Such productivity brings proportionally larger improvements to the life of the city, but often they can claim – and deserve – only tangential and attenuated links to the Olympic Games event proper. Nevertheless, investments in sewage systems, airports, major road and rail projects and, beneath that, land remediation, have regularly been brought forward in urban plans or given renewed impetus by Olympics-driven planning schedules.

The City as Focus and Vector for the Event and the Legacy

The IOC operates as a global movement and engages in networks of national and regional scale (Chappelet 2008; Finlay 2008), but it is the city that remains at the centre of the IOC's definition of its mission and activities. This city focus is compatible with contemporary geographies of the postmodern 'global city' (Sassen 1991), with 21st century Olympism neatly aligning the entrepreneurial political philosophy that has come to inform contemporary urban governance (Harvey 1989; Hall and Hubbard 1996), with de Coubertin's original ancient Olympia-inspired vision of a time-space compressed modern Olympic Games (MacAloon 1981; Fensham 1994).

The Olympic charter stipulates that the 'duration of the competitions of the Olympic Games shall not exceed sixteen days' (IOC 2007:72) (notwithstanding the Paralympics), with bidding cities advised that the 'geographical area occupied by the sports installations required for the Olympics programme should be as compact as possible'. This stipulation is partly because the 'geographical situation may also be important for post-Olympic use'.

Olympic planning decisions and outcomes (the physical structure of legacy) hinge in part on commitments to adhere to the templates of a relatively compact city-based games, with the 'spatially compact' constituting an axiomatic principle weighed against arguments in favour of utilizing extant but more distant facilities and fostering region-wide regeneration. While logistical concerns provide some rationale for this *de jure* commitment to compression, there are further considerations at the root of the approach. This is part of the impracticability (in rational terms) of some realisations of the idea of the-Games-in-the-city. It's just too compressed; needlessly so, as critics might argue.[3]

Firstly there is a commitment to ancient and modern histories of the Games, and so to Olympic traditions, or mythic 'heritage'. Secondly there is a significant aesthetic-ceremonial element enshrined in the Games format – one that lends the Games a unique experiential character (MacAloon 1982) and which is thought to depend on the compact time/space format (IOC 2003). The character of the Games as a time-space compressed global-urban experience is an important factor in thinking about 'legacy', a term which, in this context, connotes longer timescales and wider distribution of energies and productivities – a phase imaginable as a kind of decompression in the aftermath of the excitement of the 'big bang' Games. This leads to two potentially divergent commitments – to 'impact' and to 'legacy':

1. To the integrity of the Games-event as a 16 day festival, i.e. a commitment to a rigorous disciplining in the temporal and spatial framing of the Games and the assurance of an intense event experience, a spectacular event staged by the city and a re-staging of the city. This heightens the ceremonial, aesthetic, experiential and affective character of the games.

---

3 For instance, the shooting might best have taken place at the UK's National Shooting Centre in Bisley, Surrey. LOCOG opted for a temporary solution at the Royal Artillery after the International Olympic Committee expressed concerns about the distance between London and Bisley.

2. To the assertion of a long narrative conception, a lasting legacy, with widespread transformational benefits for city, region and nation, the Games as bearer and agent in vivid narratives of place-remaking, the intent to form a wide ranging historical-narrative conception of the-Games-in-the-city as catalyst for wider achievements in the long-term.

The building and infrastructure developments are relatively fixed in the urban fabric. They are, initially relatively durable, as is the space recovered by remediation and available to 'legacy projects'. The intangible legacy elements, affective, aesthetic, ethical and attitudinal 'movements' around the Games, globally and locally are not so assuredly durable.

These are potentially major elements in the transformational legacy and quite as germane to the evolving qualities of city-space as remediated land and building developments. The place of intense ideas and experience attaching to the Games (in impactful experience, memory and affective life) and linked to other attached 'human factors' is a crucial intervening element in the consolidation of longer terms 'legacy' gains. The event matters – but it is neither an end in itself nor a means to an end. Nevertheless the success of the event itself is a key stage in London's urban-developmental narrative – towards building legacies and making places.

References

Ahlfeldt, G. and W. Maennig (2009). 'Arenas, arena architecture and the impact on location desirability: The case of "Olympic Arenas" in Prenzlauer Berg, Berlin'. *Urban Studies* (Sage) 46, pp. 1343–1362.

Almarcegui, L. (2009). *Guide to the Wastelands of the Lea Valley: 12 empty spaces await the London Olympics*. London: Barbican Art Gallery.

Andranovich, G., M. Burbank and C.Heying (2001). 'Olympic cities: Lessons learned from mega-event politics'. *Journal of Urban Affairs,* 23(2), pp. 113–131.

Atkinson, G., S. Mourato, S. Szymanski and E. Ozdemiroglu (2008). 'Are we willing to pay enough to "back the bid"?: Valuing the intangible impacts of London's bid to host the 2012 Summer Olympic Games'. *Urban Studies* (Sage), 45, pp. 419–444.

Baim, D. (2009). 'Olympic-driven urban development', in G. Poynter and I. MacRury (eds.) (2009). *Olympic Cities and the Remaking of East London*. Farnham: Ashgate.

Bauman, Z. (2000). *Liquid Modernity*. Cambridge: Polity Press.

Beriatos, E. and A. Gospodini (2004). '"Glocalising" urban landscapes: Athens and the 2004 Olympics'. *Cities*, 21(3), pp. 187–202.

Berman, M. (1983). *All That is Solid Melts into Air: The Experience of Modernity*. London: Verso.

Casey, E. S. (1993). *Getting Back into Place: Toward a Renewed Understanding of the Place-World*. Bloomington: Indiana University Press.

Cashman, R. (1998). 'Olympic legacy in an Olympic city: Monuments, museums and memory', in R.K. Barney, K. B. Wamsley, S. G. Martyn and G. H. MacDonald (eds.) (1998). *Global and Cultural Critique: Problematizing the Olympic games*. Fourth International Symposium for Olympic Research. London, Ontario: University of Western Ontario, pp. 107–114.

Chappelet, J-L. (2008). 'Olympic environmental concerns as a legacy of the Winter Games'. *International Journal of the History of Sport,* 25(14), pp. 1884–1902.

Chappelet, J. (1997). 'From Chamonix to Salt Lake City: Evolution of the Olympic village concept at the Winter Games', in De Moragas, M., M. Llines and B. Kidd (eds.) (1997). *Olympic Villages: Hundred years of urban planning and shared experiences*. Lausanne: International Olympic Committee, pp. 81–88.

Crompton, J. (2004). 'Beyond economic impact: An alternative rationale for the public subsidy of major league sports facilities'. *Journal of Sport Management*, 18(1), pp. 40–58.

Feddersen, A., W. Maennig and P. Zimmermann (2007). 'How to Win the Olympic Games: The empirics of key success factors of Olympic bids', Working Paper Series of the International Association of Sport Economists, Paper No. 07–05, April 2007, available online: http://college.holycross.edu/RePEc/spe/FeddersenMaennigZimmermann_OlympicBidding.pdf (accessed December 2014).

Fensham, R. (1994). 'Prime-time hyperspace: The Olympic city as spectacle', in K. Gibson and S. Watson (eds.) (1994) *Metropolis Now: Planning and the urban in contemporary Australia*. Leichhardt: Pluto Press.

García, B. (2003). 'Securing sustainable legacies through cultural programming in sporting events'. *Culture@ the Olympics*, 5(1), pp. 1–10.

Girginov, V. and L. Hills (2008). 'A sustainable sports legacy: Creating a link between the London Olympics and sports participation'. *International Journal of the History of Sport*, 25(14), pp. 2091–2116.

Gold, J. and M. Gold (2008). 'Olympic cities: Regeneration, city rebranding and changing urban agendas'. *Geography Compass*, 2(1), p. 300.

Gratton, C. and H. Preuss (2008). 'Maximizing Olympic impacts by building up legacies'. *International Journal of the History of Sport*, 25(14), pp. 1922–1938.

Gratton, C., S. Shibli and R. Coleman (2005). 'Sport and economic regeneration in cities'. *Urban Studies*, 42(5), pp. 985–999.

Hall, T. and P. Hubbard (1996). 'The entrepreneurial city: New urban politics, new urban geographies?' *Progress in Human Geography*, 20, pp. 153–174.

Harvey, D. (1989). 'From managerialism to entrepreneurialism: The transformation in urban governance in late capitalism'. *Geografiska Annaler. Series B, Human Geography*, 71(1), pp. 3–17.

Heinemann, K. (2002). 'The Olympic Games: Short-term economic impacts or long-term legacy?' in Moragas, de M., C. Kennett and N. Puig (eds.) (2003) *The Legacy of the Olympic Games: 1984–2000*. International Symposium, 14–16 November 2002. Lausanne: International Olympic Committee, pp. 181–194.

Hiller, H. (2006). 'Post-event outcomes and the post-modern turn: The Olympics and urban transformations'. *European Sport Management Quarterly*, 6(4), pp. 317–332.

Holden, M., J. MacKenzie and R. Van Wynsberghe (2008). 'Vancouver's promise of the world's first sustainable Olympic Games'. *Environment and Planning C: Government and Policy*, 26, pp. 882–905.

Hyde, L. (2006). *The Gift: How the Creative Spirit Transforms the World*. Edinburgh: Canongate Books.

IOC (2003). The Olympic Charter, IOC Lausanne.

IOC (2009). Technical Manual on Olympic Games Impact.

IOC (2013). Olympic Legacy, IOC Lausanne.

Liao, H. and A. Pitts (2006). 'A brief historical review of Olympic urbanization'. *International Journal of the History of Sport*, 23(7), pp. 1232–1252.

MacAloon, J. (1981). *This Great Symbol: Pierre de Coubertin and the Origins of the Modern Olympic Games*. Chicago: University of Chicago Press.

MacAloon, J. (1982). 'Double visions: Olympic Games and American culture'. *The Kenyon Review*, 4(1), pp. 98–112.

MacAloon, J. (2008). '"Legacy" as managerial/magical discourse in contemporary Olympic affairs'. *International Journal of the History of Sport*, 25(14), pp. 2060–2071.

MacRury, I. (2008). 'Re-thinking the legacy 2012: The Olympics as commodity and gift'. *Twenty-First Century Society*, 3(3), pp. 297–312.

MacRury, I. and G. Poynter (2008). 'The regeneration Games: Commodities, gifts and the economics of London 2012'. *International Journal of the History of Sport*, 25(14), pp. 2072–2090.

Maguire, J., S. Barnard, K. Butler and P. Golding (2008). 'Olympic legacies in the IOC's "Celebrate Humanity" campaign: Ancient or modern?' *International Journal of the History of Sport*, 25(14), pp. 2041–2059.

Mangan, J. A. (2008). 'Prologue: Guarantees of global goodwill: Post-Olympic legacies – Too many limping white elephants?' *International Journal of the History of Sport*, 25(14), pp. 1869–1883.

Moragas, de M., C. Kennett and N. Puig (eds.) (2003). *The Legacy of the Olympic Games: 1984–2000*. International Symposium, 14–16 November 2002. Lausanne: International Olympic Committee.

Muñoz, F. (2006). 'Olympic urbanism and Olympic Villages: Planning strategies in Olympic host cities, London 1908 to London 2012'. *The Sociological Review, Special Issue: Sociological Review Monograph Series: Sports Mega-Events: Social Scientific Analyses of a Global Phenomenon*, edited by J. Horne and W. Manzenreiter, Vol. 54(s2), pp. 175–187.

Owen, K. (2002). 'The Sydney 2000 Olympics and urban entrepreneurialism: Local variations in urban governance'. *Australian Geographical Studies*, 40(3), pp. 323–336.

Preuss, H. (2004). *The Economics of Staging the Olympics*. Cheltenham: Edward Elgar.

Preuss, H. (2007). 'The conceptualisation and measurement of mega sport event legacies'. *Journal of Sport and Tourism*, 12(3), pp. 207–228.

Raco, M. (2004). 'Whose gold rush? The social legacy of a London Olympics', in Vigor, A., M. Mean and C. Tims (eds.) (2004) *After the Gold Rush: A sustainable Olympics for London*. London: Institute for Public Policy Research.

Sassen, S. (1991). *The Global City: New York, London, Tokyo*. Princeton, New Jersey: Princeton University Press.

Shipway, R. (2007). 'Sustainable legacies for the 2012 Olympic Games'. *The Journal of the Royal Society for the Promotion of Health*, 127(3), pp. 119.

Sjoquist, D. (1997). *The Olympic Legacy: Building on what was achieved*. Atlanta, GA: Research Atlanta.

Smith, A. (2007). 'Large-scale events and sustainable urban regeneration: Key principles for host cities'. *Journal of Urban Regeneration and Renewal*, 1(2), pp. 178–190.

Valera, S. and J. Guardia (2002). 'Urban social identity and sustainability: Barcelona's Olympic village'. *Environment and Behavior*, 34(1), pp. 54–66.

Zhang, L. and S. X. Zhao (2009). 'City branding and the Olympic effect: A case study of Beijing'. *Cities*, 26, pp. 245–254.

Chapter 2
How Do We Measure Legacy?

Allan J. Brimicombe

Introduction

This chapter is written in the context of the London 2012 Olympic and Paralympic Games. The author has led four major projects focused on measuring legacy: three for the International Olympic Committee (IOC) and one for the London Organising Committee of the Olympic and Paralympic Games (LOCOG). In addition, the author was appointed as Specialist Advisor to the House of Lords Select Committee on Olympic and Paralympic Legacy[1] which sought to establish whether there will be enduring regeneration and sporting legacies arising from the 2012 Games. Legacy for London 2012 has far from played itself out and whilst London 2012 can be viewed as a continuing narrative, the approaches and knowledge production that have arisen from the Games will continue to have value and currency for other cities attempting transformation through planned mega-events.

In the first chapter in this section, Iain MacRury discusses the question: 'What is Olympic legacy?'. In this chapter, legacy is taken simply to be *any net impact arising from a mega-event*. The term 'impact' refers to any change or transformation, for better or for worse, that has taken place and which is attributable to the mega-event – in other words the link, direct or indirect, to the mega-event needs to be mapped. However, the key term in the definition is 'net', that is, the impact that has occurred over and above what would have happened without the mega-event. City mega-events are rarely context-free or designed on a *tabula rasa*; rather they are superimposed on existing trajectories of historical development. Establishing a plausible counterfactual to measure net impact against is, therefore, critical to knowing what the true legacy of a mega-event is. We will therefore return to this theme later in the chapter.

The structure of this chapter is as follows. The next section discusses the growing imperative to measure legacy in the face of burgeoning mega-event costs and what we mean by 'measure'. We then move on to consider the implications of mega-event induced complexity and legacy time horizons for measuring net impacts. There then follows a critical analysis of some of the studies that have taken place in relation to London 2012 that have tried to measure legacy both in an absolute sense and in comparison to other Games cities. Finally we draw conclusions that will inform future mega-event cities and their aspirations for legacy.

Why 'Measure' Legacy?

This answer to this question may seem obvious, but it has not always been so. The first Summer Olympic Games were awarded to Athens in 1894 with the Games taking place in 1896. This initial two year lead-in to a Games has varied down the years, the longest being nine years for the 1932 Los Angeles Games. Since Sydney was awarded the 2000 Games in 1993 the pattern has settled to a seven year lead-in. The increasing length of time given over to the preparation of the Games post-World War II has reflected the growing size of the Games (exceeding 10,000 athletes in all Summer Games since Atlanta 1996) and the complexity of the facilities and infrastructure required for a Summer Games. The Olympics after all have been dubbed 'the greatest show on Earth'.[2]

---

1   http://www.parliament.uk/business/committees/committees-a-z/lords-select/olympic-paralympic-legacy
2   For example, http://www.telegraph.co.uk/sport/olympics/10187826/Boris-Johnson-on-the-London-Olympics-the-greatest-show-on-earth.html

The right to host the Games has, since 1904, been on a competitive basis and has increasingly relied on unique selling propositions that go beyond the sporting event to being economically and socially transformational, particularly since the 1992 Barcelona Games. These aspirations are expensive, even more so when individual cities (rather than groups of cities or whole countries) must compete and bear the financial burden. In addition, costs around security have escalated since 9/11 and since the 2004 Athens Games have been ten times greater (more than £1bn for London 2012) than Summer Games prior to 9/11 (based on data in Coaffee and Fussey, 2011). Given the cost attached to the prestige of the Games, it is only to be expected that from the very beginning of the modern Olympics there has been an on-going debate about whether or not money should be spent on the Games as a prestige project, as against other competing needs. The debate is not just about the building of new, often iconic, venues but other infrastructure such as transport and hotels, and whether the Games should be a catalyst for change. A turning point in this debate was the 1976 Montreal Games which had a budgetary shortfall equivalent to £2.2bn at 2012 prices. This proved catastrophic for the city's economy. More recently the 2000 Sydney Games have been shown to have made a net loss in GDP of £1.8bn at 2012 prices and that the *ex ante* valuations were over optimistic about the economic stimulus expected from the Games (Giesecke and Madden, 2011). As for the 2004 Athens Games, Greece was one of the smallest economies to have hosted the Games. There is no doubt that the scale of expenditure required, particularly for infrastructure and security, strained the economy and might be argued to have materially contributed to Greece's current economic problems. Whilst direct comparison between host city budgets is difficult because of differences in the extent to which infrastructure and other projects are accounted for within a Games budget, the trend has been for increased spending in real terms, with an increasing proportion of public funding. Thus over the last few decades the Games have evolved from the business-centred approach of the 1984 Los Angeles Games in which public spending was only 10 per cent of the budget, to the transformational legacy approaches, including that of London 2012 and Rio 2016, in which public spending predominates (see Poynter, 2009).

The International Olympic Committee (IOC) has been sensitive to criticisms around the commercialisation of the Olympic brand through media rights and sponsorship deals on the one hand and burgeoning public spending on the other. The response has been to progressively modify their mission statement in the Olympic Charter.[3] Thus, the 1991 revision codified responsible concern for the environment in the preparation for and running of the Games, to which was added in 2003 the concept of *sustainability* (some two decades after the Rio Declaration). Furthermore, specific mention of *legacy* was also introduced in the 2003 revision. The relevant parts of the IOC mission statement currently read:

> 13. to encourage and support a responsible concern for environmental issues, to promote sustainable development in sport and to require that the Olympic Games are held accordingly;

> 14. to promote a positive legacy from the Olympic Games to the host cities and host countries;

> (IOC, 2011:15)

This can be interpreted as a *post hoc* rationalisation to include within 'Olympism' some justification for the huge sums being spent and a belated formal recognition that transformational legacies have been overt aspects of bids to host the Games since the 1960s. Thus the 2016 Candidate questionnaire specifically asks for details on Olympic legacy initiatives and how they are linked to the long-term planning and objectives of the bidding city. The questionnaire also requests details on how these initiatives will be supported, financed, monitored and measured by all relevant stakeholders prior to, during and post-Games. Put simply: is there a legacy plan, can you afford to carry it through and how will you measure and evaluate both the event and its legacy? Measuring the Olympic legacy has become mandatory.

---

3   Available at: http://www.olympic.org/olympic-charters?tab=the-charter-through-time

In a positivist sense, the term 'measure' is usually associated with quantitatively assigning numbers to things in an objective way by following some valid, systematic approach and/or by reference to some standard. To be effective, these things should be both tangible and countable. In the context of legacy this leads to approaches based on quantitative indicators, whether from official statistics or surveys. But there are also intangible aspects of legacy such as whether or not one has been inspired by the Games. These require ascertaining people's subjective opinions. Thus when dealing with legacy, measurement needs to be viewed from a broader sense of reliably detecting and evidencing that some change has occurred. Social research textbooks (e.g. May, 2011; Robson, 2011) are replete with methods, from the use of official statistics, through questionnaires and interviews, to documentary research. All of these become applicable in the case of evidencing the extent to which planned legacy outcomes have occurred and equally importantly for detecting unplanned legacy outcomes.

Complexity and Time Horizons

Legacy aspirations are often stated simply for public, non-technical consumption and tend to gloss over their underlying complexity and multi-dimensionality. Thus in the case of London 2012 legacy promises, exactly what is meant by 'to transform the heart of East London' (DCMS, 2008) in measurable legacy terms? For a start 'East London' does not have any standardised (administrative) geographical definition and so where is its 'heart' and furthermore, what is intended to be transformed over what time horizon? For political reasons legacy promises are often left vague so it is easier for politicians and administrators to say this or that aspect was a success (cherry-picking legacy), or if things are not going to plan to say that it is too soon. But if we are to measure legacy, then these legacy aspirations need to be translated and decomposed into measurable components of outcomes for defined geographical extents and for defined time horizons. Each legacy thus becomes a series of thematic space-time cubes and invokes Sinton's conceptualisation (Sinton, 1978) in which regardless of the theme one dimension of the cube is fixed, another is controlled and the third is measured, as illustrated in Figure 2.1.

Of course the cross-sectional cubes can become time series data by taking a series of snap shots – a piling up of temporal layers. And there is a point here: for measuring legacy, time series data are required; we need to know trends even if it is just between two fixed points, one at the beginning and one at the end. Furthermore, the evidencing of a trend needs a baseline – a starting point. A common misconception is that the measurement of legacy only occurs after the mega-event is over and the 'show has left town'. Because the very act of preparing for a bid for some mega-event may have already set in place changes that anticipate some legacy, ideally the baseline needs to predate the bid preparation. Thus the IOC stipulates for the Olympic Games Impact (OGI) studies that the baseline must be for two years prior to being awarded the Games. In the case of London 2012, the base year is 2003, the Games having been awarded in 2005. This leads to a paradox: until the legacy aspirations have been articulated and agreed, how can the necessary baseline data collection have already taken place? So there is a certain reliance on accessible data infrastructures already being in place capable of supplying a range of statistics from which a retrospective baseline can be established. The term 'accessible' is purposefully used. In some countries official statistics are still regarded as state secrets and from the point of view of a transparent process of measuring legacy, either they might as well for all intents and purposes not exist, or where releases are made, they are unverifiable and of uncertain reliability. Western countries and organisations such the United Nations and the International Monetary Fund are espousing greater transparency and accountability through open data policies. This has seen, for example, a dramatic increase in the number of freely available, on-line releases of official statistics in the UK since 2008 (Figure 2.2). This increases the likelihood that secondary data for measuring legacy are already available and limits the need for primary data collection to be carried out by researchers.

Cities rather than countries host the Olympic Games and whilst the impact of such a mega-event may be (purposefully) uneven across the city, its impact may extend well beyond the city itself. In Figure 2.1, location features strongly because legacies happen in specific places – they are geographically rooted – but

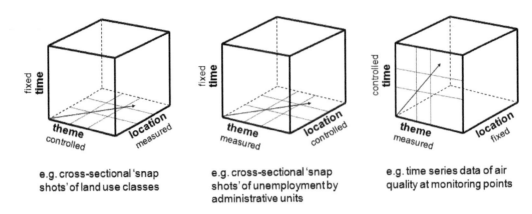

Figure 2.1   Sinton's conceptualisation of the data cube
Source: Adapted from Brimicombe, 2010.

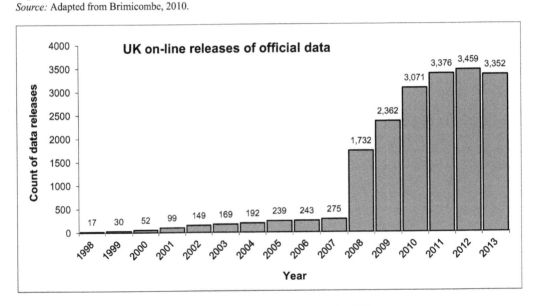

Figure 2.2   Annual number of releases of on-line data sets in the UK
Source: Based on an analysis of the UK National Statistics data release hub.

nevertheless can manifest themselves across a number of scales. The OGI studies now required by the IOC stipulate three scales of analysis: city, region and country; and of course these terms are themselves open to interpretation. Thus, in the case of London 2012 the 'city' is deemed to be the London boroughs hosting the Games venues (the Host Boroughs; initially five boroughs but extended to six in 2011),[4] the 'region' is London as a whole and the country is preferably the United Kingdom but could be Great Britain, England and Wales or England depending on availability of compatible data amongst the devolved administrations that currently constitute the United Kingdom.

Returning to the London 2102 legacy promise 'to transform the heart of East London', in order to measure the emergence of such a legacy, it needs to be decomposed into measurable units that are then evidenced and monitored from the baseline year until whatever time horizon is deemed should have elapsed for the legacy to

---

4   London Boroughs of Newham, Hackney, Tower Hamlets, Waltham Forest and Greenwich, to which was added Barking and Dagenham.

have been delivered. The six Host Boroughs (now re-branded as the 'Growth Boroughs') have, for example, interpreted this legacy aspiration as ' … a 20 year target to ensure that by 2030 Growth Borough residents will have the same social and economic chances as their neighbours across London' (Growth Boroughs, 2013), a strategy known simply as 'Convergence'. Convergence is decomposed into themes, key measures and indicators, as shown in Table 2.1. These are more complex than just tracking change year-on-year; they are supposed to converge on the same measures for London as a whole over the 20 year period, and are a moving target since the rest of London will also be changing. Thus, although over this period median earnings in the Growth Boroughs might be made to increase through a range of interventions, median earnings might rise even faster in the rest of London (as has been happening), resulting in divergence rather than convergence.

The above example, which uses quantitative indicators, makes a number of important assumptions. The first is that the collection of indicators will capture the movement towards (or away from) the legacy aspiration. The second is that the underlying data sets are available or can tractably be compiled. The third is that the data series can remain consistent over the necessary duration of time. These are fundamentally important for measuring legacy outcomes. Data series can be disrupted or discontinued due to policy changes, changes in definition of key variables and/or categories and changes in the data collection methodology. This can make earlier data incompatible with later data, such that the true effect over time becomes difficult to discern. Two examples can be made from Table 2.1 – violent crime and families receiving key benefits.

Table 2.1   Convergence Indicators

Creating Wealth Reducing Poverty	Support Healthier Lifestyles	Developing Successful Neighbourhoods
Key Measure	Key Measure	
Economically active people in employment	Life expectancy • male • female	
Indicators	Indicators	Indicators
Employment rate	Children achieving good level of development at age 5	Violent crime levels
Median earnings, full time workers living in the area	Obesity levels in school children in Year 6	Economic growth • income • jobs
Proportion of children in families receiving key benefits	Mortality rates from all circulatory diseases at ages under 75	Overcrowded households
Pupils achieving at least Level 4 in English and Maths, Key Stage 2	Mortality rates from all cancers at ages under 75	
Pupils achieving 5 GCSE grades A*–C	Recommended adult activity (3 x 30 mins per week)	
19 year olds achieving Level 2	Pupils participating in Physical Education and School Sport	
19 year olds achieving Level 3	No sport or activity	Target
Working age population with no qualifications		Additional housing units • total planned • affordable delivered
Working age population qualified to at least Level 4		

Source: Based on Appendix A in Growth Boroughs, 2013.

Crime statistics are collected by the Home Office from the 43 independent police forces in England and Wales according to the Notifiable Offences List (NOL) – mostly those offences tried at a crown court. The list does change as new offences are added in line with new legislation or in attempts to simplify it[5] – and these in themselves can cause disruption of the series. Classification and recording of these offences is governed by the National Crime Recording Standards (NCRS) and the Home Office Counting Rules (HOCR). Revisions of these cause major discontinuities in the data series such that statistics on violent crime up to 1997 are incompatible with the period 1998/99 and again with the period from 2002/03 onwards, as illustrated in Figure 2.3. Whilst NCRS and HOCR are intended to standardise classification and recording across the 43 forces, they can lack consistency even between command units within forces. Thus there could be a lack of consistency between London boroughs on how certain crimes are recorded, and we have seen documented recently the deleterious effect of a targets culture on police recorded crime (PASC, 2014). 'Violent crime' is itself not a single statutory category but a composite of a number of offences including wounding, assault with minor injury, assault without injury and robbery. A local policy to crack down on, say, robbery can lead to an increase in the number of cases being recorded and is a perverse consequence of police action. Audits of crime figures can also lead to changes in the statistics as officers pay more attention to the correct classification and counting of crimes. Thus what may seem to be a simple indicator – violent crime levels – is not so simple to interpret year on year and to do so requires contextual knowledge as well as the statistics.

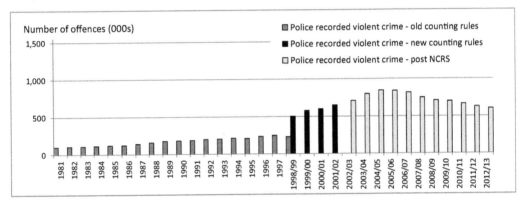

Figure 2.3   Discontinuities in the national trend of violent crime
Source: Adapted from Figure 1.2 in ONS, 2014.

With regard to an indicator monitoring the proportion of children in families receiving key benefits, a change in government policy towards benefits payments has seen the gradual phasing-in of the 'universal credit' from April 2013. This is meant to replace six key benefits from jobseeker's allowance to housing benefit. Not only does this means that there is to be a major discontinuity in the benefits statistics, but that from 2013 to 2017, when all claimants will have been moved to the universal credit, the benefits statistics will be a mixture of old and new systems, which is likely to result in considerable confusion with regard to how to interpret the statistics.

A confounder is an extraneous factor, that is, one which is not directly part of a study but which correlates with the phenomenon of interest and thus affects its behaviour. An example would be a heightened national feel-good factor detected after the London 2012 Games, which in actuality might be influenced by people's reaction to the economy returning to strong growth after a number of years of recession and austerity. In the face of confounders one has to be careful about how one draws conclusions. Returning to the case of the violent crime indicator discussed above, we reliably know from the Crime Survey for England and Wales that

---

5   The Crime Statistics Advisory Committee from time to time makes recommendations on simplifications; see http://www.statisticsauthority.gov.uk/national-statistician/ns-reports--reviews-and-guidance/national-statistician-s-advisory-committees/crime-statistics-advisory-committee-correspondence.html

violent crime has fallen by 50 per cent since 1995 (ONS, 2014) and mirrors crime decline in general across many Western countries (Zimring, 2007). It is a phenomenon that is not yet fully explained by criminologists. Thus, falls in violent crime that have been evidenced in the Growth Boroughs may have very little to do with the London 2012 Games but rather arise as a consequence of long term changes in society and the life experiences of adolescents.

A more subtle effect, of which legacy researchers need to be aware, is the potential for bias in secondary data and the reporting of legacy effects. Whilst not strictly a confounder, it can have similar effects. Take the following statement from Legacy Trust UK:

> But what do young people really think of the [London 2012] Games and what impact did it have on them? These are some of the questions that Legacy Trust UK wanted to answer whilst giving young people a voice in what they think should happen next. To do this, we commissioned specialist research agency Nielsen to undertake an independent piece of research.
>
> (Legacy Trust UK, 2013)

Legacy Trust UK is sponsored by the Department for Culture, Media and Sport, the lead government department for ensuring that the London 2012 Games are successful and that legacy aspirations are achieved. If we then go to the Nielsen web pages we find: 'The Nielsen Company is the Official Market Research Services provider to the London 2012 Olympic Games and Paralympic Games. We are a Tier Three provider and were the 20th domestic commercial sponsor announced'.[6] In other words the research agency used for the survey also seems to have a vested interest in the Games. This is not to say that the results of this survey are not valid, but that legacy researchers need a guarded, informed approach when using such evidence of a legacy.

In order to understand the net effect of a legacy (so as not to overstate it), it is important to establish what the counterfactual is. The counterfactual represents what would have happened anyway without the mega-event. One example in relation to the London 2012 Games is the upgrading of the public transport infrastructure, mostly in London, through a £6.5bn investment. Most of these works were planned anyway and would have taken place, albeit over a longer time frame, even if London had not won the Games. The net effect was to accelerate the investment so that the infrastructure was available somewhere in the region of 20 years earlier than otherwise with an immediate economic stimulus. Another example of the counterfactual is in sports participation. Evidence from Sport England (see House of Lords, 2013) indicates that between 2005 (when the Games were awarded) and 2011/12, 1.6 million more people were engaged in sport. This seems a laudable result, but it represents the *gross* outcome. Over the same period, population in England increased by 2.5 million which, with a participation rate of 36 per cent, means that population growth alone would account for about 0.9 million more people engaged in sport, leaving a *net* increase in 0.6 million and not much to show for the £450m spent on increasing sports participation.

Approaches to the London 2012 Games

The previous sections have focused on definitions, principles and pitfalls of measuring legacy. Although these have been illustrated by reference to London 2012, this section of the chapter looks more closely at how the legacy of the London 2012 Games is being measured to date. At the time of writing the author has on his hard disk nearly 200 reports and papers on all aspects of the (mostly) anticipated and (some) measured legacy of the Games. It is early days yet for all but the most immediate of legacies arising from the Games to have emerged and, given the lag of 18 months in official statistics, to have been reliably measured. Nevertheless, this volume of reports is symptomatic of the government's preoccupation with legacy and a deep-felt need to show that the Games were worthwhile and beneficial (profitable) for the whole country, not just London. This is part of a trend of democratically accountable governments increasingly seeking to justify their policies, actions and public sector spending by evidencing their merit, quality and efficacy. Thus, formal evaluation of

---

6   http://www.nielsen.com/uk/en/insights/press-room/2011-news/nielsen-the-intelligence-behind-london-2012.html

policy interventions and their outcomes have been an integral part of achieving transparency in accountability. The UK Government has a number of documents giving guidance for evaluation, impact assessment and cost-benefit analysis. Key amongst these are:

- *The Green Book* (HMT, 2003): This book provides a framework for the appraisal and evaluation of government policy and projects. The difference between 'appraisal' and 'evaluation' is in the timing. Appraisal is an assessment as to whether a proposed project or policy intervention is worthwhile and most commonly takes the form of a cost-benefit analysis. Evaluation takes place post-implementation to see to what extent the objectives of the project have been achieved and what lessons might be learnt. A supplement to the Green Book has been issued (HMT and DWP, 2011) to cover techniques in social cost-benefit analysis. An example of cost-benefit analysis for London 2012 is one carried out at the point of deciding whether or not London should make a bid for the Games (Arup, 2002). On an estimated expenditure of £1.9bn the income ranged from £1.65bn to £1.98bn, that is, from an 8 per cent loss to a 4 per cent surplus. Much of this variation arose from uncertainty in the tourism income which in many Olympic cost-benefit analyses is viewed as the key wider economic benefit of staging the Games. The decision, as we know, was to go ahead with a successful bid. But by 2005 the estimate for public sector funding had risen to £2.4bn and by 2007 to £9.3bn (Berman, 2010), primarily because the anticipated private sector involvement in the construction did not materialise. Giesecke and Madden (2011) conclude that many such *ex ante* cost-benefit valuations are over optimistic of the economic stimulus that can be expected from the Olympics.
- *The Orange Book* (HMT, 2004): This book provides a model of risk management. Any mega-event has substantial risks (financial, reputational and otherwise). If these risks are identified and monetarised at an early stage, then sufficient mitigation and financial contingency can be put in place to cover the risks. In 2007 the National Audit Office (NAO) carried out a risk assessment of the preparation phase for the London 2012 Games (NAO, 2007) and identified six areas of risk that needed to be managed. Prime amongst these was delivery against an immovable deadline; also included was planning for a lasting legacy. By 2010 a contingency of £2.2bn was in place to cover risks. These were of three types: programme contingency (£0.97bn) for the construction of the Olympic Park on a constricted site to a fixed deadline; funders' contingency (£1bn) for changes in scope and wider economic conditions; and a security contingency (£0.24bn). The post-Games audit (NAO, 2012) showed that not all the risks materialised and that an under spend of £0.38bn on the contingency was achieved. The same audit also looked at the immediate legacy benefits, which were viewed principally from the point of view of job creation (177,000 job years of employment in the construction period between 2007 and 2012; 34,500 people in Games-related employment), progress on planned legacy use of the venues and problems of governance and coordination of legacy delivery: '... . it remains the case that numerous individual organisations are delivering aspects of the legacy and that coordination of this activity remains a challenge' (NAO, 2012).
- *The Magenta Book* (HMT, 2011): This book provides further guidance on programme evaluation and complements the Green Book. The key focus is to identify 'what works', highlight good practice, identify any unintended consequences or unanticipated results, and identify value for money that can be used to improve future decision-making. 'Not evaluating, or evaluating poorly, will mean that policy makers will not be able to provide meaningful evidence in support of any claims they might wish to make about a policy's effectiveness. Any such claims will be effectively unfounded'. Meta-evaluation, the synthesis of separate smaller evaluations, is also covered in this volume, and is further discussed below in relation to London 2012.

From the preceding discussion it is clear that measuring legacy can take a number of forms – valuation, evaluation, meta-evaluation, audit – and whilst they tend to be quantitative, they can nevertheless be qualitative or a mixture of both. There is also a tendency to include historical analogues, that is, to use previous mega-events as a benchmark or as a comparator to gauge the progress or likelihood of any legacy emerging. One such study was carried out by researchers from the University of Westminster for the Royal Institution of Charted Surveyors (RICS, 2011). This study used six case studies of previous events, including the Barcelona 1992 and Sydney 2000 Olympic Games, the 1998 FIFA World Cup, and the Manchester 2002 and Melbourne 2006 Commonwealth Games, from which to draw conclusions about the likelihood of a regeneration effect from the London 2012 Games. A scorecard approach was used whereby, having identified from the case studies 15 criteria of good practice, London 2012 was rated and scored 165 out of a possible total of 200, or 82.5 per cent. What becomes clear is that for cities which plan for legacy from the moment they start planning for a mega-event, it is more likely that the legacy will come to fruition. To which we would add that the planning for legacy, rather than leaving it to chance or serendipity, ought to be accompanied by decisions at the earliest stages on how to measure legacy and to put in place the necessary data collection of time series from inception. All too often this is an afterthought. The counterfactual also needs to be assessed and documented early.

Of the quantitative indicator-based approaches, we have already discussed their use in the Convergence strategy of the London 2012 Growth Boroughs. A relatively small number of indicators are used to focus on one area of legacy: the promise to transform the heart of east London. Even so, the focus of the London 2012 Games on east London came after decades of attempts at regeneration; deprivation in east London has proved a very stubborn problem. Thus the decision to build the venues and Athletes' Village in east London as a solution to such a long-standing problem was born of either desperation or inspiration. Nevertheless, with regard to Convergence, London 2012 was intended to (re)kick-start the wider regeneration of the area with subsequent local policies and interventions aimed at sustaining the long-term outcome. Here again it is likely to be difficult to separate the legacy effect of the Games from the preceding attempts at regeneration and subsequent local policy changes and additional interventions to assure the legacy.

The IOC naturally has a much broader interest in legacy than just regeneration and has mandated, through its Host City Contracts, for Olympic Games Impact (OGI) studies and, as already discussed above, the baseline is two years prior to being awarded the Games. OGI studies need to be carried out independently of the local organising committee and the research may therefore be carried out by consultants or a university (or a consortium of both). For London 2012, the main phases of the OGI study were carried out by the University of East London (UEL), funded by the Economic and Social Research Council (ESRC).[7] At the time of writing two reports (pre-Games and Games-time) and all underlying data have been placed online at http://www.uel.ac.uk/geo-information, with a third, post-Games report due to be released by the end of 2015. The purpose of these studies is to advance the sustainability of the Games, help promote legacy and create a knowledge base that can be drawn upon by future host cities. Thus the London 2012 OGI study has already fed into Rio 2016 and Pyeong Chang 2018. OGI studies are based on an IOC Technical Manual which, despite its clear importance, is always subject to non-disclosure agreements, even though the outputs of the study are openly available. The content of the Technical Manual therefore cannot be discussed in detail here. The 2007 revision of the manual which formed the basis for the London 2012 OGI study provides the specification for the production of standardised data on 120 possible indicators in the environmental, social and economic spheres. Some of these indicators, such as So9 Health, are themselves baskets of indicators capturing many dimensions. Not all 120 indicators are expected to be reported on for all Host Cities, but an appropriate selection is negotiated between the IOC and the local organising committee and stipulated in the research contract.

Figure 2.4 gives a sample indicator from the Games-time report (UEL, 2013) – Ec26 Public Debt. The first page provides the narrative and analysis whilst the second (and any subsequent) pages are for tables, graphs and diagrams. These latter pages reflect what the team perceives to be the main story and do not necessarily show all the individual spreadsheets that underlie the indicator. Looking in detail at the first page, immediately under the title line is an indication of the scale at which this indicator is relevant. In this case it is *Country (United Kingdom)*, reflecting the availability of these statistics in relation to devolved administrations, which sometimes restricts the coverage of 'national' statistics. An indicator can be required to show coverage at 'City', 'Region' and 'Country' scales, requiring a spatial hierarchy of statistical tables to be analysed. The second section on *Data Issues* describes what the indicator is intended to measure, any pertinent technical definitions, data sources and any quality issues such as any breaks or discontinuities in the time series. It is not always possible to match intended definitions in the Technical Manual (in order to

---

7   http://www.esrc.ac.uk/news-and-events/features-casestudies/features/30529/olympic-impact.aspx

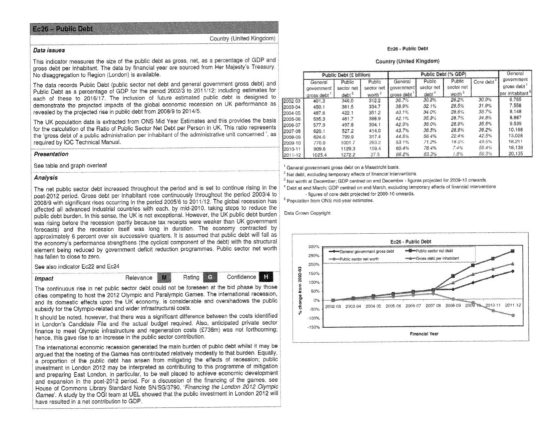

Figure 2.4   Example of OGI study output; Ec26 Public Debt
Source: From UEL, 2013: pp. 111–112.

have compatible data across host cities) and those used by governments for their statistics. For example, the term 'hospitalisation' in the Technical Manual in indicator So09 Health has ambiguity in relation to changing models of health care in the UK where many minor procedures are now carried out in clinics by a GP rather than in hospitals. Also terms such as 'ill person' for assessing morbidity are problematic. On occasion, proxy variables need to be sourced that reflect the spirit of the indicator desired in the Technical Manual. The section on *Analysis* provides an explanatory narrative of the tables and graphs provided. The final section on *Impact* evaluates the implications of the statistics for positive legacy from the Games. An innovation from the London 2012 OGI study team was to use a three-part impact coding as detailed in Table 2.2. The accompanying narrative under *Impact* provides reasoning and contextualisation for these impact codes.

The tables and graphs presented in Figure 2.4 illustrate the scrupulous attention paid to definition, sources and data copyright. This is necessary for complete transparency and avoidance of any ambiguity when comparing in-country statistics and statistics between host countries, as will happen in time. For this reason all the spreadsheets containing the original data files are accompanied by metadata, that is, data about the data. The Dublin Core Metadata Element Set (ISO 15836:2009) has been adopted for the London 2012 OGI study so that third-party users of the datasets can judge their provenance, value, reliability and suitability for the purpose for which they will be used. Finally, it is worth noting from the table in Figure 2.4 that the most recent available data (2011/12) is at least a year old at the date of the OGI study's publication (2013) and reflects the inevitable lag in the release of official statistics whilst they go through collation and quality control processes before being released. Thus most legacy studies using official statistics are looking back rather than necessarily knowing the situation right now.

Table 2.2   Impact coding of indicators for a Games effect

Relevance The considered degree to which the data informs the causality of a Games effect vis-à-vis legacy promises.	H	High
	M	Medium
	L	Low
Rating The level of impact that is judged to have taken place over the data period, given relevant context.	G	Green (positive impact)
	Y	Yellow (small or indeterminate impact)
	R	Red (negative impact)
Confidence The level of confidence with which the conclusions concerning impact can be derived from the data.	H	High
	M	Medium
	L	Low

Table 2.3   Components of the Dublin Core Metadata Element Set (ISO 15836:2009)

Label	Definition
Title	name given to the resource
Creator	entity primarily responsible for making the resource
Subject	topic of the resource
Description	account of the resource
Publisher	entity responsible for making the resource available
Contributor	entity or entities responsible for making contributions to the resource
Date	point or period of time associated with an event in the lifecycle of the resource
Type	nature or genre of the resource
Format	file format, physical medium, or dimensions of the resource
Identifier	unambiguous reference to the resource within a given context
Source	related resource from which the described resource is derived
Language	language of the resource
Relation	related resources
Coverage	spatial or temporal applicability of the resource, or applicable jurisdiction
Rights	information about rights held in and over the resource

The OGI study, and other similar types of studies, show change in relation to a baseline, but do not consider the counterfactual. To establish a coherent counterfactual for each and every indicator, when there are many, would be a daunting task. The counterfactual is more easily achieved where there is a single or small group of indicators, such as when valuing the contribution to GDP of a mega-event. The meta-evaluation study[8] of the London 2012 Games carried out by the Department for Culture, Media and Sport (DCMS) provides one such example (DCMS, 2013: Annex A). The direct spending on the preparation for the Games by the public sector – the Public Sector Funding Package (PSFP) – totalled just over £8.9bn. This included the land

---

8 A meta-evaluation is an over-arching synthesis of the findings of individual project-level evaluations in order to provide a comprehensive understanding of outputs and impacts associated with a mega-project.

purchase, infrastructure and venue construction. The economic calculation of the *gross* GVA[9] impact of this spending for the period 2007–2012 is £11.5bn. Regarding the counterfactual for calculating the *net* impact, the report states: 'The modelling compares the impact of the Olympics with the counterfactual assumption that the Olympics weren't awarded to London and therefore there was no construction or operational spending. There is no counterfactual assumption related to spending the public money on anything else' (DCMS, 2013: 124). This is a politically convenient assumption for the government to make because it means the gross is the net. Some displacement was accounted for (movement of production from other parts of the economy to the Games preparation) giving a net GVA impact of about £10bn and therefore the preparation for the Games made a surplus! This is not really tenable or believable. About £3bn of the PSFP came from sources such as the National Lottery and would have circulated in the economy and contributed to GVA even if London had not won the Games – so the counterfactual cannot be zero. The other £6bn of central government spending, a miniscule amount of total government spending, may well have been spent on other projects in the boom years prior to the recession. A more realistic counterfactual would be to model the value of PSFP as government consumption (spending on goods and services). The net impact would then more properly reflect the difference between spending on consumption versus spending on infrastructure and thus value the true legacy.

Conclusions

A number of lessons can be drawn from our experience of measuring the legacy of the London 2012 Games. The first is that the concept of legacy is inextricably bound up with the notion of additionality; we are looking to measure the net effect wherever we can and thus by definition, legacy is net impact rather than just change *per se*. Governments, sponsors and other stakeholders are always looking for a 'good news' story from mega-events: that the enormous investment has resulted in a positive legacy. Over zealousness in this regard can easily lead to 'legacy inflation' once the event is over. Thus, every new piece of inward investment gets branded as legacy even if that investment would have taken place anyway regardless of the mega-event, or the fact that the decision to invest had nothing to do with the mega-event itself.

It is clear that if planned legacies are to succeed, then they must be incorporated into the mega-event planning from the beginning. Legacy does not succeed as an afterthought. The question of how to measure and evaluate that legacy also needs to be given careful thought at the earliest possible stage. The data infrastructures and their financing need to be put in place, the relevant variables need to be identified and their capture designed. London 2012 was fortunate in that government policies towards transparency and open data meant that many of the right data were being released at the right quality. Some continuity of environmental variables at local levels has been problematic because many of the governance structures created to prepare for and run the mega-event quickly vanish afterwards.

Finally, it is important not to try and measure longer-term legacy too soon. Some short-term gains can be expected but many of the more profound transformations are likely to take fifteen to twenty years to fully emerge. Problems then arise in keeping data series intact without discontinuities over such long periods. Also, fresh initiatives to bolster and guarantee legacy serve to complicate assessments of the true legacy emerging from the mega-event itself, though it can be argued that these subsequent initiatives are therefore part of the legacy. But in time, the cumulative net impact approaches its maximum and the new normal is established. Measuring the legacy can stop.

Acknowledgements

The author would like to thank colleagues at the University of East London and University College London who worked on the OGI studies, and the ESRC for their funding of the OGI studies.

---

9   Gross Value-Added (GVA) measures the economic contribution of each producer and contributes to the calculation of Gross Domestic Product (GDP); GVA + taxes – subsidies = GDP.

References

Berman, G. (2010). *Financing the London 2012 Olympic Games*. House of Commons Library Standard Note SN3790. Available from: http://www.parliament.uk/briefing-papers/SN03790.pdf

Brimicombe, A.J. (2010). *GIS, Environmental Modeling and Engineering*. Boca Raton, FL: CRC Press.

Coaffee, J. and P. Fussey (2011). 'Olympic security', in J.R. Gold and M.M. Gold (eds.) (2011). *Olympic Cities: City agendas, planning and the world's games*. London: Routledge.

DCMS (2008). *Before, During and After*. London: Department for Communities, Media & Sport.

DCMS (2013). *Meta-Evaluation of the Impacts and Legacy of the London 2012 Olympic Games and Paralympic Games; Report 5: Post-Games Evaluation, Economic Evidence Base*. London: Department for Communities, Media & Sport.

Giesecke, J.A. and J.R. Madden (2011). 'Modelling the economic impacts of the Sydney Olympics in retrospect – game over for the bonanza story?' *Economic Papers*, 30, 2, 218–232.

Growth Boroughs (2013). *Convergence: Annual Report 2012–2013*. Available at http://www.growthboroughs. com/convergence

HMT (2003). *The Green Book: Appraisal and evaluation in Central Government*. London: Her Majesty's Treasury. Available from: https://www.gov.uk/government/publications/the-green-book-appraisal-and-evaluation-in-central-goverment

HMT (2004). *The Orange Book: Management of risk – principles and concepts*. London: Her Majesty's Treasury. Available from: https://www.gov.uk/government/publications/orange-book

HMT (2011). *The Magenta Book: Guidance notes for policy evaluation and analysis*. London: Her Majesty's Treasury. Available from: https://www.gov.uk/government/publications/the-magenta-book

HMT and DWP (2011). *Valuation Techniques for Social Cost-Benefit Analysis: Stated Preference, revealed preference and subjective well-being approaches*. London: Her Majesty's Treasury and Department for Work and Pensions. Available from: https://www.gov.uk/government/publications/valuation-techniques-for-social-cost-benefit-analysis

House of Lords (2013). *Keeping the Flame Alive: The Olympic and Paralympic legacy*. London: House of Lords Select Committee on Olympic and Paralympic Legacy.

IOC (2011). *Olympic Charter*. Lausanne: International Olympic Committee.

Legacy Trust UK (2013) *London 2012: Have we inspired a generation?* Available from http://www. legacytrustuk.org/info/About_Publications

May, T (2011). *Social Research: Issues, Methods and Process*. 4th Edition. Maidenhead: Open University Press.

NAO (2007). *Preparations for the London 2012 Olympic and Paralympic Games – Risk assessment and management*. London: National Audit Office.

NAO (2012). *The London 2012 Olympic Games and Paralympic Games: Post-Games review*. London: National Audit Office.

ONS (2014). *Focus on: Violent Crime and Sexual Offences, 2012/13*. London: Office of National Statistics.

PASC (2014). *Caught Red Handed: Why we can't rely on police recorded Crime statistics*. London: House of Commons Public Affairs Select Committee.

Poynter, G. (2009). 'The evolution of the Olympic and Paralympic Games 1948–2012', in G. Poynter and I. MacRury (eds.) (2009). *Olympic Cities: 2012 and the Remaking of London*. Farnham: Ashgate.

RICS (2011). *The 2012 Games: The Regeneration Legacy*. London: Royal Institution of Chartered Surveyors.

Robson, C. (2011). *Real World Research*. 3rd Edition. Chichester: Wiley.

Sinton, D.F. (1978). 'The inherent structure of information as a constraint to analysis: Mapped thematic data as a case study', in G. Dutton (ed.) (1978). *Harvard Papers on Geographic Information Systems Volume 7*. Graduate School of Design, Harvard University.

UEL (2013). *Olympic Games Impact Study – London 2012: Games-time Report*. University of East London. Available at http://www.uel.ac.uk/geo-information/London_OGI2/index.htm

Zimring, F. (2007). *The Great American Crime Decline*. New York: Oxford University Press.

Chapter 3

The Economic Power of the Olympic Brand and the Legacy of London 2012

Alvaro de Miranda

Introduction

Mega spectacles such as the Olympics have immense economic power. It is not only that they can attract public investment on a scale such as the £9 billion that London 2012 cost, a sum greater than the GDP of many countries with a sizeable population, but their commercial value to a number of major private economic interests is also considerable. In the four year period that led up to London 2012, the revenues of the International Olympic Committee (IOC) reached approximatedly US$5 billion (approximately £3.3 billion) (Mickle 2012). According to the IOC itself, 47 per cent of this revenue came from the sale of the broadcasting rights to the Summer and Winter Games, 45 per cent from commercial sponsorship deals and three per cent from licensing of the Olympic logo. Only 5 per cent came from the sale of tickets (IOC 2013).

Following his success in the London 2012 Games, Olympic multiple gold medal sprinter Usain Bolt's earning power from sponsorship was predicted to rise to US$23 million a year (approximately £15.3 million) (Sports Business Daily 2012). In September 2013 he renewed his sponsorship deal with sports equipment brand Puma until after Rio 2016 for a reputed annual fee of US$10 million (Sports Business Daily 2013). London 2012 heptathlon gold medal winner Jessica Ennis (now Ennis-Hill) signed multiple sponsorship deals with British and global brands which included Santander Bank, Adidas, Olay, Jaguar, British Airways and private health insurer Pruhealth. Estimates of her annual earnings from these deals range from £2 to £3 million (Burn-Callander 2012).

The Olympic Games is a brand of such economic power that it has world megacities and their national governments willing to advance many billions of pounds of taxpayers' money in competing with each other to become associated with it. It is also the brand that adds considerable market value to its medal winning athletes. It is a truly global brand whose ownership is vested in the International Olympic Committee. The IOC and its National Committees protect their ownership of the trademark zealously. Any organisation associating the word Olympic or the Olympic logo to itself, however innocently, has to contend with legal threats from the IOC or its National Committees (Raustiala and Sprigman 2012). The IOC has such muscle that it can have the national authorities of the Olympic Games sites policing its ownership of the brand and of its sponsors. During London 2012 there were numerous cases of the British authorities warning off small traders and even individuals from referring to the Olympics or the Olympic logo in their literature or merchandise. An 81 year old grandmother who knitted a white shirt for a doll to be sold in a fund raising activity in Downham Market and embroidered it with 'GB 2012' and the five Olympic rings was warned by UK Trading Standards, the government agency that enforces commerce laws, that selling her doll for charity would be against the law. An east London coffee bar was forced to drop the O from its name and call itself Café Lympics (Boudway 2012). During the Games themselves stewards prevented spectators from entering the Olympic park whilst carrying branded goods other than those from the official sponsors.

The IOC has even sought and been given a trademark on the number '2014' by the UK's Intellectual Property Office as the trade mark of the 2014 Winter Olympics (Techdirt 2013). The allure of the brand is such that attempts to associate companies to the Olympic brands and bypass the IOC's policing efforts are also numerous. Amongst the more ingenious ones was that of a local Utah brewing company in the US, Schirf Brewing, which advertised its product during the 2002 Salt Lake City Winter Olympic Games as 'The Unofficial Beer, 2002 Winter Games' (Boudway 2012).

Two obvious questions arise from all this: Where does the power of this brand come from? What are its implications for host cities and countries? What follows is an attempt to answer these questions.

Understanding the Economic Power of the Olympic Brand

The economic power of the Olympics brand is based first and foremost in its unique ability to capture the public's imagination and to inspire people throughout the world. Some of the Olympics' appeal as a spectacle is shared with other mega-events which have international prominence in modern society. These include the FIFA World Cup, the World Expos and even the Rio Carnival. Insofar as all these globally visible events attract the attention of the world they have the common appeal of large scale spectacles. Such spectacles have a long history of appealing to the public and of having international prominence. 'Bread and circuses' were reputed to be the twin main demands of the Roman plebs. However, sports mega-events have a particular allure which goes well beyond what is shared with other large scale spectacles. The attraction that is specific to sport is shared by the Olympics and the FIFA World Cup, but the Olympics offer something even beyond that.

The Power of Myth

The Olympics have a mythic dimension that captures the public's imagination in unique ways. According to Burke (1947), myth acts essentially on the imagination. It has the power to create a transcendental feeling of unity between being and the universe similar to the effect of poetry. Roland Barthes (1956) argues that in order to be effective, myth has to create 'euphoric clarity', a simple and clear image free from contradictions. Myth, therefore, according to Barthes, has to be apolitical as politics involves a struggle between competing interests and destroys clarity. Myth creates in the imagination an ideal world, a utopia in which the individual feels him or herself realized, transcending his or her real, unhappy existence; the more unhappy the real existence, the greater the power of myth over the imagination. Myth has thus the capacity to make the individual forget reality and live, at least temporarily, in an idealized world in which he or she is a legend. The mythic power of the Olympic Games dates back to its origins as a festival in honour of Zeus in 776 BC in its association with Greek mythology.[1] In modern days this unique ability of the Olympic Games to capture the public imagination has been increasingly harnessed for commercial purposes.[2] The advent of digital communications technologies with a global reach has enabled its mythic power to enter most homes, offering multinational corporations exceptional marketing opportunities.

Aware of the power of the Olympic myth, the IOC has been quick to claim its ownership by treating it as a brand. The ability of the Olympics to create the transcendental feeling of unity between being and the universe, and the 'euphoric clarity' of its ideal world lies in the core values it is said to promote. Protection of the public image of these core values is therefore essential for the protection of the Olympic brand. The core values underlying the Olympic brand include: (1) hope for a better world through involvement with sport; (2) the inspiration to achieve personal dreams through the lessons of athletes' sacrifices, striving and determination; (3) friendship and fair play; (4) joy in the effort of doing one's best. The IOC itself conducted extensive research in 11 countries on the attributes of the Olympic Games that were identified by the public as the most valuable and found these to be multiculturalism, globality, participation and fair competition (Madrigal et al. 2005).[3]

---

  1   In its modern version, de Courbertin, the founder of the modern Olympic Games, re-asserted the 'myth' as one modelled on the achievements of the English gentry and contrasted this with the French, defeated in 1871 in the war with Prussia, who he felt lacked the moral fibre to protect the Republic (Weber 1970). For an interesting discussion of the evolution of the Olympian myth from its inception up to the present day, see McDowall (2014).

  2   See, for instance, Shaw (2012) for a description of how this was done in the case of the Vancouver 2010 Winter Olympics.

  3   The explicit adoption of the core values by the IOC was part of an attempt to clean up its image following the corruption scandal over the bid for the 2002 Salt Lake City Winter Olympics. This is discussed later on in this chapter.

The commercial power of the Olympic brand is due to firms wishing to harness the extraordinary mythic quality of the Olympics to their own brand image by association. Sponsorship of the Olympics, which keeps competitors at bay, is one of the main ways of achieving this. However, there are perhaps even more secondary marketing opportunities that arise from advertising to the global media audience of the Olympics. Such advertisements often play on the mythic quality of the Olympics and associate its core values to the product and brand being advertised. This, in turn, provides world commercial media companies with a major incentive to bid for the exclusive broadcasting rights of the Olympics. They expect to more than cover the fees paid to the IOC through their own advertising revenues which come not only from the broadcasting of adverts during the Games themselves, but also from the effect on the general advertising market of the enhanced audience figures which the Games provide.

For politicians too, the mythic qualities of the Olympics are extremely attractive. Those who are associated with and promote the Games hope to acquire the aura of rising above politics by making use of the apolitical character of the myth. It enables them to present what they are promoting as the solution to a host of social problems they are unable to resolve through the contested operation of politics. This is particularly useful at a time when politicians have become largely discredited and seen as self-interested and corrupt in the eyes of the public. They hope that association with the altruistic core values of the Olympics will help cleanse their image and simultaneously divert attention from the reality of the social and economic problems they are unable to deal with, by promoting the myth that the Games will provide the solution.

There is another way in which the Olympic myth proves useful to both capital and politicians and which it shares with sport in general. In the age of neoliberal ideology, the aim is to shift responsibility for failure from society in general and the powerful in particular to the failing individual. Sport in general and the Olympics as spectacle are all about the celebration of success through individual effort.[4] Those who fail to make the grade do so because they did not try hard enough, not because of structural barriers that exist within society which prevent disadvantaged individuals whether through class, ethnicity or gender, from achieving success. The role model of individuals from disadvantaged backgrounds who achieve success in sport is paramount in supporting the argument that if they did it, then anyone can and disadvantage is not the fault of the social structure. This is the major reason why Jessica Ennis-Hill, the girl from a Sheffield mixed race working class background, Usain Bolt, the Jamaican village boy, or David Beckham, the boy from Leytonstone, are able to command such enormous brand value themselves. The multi-ethnic, multicultural and multi-class image which the mythic ideal world of the Games projects, and which the IOC correctly cherishes for economic reasons, creates an impression that the Olympics are a reflection of the real world, a world in which everyone can succeed if they try hard enough and is therefore devoid of structural injustice.

The economic power of the Olympics brand is thus immense. It can mobilise billions from private capital and from the public purse. However, this power relies essentially on its brand image remaining untarnished, a task which is made increasingly difficult by its growing association with a business world dominated by the search for profit. In this contradiction lies the central problem facing the IOC and its sponsors.

The Tarnishing and the Cleansing of the Olympic Brand and the Legacy of London 2012

In 1998, the IOC experienced a major threat to its brand image in the shape of a corruption scandal surrounding the organisation of the 2002 Salt Lake City Winter Games. The scandal broke when a Salt Lake City (Utah) television station, KTVX, reported that the Salt Lake Olympic Organizing Committee (SLOC) for the Olympic Winter Games of 2002 had offered a scholarship for the daughter of the IOC member for Cameroon to attend the American University in Washington. It later transpired that this scholarship was part of a wider SLOC scheme to offer scholarships to the family members of the IOC in an attempt to win their vote for Salt Lake City as the venue for the 2002 Winter Olympics. Thirteen people had been offered scholarships from the Salt Lake City bid funds or from SLOC, at least six of whom were close relatives of IOC members. A member of the IOC subsequently revealed that seven to 10 per cent of IOC members had received or requested bribes from bid cities and agents were claiming to be able to deliver

---

4   Team sports have a relatively low profile in the Olympic Games.

IOC votes in exchange for a fee (Mallon 2000). An avalanche of further media revelations ensued and the IOC was engulfed in a worldwide corruption scandal. The scandal affected also the preparations for the 2000 Sydney Summer Olympic Games when in January 1999 Australian Olympic Committee president John Coates released documents revealing that he, and other officials, had been involved in extensive vote buying in 1993 to secure Sydney's bid for the Games (Phillips 1999). The scandal greatly alarmed the IOC, particularly because of its concern that it might tarnish the image of the Olympics brand and affect its ability to procure commercial sponsorship and advertising for the Games (Payne 2006). It led to several major changes in the way the IOC operated, starting with the formation of two new commissions to help reform its structure and that of the Olympic Movement to ensure that such problems would never occur again. These were the Ethics Commission and the IOC 2000 Commission. Ten IOC members were expelled and another 10 sanctioned. The scandal rumbled on during the bid period for the 2008 Summer Olympics which came to an end in 2001 with the announcement that Beijing had won (ibid.). It was also at the time that the IOC was trying to deal with the aftermath of the corruption scandal that the concept of legacy began to acquire increasing importance in IOC thinking. In 2001, the IOC started working towards the development of a framework for the collection and analysis of economic, social and environmental data from future Olympic Games that became known as the Olympic Games Global Impact Study (OGGI) and, later, the Olympic Games Impact Study (OGI).

The start of the bid period for the 2012 Olympic Games came at a time when the IOC badly needed to cleanse its image and that of the Olympics brand. Also, the escalating costs to the public purse of the infrastructure required to support the Olympic Games as mega-spectacle increased the pressure on host city and national policymakers to provide a justification to the electorate for the spending of such huge sums. The increase in the costs of staging the Games had been prompted by the realisation that if the Games became more spectacular, their capacity to attract a wider global audience, now made possible by the new digital and satellite technologies, would increase and, with it, the value of the Olympic brand to commercial partners, a phenomenon which the Games share with Hollywood blockbuster movies. The concept of the legacy of the games thus became increasingly important in the minds of both the IOC and the city and national policymakers sponsoring bids for the Games (Poynter 2006).

London perceived all this in its bid for the 2012 Olympics and cleverly combined all the required elements for the bid to succeed. The presentation of the bid was orchestrated to stress all the mythic elements of the Olympics and empathised strongly with the core values of the IOC. Their inspirational character was emphasised throughout. The bid was structured around the concept of legacy with the dual emphasis on the sports legacy and the social and economic benefits that the Games would bring to the deprived and multi-ethnic and multicultural population of east London. The sports legacy would be to inspire a new generation of young children to participate in sport. At the presentation of the bid to the IOC Commission in Singapore on 6th July 2005 by a top level British delegation that included Princess Anne, Tony Blair, London Mayor Ken Livingstone, Sebastian Coe and David Beckham, London's relationship to the inspirational power of the Olympics was the key theme. This was summarised in the few poetic sentences with which Coe began his speech:

'To make an Olympic champion it takes eight Olympic finalists. To make Olympic finalists, it takes 80 Olympians. To make 80 Olympians it takes 202 national champions, to make national champions it takes thousands of athletes. To make athletes it takes millions of children around the world to be inspired to choose sport.'

(Mason 2012).

The same sentences introduce the short bid presentation video which was shown to the IOC. They are narrated over pictures of black children filmed in sepia colours and soft focus. Each sentence is illustrated by images depicting teenage children preparing to participate in a race and then actually taking part in it, ending with a close up of the facial expression of joy of the winner. The scene then moves, in apparent flash-back, to what seems to be a run-down African township, where a group of poor, very young black children, some barefoot, implicitly the same children that earlier in the film were seen taking part in, and one of them winning, a race, but at a younger age, sitting idly on the steps of a local store playing by throwing stones across the dirt road over roaming chickens, at a pile of second hand tyres. A police car slowly drives by and the police driver looks suspiciously at the children. A television hanging on the side of the store starts broadcasting the 2012 Olympics 100 metres sprint final. The children turn to the television and watch transfixed as the commentator shouts, 'It's Nigeria's Tony Neery, the Olympic 100 meters champion' and the film ends.[5]

The intended message was that if the Olympics were awarded to London, London would ensure that they would inspire children to fight their way out of poverty, idleness and possibly even a life of crime to success through sport. The theme was taken up again by Sebastian Coe in the speech that immediately followed the film. References to the inspirational nature of the Games abounded in his presentation. The mythic dimension was developed further through Coe's reference to the three principles which underpinned the bid, the first of which was 'to provide a magic experience'. The filmed address to the session by Tony Blair contained a quote from someone to whom he referred as an inspirational figure, Nelson Mandela, to the effect that there was no better place than London to hold 'an event that unites the world'. He finished his address with the statement that 'London will inspire young people throughout the world and ensure that the Olympic Games remain the dream for future generations'.

The fact that poverty would still be there was not considered by the film. However, this aspect was addressed through the second major promise made by London, that the Games would act as a major catalyst for the economic regeneration of the deprived area which was east London and thus benefit the dispossessed. The same mythic language permeated the promises made.

The Meaning of Legacy for London 2012

In the context of the foregoing discussion, the concept of legacy in relation to London 2012 should be understood not merely at face value, but also as a rhetorical device to win the Games for London and to sell the Games to the public at large, itself a necessary condition for winning the Games. The mythic dimension of the concept is designed to capture the public imagination and win it over to the bid. Without this, the necessary investment from the public purse could not be won and the bid would not be acceptable to the IOC. In order to win the 2012 Games by impressing the IOC Commission in Singapore, the London bid placed the main emphasis on the inspirational nature of the Games and their capacity to enthuse all young people to participate in sport. The subtext, as illustrated by the promotional video, was that sport has the power to free young people from poverty, anonymity and even crime by inculcating in them the Olympic core values. London promised that this sports legacy would be felt in London and throughout England, where specific targets were eventually set, but also worldwide, as implicit again in the video.

The sports legacy promise, and the way it was presented, chimed with the second main theme of legacy, that of the regeneration of east London. This was designed more to appeal to policymakers and to the British public in order to win their support and leverage the enormous amount of investment required from the public purse. The synergy between the two was achieved by the promise that the regeneration of east London would benefit a deprived, multi-ethnic local population. This was symbolically underlined by taking to Singapore the 30 young people from east London who would ooze enthusiasm for the Games. It also underlined the Olympic myth of bringing the world together. East London is, after all, a microcosm of the world in its multi-ethnicity.

---

5   This film in style and atmosphere bears a remarkable resemblance to a well known series of Nike television commercials featuring the black US star basketball player Michael Jordan, that all end with the slogan 'Become Legendary!'. In these advertisements the apparent message is that success comes from individual effort. However, the implied but very understated (and in one case even explicitly denied) subtext is that the route to become a legend is to own Nike sports equipment. At the time of writing, the Nike Michael Jordan adverts are widely available on YouTube. The presentation by the London delegation to the IOC Commission in Singapore, including the short film that I have described, was accessed at: http://news.bbc.co.uk/media/avdb/news_web/video/9012da6800177d8/bb/09012da680017a08_16x9_bb.asx on 31 October 2013.

Was the Myth Ever Credible?

The mythic nature of the promises to use the Olympics to inspire young people in order to increase their participation in sport is underlined by contrasting it with the reality. Declining participation in sport by people of virtually all ages is a phenomenon that has been affecting many countries in recent years. In the UK it has been taking place since 1990 and is most marked in young people.[6] Maarten van Bottenburg, Professor of Sport Studies at Utrecht University and Research Director of the Mulier Institute, tracked the change in sports participation in the UK amongst various age groups between 1987 and 2002 and showed that it had declined for most age groups. The most marked decline was for the 20 to 24 age group where participation fell from a maximum of 71 per cent in 1990 to 60 per cent in 2002. Sports participation amongst the 16 to 19 age group experienced a similar decline from 81 per cent to 71 per cent. Only the over 70s maintained their sports participation rate (Van Bottenburg et al. 2005).

The decline has coincided with a large rise in obesity and this has raised concerns that the two may be linked. The factors influencing sport participation have been the subject of research and have been found to be complex. However, the clearest determinant of sport participation is income and social class for all age groups. This was identified in a report containing academic review papers commissioned by Sport England (Rowe, Adams and Beasley 2004) as contextual analysis to inform the preparation of the framework for sport in England. The decline in participation is most marked in the lowest social group. This is also the group which displays also the greatest rise in obesity.

In addition, the authors pointed to the fact that '(t)he demographics and many of the cultural drivers are pushing us towards a more sedentary rather than active nation'. Studies have further shown that people regard lack of time as a significant barrier to participation in sport. The Sport England study notes that working hours and the intensity of work have increased. More recent research and reports confirm these findings. The 2013 report by Public Health England (Roberts et al. 2013) confirmed the strong class and income bias of participation in sport. The report also stressed the importance of the built and natural environment (e.g. parks, paths, pavements) on how people are able to be physically active. People from lower socio-economic groups have less access to these features. It pointed to affordability of sports facilities as another important factor and once again, the importance of lack of time is underlined. The report states that the Health Survey for England (2007) showed 45 per cent of men and 34 per cent of women thought work commitments were a barrier to being active. This means that people from lower socioeconomic groups may have a lack of discretionary time to take part in physically active leisure pursuits – even if they are free. The participation of young people of school age is also strongly related to class and environment (Dagkas and Stathi 2007; Edwards 2011).

The underlying justification for the approach to increasing participation in sport taken by the London 2012 bid is the assumed demonstration and inspirational effect of sporting success. The bid presentation in Singapore used an approach that the academic literature calls the 'sport pyramid metaphor', the assumption that success in elite sports increases mass participation. This is also used to justify the concentration of state investment in supporting elite sport. However, there is little evidence that this metaphor is generally valid (Colter 2004; Girginov and Hills 2008; De Bosscher et al. 2013). Where it does occur, its effect tends to be temporary. The notion that support for elite competitive sport will increase mass participation by young people is further undermined by some research evidence from schools which shows that concentration of support on excellence in competitive sport may have the opposite of the intended effect, as it can act as a

---

6   The fact that these trends were well known at the time that the London 2012 bid was made, with its promises to increase sports participation, has been recognised by a recent House of Lords report on the legacy of London 2012. The report states: 'In its 2007 report, *London 2012 Olympic Games and Paralympic Games: Funding and Legacy*, the Commons Culture, Media and Sport Select Committeeconcluded that "no host country has yet been able to demonstrate a direct benefit from the Olympic Games in the form of a lasting increase in participation". Systematic reviews of literature both academic and policy-related further demonstrated this point. The challenge therefore is a lofty one'. (House of Lords 2013: 25). The report points to the fact that the early evidence available for the post-Games period does not give rise to optimism that the promise will be kept and blames the level of funding for sports clubs, the lack of coordination between the grassroots level sports organisations and the organisations responsible for high performance sport, and the lack of a clear legacy plan for capturing the enthusiasm of the Games within all sports (ibid. 27).

deterrent for the less able to participate because they feel excluded and inadequate. A more general emphasis on physical activity that is not necessarily competitive may be more effective (Allender, Cowburn and Foster 2006). Despite the research evidence and the promises of the sporting legacy of the 2012 Olympics, the Conservative/Liberal Democrat coalition government was continuing to authorise the sale of school playing fields as a form of mitigating the inadequate funding of schools (Hope 2013).

The promise of a lasting social legacy benefiting the deprived population of east London through the regeneration of its economy that the Olympic Games would bring, which constituted the other main plank of the London 2012 bid, was also based on dubious historic evidence. There is little evidence that previous Olympics have brought much in the way of benefits to such social groups (Poynter 2006; Minnaert 2012). It could be that London 2012 would prove to be an exception and there have been undoubtedly genuine efforts on the part of key actors in the London Legacy Development Corporation (LLDC) to try to ensure that London achieves the exceptional. However, the playing field is far from level. Whilst the budget for the 2012 London Olympics was elicited from the public purse with few or no strings attached, the LLDC's brief requires that its post-Games developments provide a commercial return, to make up in part for the public expenditure on the Games. This is likely to place severe limitations on its ability to bring benefits to the local disadvantaged population. The first casualty may be the target of providing 35 per cent affordable housing in the Olympic Park housing developments, which may be impossible to achieve, even after the coalition government has introduced a new, much broader, definition of 'affordable'. It defines homes being sold at up to 80 per cent of market rates as still being 'affordable' (Bernstock 2014).

An early post-Games test of the way the interests of the local disadvantaged population were viewed when they came into conflict with other perceived economic development priorities of the area was provided by the proposal to demolish the Carpenters Estate on the edge of the Olympic Park, a run-down Newham Council housing estate, and build on its site a new campus of University College London (UCL). The creation of the UCL campus had been part of the regeneration strategy which envisaged the setting up of a ′world class′ university campus in the Olympic Park area as a contribution to bringing east London into the 'global knowledge economy' (de Miranda 2014). The residents of the Carpenters Estate, supported by UCL students and staff, fought the demolition proposals on the grounds that their needs and interests had not been properly considered, a claim that was supported by a research report on a study undertaken by staff and students on the UCL M.Sc.in Social Development Practice (Frediani et al. 2013). The residents' struggle against the proposal ended in victory when negotiations between Newham Council and UCL over the estate finally broke down in March 2013. The future of the Carpenters Estate is now uncertain.

Conclusion

I have sought to show that the promises made, that the legacy of London 2012 would bring social and economic benefits to the deprived population of east London and greater participation in sport by the population of Britain (and, indeed, of the world), in particular that of young people, should not be taken at face value, but rather understood as a rhetorical device to leverage the immense investment of public funds that the holding of the modern Olympic Games require. The Games function primarily as a spectacle which attracts the attention of the world, facilitated in modern times by the widespread availability of digital communications technologies. As such, they provide an unparalleled opportunity for the commercialisation of products and for the dissemination of brands.

I have argued that the unique effectiveness of the Olympic Games as a commercial vehicle relies primarily on their mythic power. This enables the Games to capture people's imagination by creating a vision of a utopian world, free from conflicts and divisions, in which everyone is able to succeed through their own efforts. In order to be effective in this way, the image of Olympic Games has to remain pure and the Games themselves have to be perceived as dedicated to altruistic ideals and to the construction of a better world, however much this might not correspond with reality. The tarnishing of the image would undermine the ability of the Games to capture people's imagination and thereby their ability to attract commercial sponsorship. This was the threat that the IOC faced when the Salt Lake City Winter Olympics scandal broke and which the London 2012 promises helped it to overcome.

As a rhetorical device, with little real evidence to support them, the specific promises of London 2012, that the main benefits of the huge investment of public resources would accrue to the disadvantaged population of east London, are unlikely to be fulfilled, and will in a short time be largely forgotten as public and media attention turns to Rio 2016. This is not to say that east London would have been better off without the Games. Without them, it is likely that the area would have remained unnoticed and its deprived population would have been denied even the few crumbs that they will have been able to collect from the table of the Olympic feast. The Olympic bid, and the government promises that followed its success, at least provided those genuinely committed to improving the position of the deprived population of east London with a rhetorical weapon of their own.

References

Allender, S., G. Cowburn and C. Foster (2006). 'Understanding participation in sport and physical activity among children and adults: a review of qualitative studies', *Health Education Research*, 21 (6), pp. 826–835. http://her.oxfordjournals.org/content/early/2006/07/20/her.cyl063; accessed 11 November 2013.

Barthes, R. (1972, orig. 1956). *Mythologies*. New York: Noonday.

Bernstock, P. (2014). *Olympic Housing*. Farnham: Ashgate.

Burke, K. (1947). 'Ideology and Myth', *Accent Magazine*, 7, Summer, pp.195–205 reproduced in J.R. Gusfield (ed.) (1989) *Kenneth Burke: On symbols and society*. Chicago and London: University of Chicago Press.

Van Bottenburg, M., B. Rijnen and J. van Sterkenburg (eds.) (2005). *Sports participation in the European Union: Trends and differences*. http://ec.europa.eu/health/ph_determinants/life_style/nutrition/platform/docs/ev20050525_rd02_en.pdf; accessed 14 May, 2014.

Boudway, I. (2012). 'Don't mess with the Lord of the Olympic Rings'. *Bloomberg BusinessWeek*, June 14. http://www.businessweek.com/articles/2012–06–14/dont-mess-with-the-lord-of-the-olympic-rings; accessed 30 May 2014.

Burn-Callander, B. (2012). 'Jessica Ennis Strikes Gold (In More Ways Than One)', *Management Today*, August 6. http://www.managementtoday.co.uk/features/1144376/jessica-ennis-strikes-gold-in-ways-one; accessed 30 May 2014.

Colter, F. (2004). 'Stuck in the blocks? A sustainable sporting legacy', in Vigor, A., M. Mean and C. Tims (eds.) (2004). *After the Gold Rush: A sustainable Olympics for London*. London: IPPR and Demos.

Dagkas, S. and A. Stathi (2007). 'Exploring social and environmental factors affecting adolescents' participation in physical activity', *European Physical Education Review*, 13(3), pp. 369–384.

De Bosscher, V., P. Sotiriadu and M. van Bottenburg (2013). 'Scrutinizing the sport pyramid metaphor: An examination of the relationship between elite success and mass participation in Flanders', *International Journal of Sport Policy and Politics*, 5(3), pp. 319–339.

de Miranda, A. (2012). 'The real economy and the regeneration of East London', in Poynter, G., I. MacRury and A. Calcutt (eds.) (2012). *London After the Recession: A fictitious capital?* Farnham: Ashgate, pp. 161–176.

Edwards, M.J. (2011). 'The Impact of School Sports Partnerships on Primary Schools: An in-depth evaluation', Ph.D. Thesis, Durham Theses. http://etheses.dur.ac.uk/3294/2/Mark_Edwards_PhD_thesis.pdf; accessed 08 November 2013.

Frediani, A. A., S. Butcher and P. Watt (eds.) (2013). 'Regeneration and well-being in East London: Stories from Carpenters Estate', MSc Social Development Practice Student Report, March. http://www.bartlett.ucl.ac.uk/dpu/programmes/postgraduate/msc-social-development-practice/in-practice/london-based-fieldwork/carpentersreport; accessed 3 April 2014.

Girginov, V. and L. Hills (2008). 'A sustainable sports legacy: Creating a link between the London Olympics and sports participation', *The International Journal of the History of Sport*, 25(14), pp. 2091–2116.

Hope, C. (2013). 'One school playing field sold off every week since Coalition was formed', *The Telegraph*, 13 December. http://www.telegraph.co.uk/education/keep-the-flame-alive/10516870/One-school-playing-field-sold-off-every-three-weeks-since-Coalition-was-formed.html; accessed 31 December 2013.

House of Lords (2013). 'Keeping the flame alive: The Olympic and Paralympic Legacy', *House of Lords Select Committee on Olympic and Paralympic Legacy Report of Session 2013–14*, London: The Stationery Office Ltd. Online at http://www.publications.parliament.uk/pa/ld201314/ldselect/ldolympic/78/78.pdf; accessed 31 May 2014.

IOC (2013). 'Revenue Sources and Distribution', http://www.olympic.org/ioc-financing-revenue-sources-distribution?tab=sources; accessed 30 May 2014.

Madrigal, R., C. Bee and M. La Bouge (2005). 'Using the Olympics and the FIFA World Cup to enhance global brand equity', in Amis, J. M. and T. B. Cornwell (eds.) (2005). *Global Sports Sponsorship*, Oxford: Berg.

Maldon, B. (2000). 'The Olympic bribery scandal', *Journal of Olympic History*, May, pp.11–27. http://library.la84.org/SportsLibrary/JOH/JOHv8n2/johv8n2f.pdf; accessed 29 October 2013.

Mason, C. (2012). 'Who's who: Political credit from London Olympics', *BBC News*. http://www.bbc.co.uk/news/uk-politics-19179711; accessed 31 May 2014.

McDowall, C. (2014). 'Olympians & Gentlemen – Freedom, liberty, honour & celebrity', *The Culture Concept Circle*, February 2. http://www.thecultureconcept.com/circle/olympians-gentlemen-freedom-liberty-honour-celebrity; accessed 30 May 2014.

Mickle, T. (2012). 'IOC generates a record $5 billion during last quadrennium', *Sports Business Daily*, July 24. http://www.sportsbusinessdaily.com/SB-Blogs/Olympics/London-Olympics/2012/07/ioccommission.aspx; accessed 30 May 2014.

Minnaert, L. (2012). 'An Olympic legacy for all? The non-infrastructural outcomes of the Olympic Games for socially excluded groups (Atlanta 1996–Beijing 2008)', *Tourism Management*, 33(2), pp. 361–370.

Olympic Review (2006). 'What is the Olympic Games global impact study?' *Olympic Review*, June. http://www.olympic.org/Documents/Reports/EN/en_report_1077.pdf; accessed 11 November 2013.

Payne, M. (2006). *Olympic Turnaround*. Westport, Connecticut: Greenwood Publishing Group.

Phillips, R. (1999). 'Sydney revelations deepen Olympics corruption scandal', *World Socialist Web Site*. https://www.wsws.org/en/articles/1999/01/olym-j30.html; accessed 31 May 2014.

Poynter, G. (2006). 'From Beijing to Bow Bells: Measuring the Olympics Effects', *Working Papers in Urban Studies*, London: London East Research Institute. http://www.uel.ac.uk/londoneast/research/FromBeijingtoBowBells.pdf; accessed 30 October 2013.

Roberts, K., N. Cavill, C. Hancock and H. Rutter (2013). *Social and Economic Inequalities in Diet and Physical Activity*. London: Public Health England.

Rowe, N., R. Adams and N. Beasley (2004). 'Driving up participation in sport: the social context, the trends, the prospects and the challenges', in *Driving up Participation: The challenge for sport*, Sport England. http://www.sportengland.org/media/142505/driving-up-participation-the-challenge-for-sport-2004-.pdf; accessed 8 November 2013.

Shaw, C. (2012). 'Myths, heritage and the Olympic enterprise', in Lenskyl, H.J. and Wagg, S. (eds.) (2012). *The Palgrave Handbook of Olympic Studies*. Basingstoke: Palgrave Macmillan, pp. 289–303.

Sports Business Daily (2012). 'Usain Bolt heads the list of Olympians most likely to cash in following London', *Sports Business Daily*, August 14. http://www.sportsbusinessdaily.com/Daily/Issues/2012/08/14/Olympics/Athlete-Mkting.aspx; accessed 30 May 2014.

Sports Business Daily (2013). 'Usain Bolt signs $10M extension with Puma to Run through 2016 Rio Olympics', September 25. http://www.sportsbusinessdaily.com/Global/Issues/2013/09/25/Marketing-and-Sponsorship/Bolt-Puma.aspx; accessed 30 May 2014.

Techdirt (2013). 'The International Olympic Committee has already staked a trademark claim on the number "2014"', *Techdirt*. http://www.techdirt.com/articles/20130115/13593121691/international-olympic-committee-has-already-staked-trademark-claim-number-2014.shtml; accessed on 30 May 2014.

Weber, E. (1970). 'Pierre de Coubertin and the introduction of organised sport in France', *Journal of Contemporary History*, 5(2), pp. 3–26.

PART II
Urbanism: Space, Planning and Place-making

Part II – Introduction
Urbanism: Space, Planning and Place-making

Gavin Poynter

In the closing pages of the 'Wealth of Nations', Adam Smith recorded the historic importance of the city as that which represented society's progress from subsistence to higher forms of social life, including its 'conveniences and elegancies' (Smith 1977: 502). Almost two centuries later, Lewis Mumford noted that the city provided not merely the conditions for society's material improvement but that its very forms were simultaneously a product of the mind and also conditioned it:

> 'Mind takes form in the city; and in turn urban forms condition mind. For space, no less than time, is artfully reorganized in cities; in boundaries, lines and silhouettes ... the city records the attitude of a culture and an epoch to the fundamental facts of its existence'. (Mumford 1995: 23)

The city was not simply a physical place but also an imagined or symbolic space that was reflective of the dominant culture of its time.

For Walter Benjamin, Paris, the capital of the nineteenth century, exemplified the relationships that may arise between the physical place and the imagined space of a city. The bourgeoisie's own imagination of the city's form being reflected in and given shape to by the architect Haussmann's 'long perspectives of streets and thoroughfares' in which the dominant class could feel secure; their construction achieving simultaneously the confinement of the working classes to their own districts and the new wide thoroughfares providing routes along which the military could quickly move from their barracks to enter the districts if insurrection or social unrest was threatened. For Benjamin, this nineteenth century new order, ushered in by industrialization, was also safeguarded within the city and nation by events – world and national exhibitions:

> 'World exhibitions are the sites of pilgrimages to the commodity fetish ... .They open up a phantasmagoria that people enter into to be amused. The entertainments industry facilitates this by elevating people to the level of commodities'. (Benjamin 1995)

These promoted the new technologies and commodities of the great nations but also served as entertainment for the masses for whom the exhibition was designed to ensure they were in awe of luxury and enjoying their 'own alienation from themselves and others' (Benjamin 1995: 50).

These views, expressed by eminent social theorists and written over the course of almost three centuries, simply affirm the enduring significance of the themes addressed in this section. The city is now the form of social organization in which, in this century, the majority of the world's population will live. The pace of urbanization demands that we re-visit the analysis of the urban economy and its transformative effects upon the city; the city is also rapidly changing in a highly internationalised economy in which time and space are compressed and new patterns of physical and virtual relations between global metropolises have emerged. Lastly, and directly related to the theme of this section of the book, the great exhibitions particularly popular in the latter part of the nineteenth and early twentieth century (Roche 2000) have been augmented or, indeed, surpassed in their significance by the emergence of the mega-event and, especially in these times, the sports mega-event.

By hosting such events the city seeks simultaneously material and symbolic transformation; changes in the purpose(s) and structure of place along with changes in the way that the city is imagined or in contemporary terms, its image is branded, received or perceived by the outside world. For the communities and citizens of the host city, the sporting festival may enthrall but it may also presage a social legacy that resonates, in contemporary ways, with Benjamin's critique of the great exhibitions. At the very least, the sporting mega-event relies for its success upon the enthusiasm and support for it demonstrated by the citizens of the host city and nation. The close association of a nation's political elite with the event necessarily highlights the elite's relationship with its own citizens. The former may seek to capture for itself a little of the reflected glory of a successful event or it may experience widespread public criticism and social unrest if the event and its promised legacies are received by citizens as an opportunity to raise social and political grievances. In their analyses of the complexities of these urban developments and the social and political contexts in which they occur, authors over recent years have embraced what has been called the 'spatial turn' in the social and human sciences.

The Spatial Turn

If place is bounded and known, at least to those who visit, work or live in it, then space is, arguably, less clearly defined and may be interpreted in a variety of ways – materially as a territory, as an organizational/ administrative form, as an often contested memory or as an imagined, planned or designed, future. In his 'Production of Space', Henri Lefebvre (Lefebvre 1991) provides a conceptual frame for understanding space. For him, space contains a multitude of inter-connections that constitute the social relations of capitalism. These may be presented as a conceptual triad of spatial practice involving the processes of production and reproduction; representations of space which the relations of production impose overtly to provide and secure the prevailing order and, lastly, representational spaces that contain a symbolic, often hidden, character or, as he calls it, the underground side of social life. This hidden character arises, according to Lefebvre, from the special position of labour whose exchange with capital creates abstract labour and, in turn, abstract space, a world of commodities supported, for example, by the state and networks of businesses and banks (Lefebvre 1991: 33–53).

Doreen Massey sought to uncover this hidden side of social life, to unearth some of the influences on 'the hegemonic imaginations of space' (Massey 2005: 17) in order to focus on the potential offered by the negotiated, disputed social relations that lie at its core. In providing a passionately argued case for re-imagining space, Massey discusses several of the key authors that have informed the 'spatial turn'. In proposing that space presents us with the social *in the widest sense*; she argues that the re-imagining of space is central to challenging those dominant discourses that fix or bind the local to 'place' and attach the concept of space to the immutable laws of the market, through the process of globalization. By focusing on the relationships between place and space, Massey challenges the counter-positioning of the local and the global by seeking to affirm the significance of their being mutually constituted (Massey 2005: 184). For Massey, therefore, the conceptions of place and space should not be conceived as binaries such as real (place), abstract (space) but as inter-acting and socially constructed.

Space, Place and Sports Mega-events

Whilst this briefest of discussions of the 'spatial turn' does little justice to the richness and diversity of the writings of those who have contributed to its development from different disciplines and theoretical perspectives, it is perhaps possible to argue that the process of urban change often catalyzed by hosting the sports mega-event reveals in many ways what Massey has referred to as the mutual constitution and re-constitution of place and space.

The sports mega-event is materially bounded, fleetingly, to a place but it is also used to project an imagination of that place to a global audience and, in so doing, the host city and sometimes nation, seeks to assert or re-affirm its longer term status or position in the world. In the course of that longer term, or legacy phase, the images of place may translate into its material re-making via, for example, inward investment and increased tourism. Secondly, in Lefebvre's terms, through hosting the mega-event, the practices and symbolic representations of space may be reconceived or reaffirmed in accordance with the interests of the prevailing order via the creation of new patterns of land ownership, occupancy, purpose and security in what might previously have been considered as an urban area of relatively low economic value (Fussey et al. 2012; Giulianotti and Klauser 2010).

Finally, the sports mega-event provides insights into the hidden character of the social life of the city not least when the prevailing interests reflected in the practices and symbolic representations highlighted and instigated by the mega-event are challenged by its own citizens (as in Rio, Istanbul or other cities). In this sense, whether the urban transformations catalyzed by the mega-event are perceived by state, enterprises or citizens to have positive or negative effects, their spatial dimensions are, in Massey's terms, an on-going open production often long after the event is over. The chapters in this section take up several themes related to the debates that have arisen from the spatial turn. They examine three mega-event cities – London, Rio de Janeiro and Turin – to explore the relationships between planning, place making and different conceptions of space.

References

Benjamin W. (1995). 'Paris: Capital of the Nineteenth Century', in Kasinitz, P. (ed) *Metropolis Centre and Symbol of our Times*. London: Macmillan, pp. 46–57.

Fussey P. J., Coaffee, G. Armstrong, and D. Hobbs (2012). *Securing and Sustaining the Olympic City*. Farnham: Ashgate.

Giulianotti, R. and F. Klauser (2010). *Journal of Sport and Social Issues*, February, vol. 34: no. 1, pp. 49–61.

Lefebvre, H. (1984). *The Production of Space*. Oxford: Blackwell.

Massey, D. (2005). *For Space*. London: Sage.

Mumford, L. (1995). 'The Culture of Cities', in Kasinitz, P. (ed) *Metropolis Centre and Symbol of our Times*. London: Macmillan, pp. 21–29.

Smith, A. (1977). *The Wealth of Nations*. Harmondsworth: Penguin Books Ltd.

Chapter 4

Urban Designs on Deprivation: Exploring the Role of Olympic Legacy Framework Masterplanning in Addressing Spatial and Social Divides

Juliet Davis

Introduction

'Within 20 years the communities who host the 2012 Games will have the same social and economic chances as their neighbours across London'.

This promise was articulated by the Five Olympic Host Boroughs in their Strategic Regeneration Framework (SRF) of 2009. It represents their understanding of the fundamental challenge of Olympic-led regeneration in east London. The task is conceived as one of reaching a 'convergence' of deprivation levels between the host boroughs and the rest of London, in recognition of existing disparities represented across multiple indicators. We are told that it is to be achieved, at least in part, by ensuring that large-scale, post-Olympic Games development is designed to adequately address long-standing social issues directly associated with its built environment context and morphology and by further bringing lasting 'economic benefits to the residents in the area'.

This chapter, is concerned with how the discourse of 'convergence' has become intertwined with urban planning and design for the Olympic legacy, and how this is said to constitute a basis for bettering existing East End lives. I do this by investigating the relationship between core themes of the convergence agenda and key aspects of the spatial strategy which began life as the 'Legacy Masterplan Framework' in 2008, and which has more recently resulted in the 'Legacy Communities Scheme' (2011). In the process, I am interested in considering how, beyond the much criticised comprehensive redevelopment of the Olympic site following compulsory purchase in 2006, the processes of gradual development and build-out of the masterplan framework are being positioned to engage with the existing realities and evolution of its local social and spatial contexts, and the place of 'convergence' in this.

The aim of using town planning and urban design to address the deprivation of locales in the East End is not new. The chapter begins by sketching key strands of historical debate on the use of renewal strategies predicated on redesign and redevelopment to address deprivation in east London in different eras over the last 100 years. In general, these strategies have been shown to create more issues than they have solved, for example by disrupting established patterns of social life or stimulating social change through engineered processes of displacement. Some critics suggest that this may be due not only to the relative bluntness of redesign and redevelopment strategies as instruments for intervening in the complex realities of lived urban places, but also to the ways of assessing deprived or declined places and of, in a connected way, construing and constructing their needs in spatial terms.

In response, the chapter goes on to look at how the concept of 'convergence' is framed, particularly in terms of a discourse of growth-focussed and change-oriented regeneration. The focus of this section is the host borough's 'convergence' strategy, as articulated in the SRF (2009) and later 'Convergence Action Plan 2011–2015', though I also take into account the SRF Progress Report 2009–2011 and Convergence Framework Annual Report 2011–2012. I look at the relationship between ways of measuring the deprivation gap within these documents and, specifically, ways of defining the scope of masterplanning. I, of course, recognise in doing so that the SRF does more than only provide the means for connecting socio-economic and physical development. However, I argue that the scope of masterplanning to address socio-economic realities can be seen to be constrained by the ways in which deprivation is construed, measured and represented. In addition, whilst an important emphasis of the SRF is on framing a regeneration process within which gradual convergence and anticipated, steady economic growth and development are seen to be connected, it is unclear to what extent it is genuinely targeted at existing deprived communities, or relies on the attraction of incoming, wealthier residents.

I then look at the connection between the 'convergence' strategy and some of the key spatial principles and parameters established for future development within the Olympic Park Legacy Company's (now the London Legacy Development Corporation [LLDC]) 'Legacy Communities Scheme (LCS) (2011) and the Greater London Authority's 'Olympic Legacy Supplementary Planning Guidance' (OLSPG) (2012), also building on my knowledge of the 'Legacy Masterplan Framework' which precedes them. I focus on how 'convergence' goals are expressed through the envisaged urban form of connected, mixed-use and compact development. I argue that whilst much emphasis is placed on connecting the site's offer of amenities to existing areas, it is apparent that 'convergence' is predominantly to be achieved through the planning of resources scaled, located and designed to suit future, imagined communities. This raises important questions, not only for urban design and urban form, but for the governance of urban form for the future.

Redeveloped Places and Displacement in East London: The Historical Myth that Lives are Bettered by Cataclysmic Change

In 1961, Jane Jacobs wrote that 'money is employed cataclysmically instead of for gradual, steady street and district improvement, because we thought that cataclysms would be good for our slum dwellers – and a demonstration to the rest of us of the good city life' (Jacobs, 1972[1961]: 401). Whilst official Olympic-led regeneration discourse in London does not, of course, refer to slums or slum-dwellers, Jacobs' conceptualisation of the relation between the use of drastic spatial change and social reform in mid-20th century New York is relevant to the context of London's post-Olympic legacy planning, as it is to the longer history of 20th and 21st century redevelopment in the East End.

The East End provides a catalogue of 'cataclysmic' redevelopments produced over the course of the past century. Developments such as the Boundary Estate in Shoreditch and the St. Vincent Estate in Limehouse reflect the blacklisting of poor neighbourhoods in the 20th century for 'slum clearance', whilst the modern urban fabric that typifies much of inner east London attests to the wide-spread reconstruction and renewal programmes that were initiated alongside slum clearance following the Second World War (Malpass and Murie, 1999). The redevelopment of the Docklands from the early 1980s has been viewed as paradigmatic for many of the neoliberal policies of property-led yet pump-primed regeneration geared toward economic restructuring and global city promotion, in the wake of industrial decline (Fainstein, 1994; Foster, 1999; Hall, 1998). Some sites of early and mid-20th century redevelopment have, in more recent times, become targets of renewal again, including numerous 20th century council estates now viewed as failed modern visions (Campkin, 2013; Watt, 2009, 2013) and 'first wave' post-industrial redevelopment in London's Docklands (Carmona, 2009) and the Lea Valley. Examples of this feature in the fringes of the Olympic site, most notably in the case of the redevelopment of the Trowbridge Estate in Hackney Wick following its memorable demolition from 1985 and the current dismantling of the Carpenters Estate in Stratford. The Olympic site's mega-event-led redevelopment from 2006 for long-term regeneration may be one of the latest redevelopment sites in this list, but it is also one of the largest single *tabula rasa* sites to date, intended to deliver lasting social change and impact.

Many of these historical redevelopments and the urban policies which have informed them have been predicated on aims of addressing deprivation and/or decline, from the deprivations of clean air, water and sanitation, as well as of health and employment experienced by the 'outcast London' of the 19th century (Steadman-Jones, 1971) to the multiple deprivations produced in the context of the decline of London's port and historical industries from the 1970s (Cattell, 1997; Lupton, 2003). However, despite the dramatic spatial effects of comprehensive redevelopment, their impact on the economic and social welfare of previously established local groups has been often said to be more limited. The effects of some projects have been found to be the reverse of what was stated or intended, including the displacement of existing social groups and the regeneration of their places for the benefit of others – the developers, owner-occupiers and or gentrifiers, who become able to realise the capital accumulation potential of demolition sites. This, in turn, has created scope for the theorisation within urban scholarship of these projects and processes in terms including uneven development, inequality and social polarisation, and accumulation by dispossession, in the context of neoliberalism and capitalist development dynamics.

For example, when the Boundary Estate first opened in 1900 to provide social housing for the 'deserving poor' of Shoreditch, the rents demanded by the local council were too high for the displaced residents of the former Old Nichol (Yelling, 1986[2007]: 87). In Docklands, although development was predicated on the notion that capital investments in development and property would 'trickle down' to poorer communities (Imrie and Thomas, 1999), commentators are generally agreed that 'leverage planning' (Carmona, 2009) did little to advance the prospects of the established communities of the Isle of Dogs or neighbouring Poplar (Fainstein, 1994, 2005; Foster, 1999; Hall, 1999; Hamnett, 2003). To this day, the line between the Isle of Dogs and Poplar along the six lane highway of the A1203 marks a staggering drop from the 72nd to between the first and tenth percentile of most deprived places in Britain.[1] The generic urban form of high-rise office and shopping mall development, and the superficial eclecticism of 1980s (Harvey, 1988) and early 1990s postmodern architecture have, in turn, been interpreted as further affronts to local urban history, spatial context and culture (Fainstein, 1994, 2005). In planning terms, Docklands has been viewed as a missed opportunity to develop instruments for harnessing private development capital to explicitly address existing local development needs (Hamnett, 2003: 244–245; MacRury and Poynter, 2009: 95–97).

Analyses of the early legacies of 2012 Olympic planning point to similar issues of disadvantage resulting from rather than being addressed by redevelopment (Raco and Tunney, 2009; Davis and Thorley, 2010; Hatcher, 2012). Whilst Raco and Tunney focus on the politics of looming redevelopment on the sizeable cluster of 200 small businesses operating within the site up until 2007, Davis and Thornley, and also Hatcher, focus on some of the negative impacts of dispersed relocation on established social groups self-defined as 'communities'. More recently, Watt has explored the experiences of social groups at the Carpenters Estate threatened by immanent 'decanting' to make way for redevelopment, a process seen to exemplify Harvey's conception of the dialectical 'accumulation by dispossession' dynamic of capitalist economic development. These studies also highlight the importance of analysing the tools by which redevelopment is carried out – from planning policy to compulsory purchase to masterplans for 'new' areas. Beyond a troubled history of demolition and relocation, this chapter sets out to explore the potential of legacy masterplanning to more effectively impact on established life at the Olympic site's fringes.

In so doing, it is important to recognise what Lees (2003: 64) refers to as the 'story lines […] used for constructing arguments' that legitimate regeneration. Motivations for cataclysmic redevelopment and reconstruction are strongly informed by discourses of 'decline' and deprivation. As Beauregard (2003: 21) argues, 'discourse is not merely an objective reporting of an incontestable reality but a collection of contentious interpretations. The 'real world' provides material for discourse, but these understandings are then mediated socially through language'. In the context of masterplans, they are, of course, further and complexly mediated through urban analytical and design conventions as well as creativity to produce urban form and architecture. The tools of urban analysis are particularly crucial as different approaches – from land-use analysis to the close observation of the relations between social life and urban places – have the potential to yield quite different conceptions of the roles and the goals of intervention.

The effects of interpretations of decline and deprivation on the fabric of place are powerfully suggested by Campkin in his book *Remaking London*, within which emphasis is placed on the connection between the discourses of 'dirt and disorder' associated with poor places and a history of redevelopment 'clean ups' traced up to the 2012 Olympics. A similar point is made by Yelling (2007 [1987]: 51–64) in looking at the relationship between Charles Booth's mappings of poverty in the 1890s, and the sites of London County Council (LCC) slum clearance over subsequent decades. In their famous ethnographic study of post-War Bethnal Green, Young and Wilmott (1957) argued that whilst official analyses of the conditions of dwellings in this working class area flagged up important issues for planning, they concurrently failed to recognise the value of the human fabric of places and, in these terms, the social and cultural dimensions of lived urban

---

1 http://opendatacommunities.org/deprivation/map

form. In Cattell's 2012 study of residential locales distributed around the Olympic site, the need for qualitative research methodologies to understand the complex realities of deprivation is emphasised. These are able to reveal the complex 'real world' of both social and material deprivation – a relation which is not always constituted or experienced in the same way – as well as some of the ways in which people develop their own strategies for coping with real and perceived hardships. This work provides a basis on which to consider the limitations of quantitative deprivation measures which predominate in official representations of the goals of Olympic-led regeneration.

'Cataclysmic' redevelopment has a problematic history in east London of failing to realise benefits for and actually within existing places deemed as poor. So how realistic is 'convergence' given the redevelopment context of the Olympic site? This question frames the primary concern of the analytical sections that follow: the relationship between ways of understanding deprivation and the priority of convergence, and ways of realising convergence spatially through a masterplan.

Convergence: A Concept and a Strategic Spatial Goal

In this first section, I provide a brief overview of the concept of 'convergence' as a way of conceptualising the challenge of Olympic legacy regeneration, and as a way of framing the goals of design and development.

The stated aim of the SRF is to establish 'an Olympic legacy vision for the area which goes beyond sport' in order to 'ensur[e] a better future for our boroughs and the people who live in them' (Five Host Boroughs, 2009: 1). The idea of a 'better future' is predicated on the recognition that the host boroughs concentrate the highest levels of disadvantage in the 'greatest cluster of deprivation' anywhere in England and Wales (Five Host Boroughs, 2009: 11).

A 'deprivation' gap between the host boroughs and other parts of London is seen to show up 'on almost every indicator available' (ibid. 2009: 11) including the Indices of Deprivation, London Labour Market Indicators and a variety of other census-based and annually collected public sector data. The Indices of Deprivation for 2007 showed that 20 per cent of the super output areas in the Host Boroughs ranked amongst the five per cent most deprived in England and Wales, and that approximately 40 per cent of those in the vicinity of the Olympic site fell into this category (Figure 4.1). In the SRF, the gap is portrayed in terms of the following key headlines:

- **Employment:** 6.2 per cent less people employed
- **Violent Crime:** 17 per cent more
- **Overcrowding:** 250 per cent – 540 per cent more
- **Adult skills:** 6 per cent more adults with no qualifications
- **GCSE:** 8 per cent lower attainment
- **Mortality:** 15 more people per 100,000 population die prematurely
- **Childhood obesity:** 25 per cent of children obese by year six – above the London average.

'Convergence' represents the goal of closing these gaps over 20 years. Whilst their presentation in these quantitative terms is compelling and creates a set of targets for intervention, it also serves to abstract the reality of deprivation as experienced – potentially in quite different ways – in the Host Boroughs' neighbourhoods. Whilst, following Beauregard, these numbers reflect the 'real world', they do not capture anything about the kinds of resourcefulness, resilience, resistance and productivity that Cattell (2012), for example, highlights as features of the multicultural east London neighbourhoods she encountered through qualitative research. What alternative implications could such qualitative research have for legacy spatial strategy? Arguably, the lack of qualitative understanding has implications for the evaluation of convergence over time as well, as not all population data is effective at capturing the degree to which a changing level of performance genuinely reflects change in the lives of deprived individuals.[2]

---

2   This is an issue that is recognised in the Initial Impact Assessment which forms an Appendix to the SRF – see pp. 63–64.

Figure 4.1   Site wide indicative phasing plan
Source: London Legacy Development Corporation.

Deprivation is connected to a story of decline resulting from 'post-war destruction and more recent de-industrialisation' within which subsequent development fails to deliver 'benefits to existing residents' (Five Host Boroughs 2009: 19). In the main, this is because its architecture is of poor quality, and its built environment blighted by fragmentation and a lack of open space and other amenities. Such an environment, it is suggested, is unable either to keep those residents who become able to leave it behind or to attract population, becoming a place of last resort. Within this narrative, the Olympic site, legacy plans and other 'large development plans' in and around the Lea Valley represent the opportunity to reverse decline through the economic restructuring of the area and the associated restructuring of the urban landscape to render it a place of choice for both existing and new residents. In these terms, the counterpoint to the decline story is one of massive change and growth, in which contemporary urban planning and design and architecture are positioned as 'key' to future convergence. The regeneration challenge is presented as one of being able to harness the investment associated with large-scale development for creating the resources needed to stabilise the existing population, address its deprivation, and at the same time attract 250,000 new residents. The regeneration process is conceptualised in these terms as a 'virtuous growth cycle', predicated on the management of property-led redevelopment through government intervention in the provision and management of place-based public resources.

The convergence strategy is presented in the SRF in terms of seven key goals or outcomes, which address the key components of the deprivation gap. These are said to be addressed through a variety of interventions from 'hard' physical transformations to 'softer' local governance and service provision-based strategies. Table 4.1 opposite summarises these outcomes alongside an analysis of their implications for legacy masterplanning.

A number of key points arise from this analysis. First, it is hard to ignore the language used to articulate the envisioned convergence process through physical regeneration. For all that this is a strategy geared toward improving the lives of established residents, it is also one in which deprivation is to be 'eradicated' and 'designed-out', in order to make places 'better' or 'good', 'decent', 'attractive', 'clean' and 'world-class' rather than 'holes', from the privileged perspectives of local leaders. Though the nature of good, clean and attractive spaces is not defined within the document, the description of physical regeneration in these terms serves to link disadvantage with aesthetic as well as moral predispositions in ways that recall Campkin's (2013) analyses of the functions of 'dirt' in historical urban renewal discourses and processes. In this context, the challenge of urban design becomes one of addressing a constructed image of urban deprivation rather than its complex realities or disparate experience. Significantly, the SRF suggests that the purpose of this is not only to transform perceptions of the Olympic site's urban context, but also to be able to market the 'add[ed] value' (SRF 2009: 21) of a regenerated image.

Second, it is clear that a number of spatial categories and strategies, including scale, location and connection, are seen as facilitators of convergence. Although the table does not highlight the weight given to different categories, within the report it is clear that connectivity is primary. However, throughout the report, the significance of connectivity for achieving convergence outcomes varies. Its purpose, in a practical sense, is said to be to deal with historical spatial fragmentation in the urban morphology of the Lower Lea Valley. At the same time, connectivity is a metaphor for 'convergence' as historical gaps in the urban fabric are seen to coincide with evidence of the deprivation gap, and the strategy of reconnection is constructed not only to facilitate more legible, convenient, safe and walkable routes between existing local places, but also to create the infrastructure for capital flows of leveraged investment. Building standards and environmental 'high quality' play similarly ambivalent roles with respect to convergence, being at once necessary for improving current living conditions and images of anticipated urban orderliness and prosperity.

Third, it is significant that convergence outcomes are to be realised over 20 years of gradual transformation through legacy masterplanning and incremental development. This is the timescale envisaged for delivering the full build-out of the legacy site and for realising each of the above masterplanning objectives in an economically viable way. The major significance of this for this chapter is that it represents a reaction on the part of the SRF's authors in general to the critical appraisal of redevelopment in the past and, in turn, suggests scope for social impacts to be managed through a carefully timed process of public and private investment. The idea, however, that existing residents will benefit from community infrastructure, new homes

Table 4.1   Convergence outcomes and their masterplanning implications

	Promised Convergence Outcomes	Masterplanning: Addressing convergence
1	Creating a coherent and high-quality city within a world city region (to address negative local levels of satisfaction with their areas rather than specific deprivation indicators)	• 'World-class'/ design-led development of the Olympic Park • Transformation of the valley from a 'hole in our urban fabric' to a 'heart for the area' • Enhancing local and wider connectivity • 'High standard' development and public realm 'providing adaptable, generous, long-lasting places to live and work'
2	Improving educational attainment, skills and raising aspirations	• Modernising and reinforcing schools infrastructure • Better physical environments for learning.
3	Reduce worklessness, benefit dependency and child poverty	• Social infrastructure
4	Homes for all	• New homes to relieve pressures leading to crowded households (8,000 – 10,000 within the Olympic Park) • 'High quality' homes and a diverse housing offer to encourage people to stay – stabilising communities.
5	Enhancing health and well being	• Provision of 'quality' green space and other local amenities • Maintaining housing standards to reduce health issues associated with life in buildings • Community infrastructure of schools, health, wellbeing and social care services including a 'world-class' polyclinic. • Provision of options for walking and cycling • Land use planning principles for ensuring a provision of healthy food outlets near schools and on high streets
6	Reduce serious crime rates and anti-social behaviour	• No specific implications. Emphasis is placed on policing.
7	Maximising the sports legacy and raising participation levels	• Transforming retained Olympic venues to provide a 'great sporting offer' to residents

and amenities is predicated on their being able to stay in place. Stabilising communities over 20 years, in the light of the dynamics of apparently ever increasing London property values (Whitehead and Travers, 2011), changing housing policy and local service provision, and the limitations of planning policies on the nature of housing provision (Bernstock, 2014), as well as the array of economic and factors which inform population mobility in London (see for example, Travers, Tunstall and Whitehead, 2007) is clearly highly challenging.

In the 'Convergence Framework: Annual Report 2011–2012' and the 'Convergence Action Plan 2011–2015', the above outcomes are reconfigured and located under three primary themes, each denoting a number of performance indicators and targets:

• Creating Wealth and Reducing Poverty
• Supporting Healthier Lifestyles
• Developing Successful Neighbourhoods.

Broadly, the implications of these for masterplanning are the same as in the 2009 document. However, there are also some notable differences: 6,870 new homes rather than 8,000–10,000 in the Olympic Park are now envisaged, which represents lower density development yet at the same time provides less scope for dealing with issues of affordability and overcrowding through housing provision. An initial wave of social infrastructure legacy development in the form of the E20 Health Centre and Chobham Academy (with 1,800 pupils aged between 3 and 18) was realised concurrently with the Olympics. The offer of the Olympic Park and transformed Games venues is described in terms of convergence outcomes in far more detail – there are '29 play spaces, plazas, canal paths, roof gardens and cycle paths' – and 'public realm improvements' at the Park's fringes are also mentioned. More emphasis is placed on how the Olympic legacy can exemplify the concept of 'healthy urban planning' and its translation into spaces such as 'fitness trails, activity trails [and] marked walking routes' that encourage physical activity. Healthy planning is said to involve focus both on improving levels of physical health, and on the role of social capital in health and well-being outcomes. It is important to note that in the context of the Olympic site, opportunities for promoting social capital are created on the basis of generalised conceptions of how to do so, rather than based on all social groups and their existing spaces of interaction.

Emphasis is also placed on social capital, as well as on 'building individual and community resilience so residents are ready to benefit from economic growth', and 'promoting aspirations' under the 'Creating Wealth and Reducing Poverty' theme. Whilst the focus on social strategies in these terms marks an interesting departure from the earlier emphasis on statistics and is potentially interesting for urban design, the emphasis on trivial 'attractiveness' under the other themes suggests a disappointingly marginal role for design in addressing local social realities directly. On a more positive note, the two reports make clear that physical intervention strategies will only facilitate convergence if their offer is managed and participation is sought in their design and realisation.

It is notable that whilst convergence targets are made more explicit in numerical terms in the 2011–2015 documents, more of the onus on determining suitable strategies is explicitly placed onto the private sector. Developers 'submitting major planning applications are expected to demonstrate how their schemes will help achieve Convergence' (The Six Host Boroughs, 2012: 7). In turn, it is argued that as the costs of deprivation to the public sector are high yet local authority spending has to be drastically scaled back over the coming years, the identification of private investment routes to convergence outcomes in terms ranging from building standards to amenities and social infrastructure is vital (The Six Host Boroughs, 2011: 3). This points both, to the dependence of convergence on the capacity of local leaders to manage the offer of development in an evolving economic climate and to significant levels of uncertainty concerning its future feasibility.

Convergence by Design

So what does this mean for legacy master planning itself, in terms of the Legacy Communities Scheme's (LCS) Development Specification and Framework (DSF)? The LCS outline planning application of September 2011 overlaid part of the area that had been explored by the Legacy Masterplan Framework (LMF), as well as the 'Olympic Facilities and Transformation' masterplans which received outline planning permission in 2007. Initiated in 2008, the LMF was commissioned by the London Development Agency from a consortium of urban design practices – Allies and Morrison, EDAW (now AECOM) and KCAP. The LMF was to be a design-led medium for mediating dynamically unfolding processes of spatial and socio-economic growth and change to create a piece of city over a period of 20 years and more and thus far more than a conventional masterplan (Figure 4.2).

This approach informs the LCS, which, championed by the same firms, covers 64.48 hectares established for post-Games mixed-use development (Figure 4.2). In general, this area corresponds to the sites of the Games temporary venues and facilities. The urban form principles established in the LCS, which now has planning approval, are further supported by the development principles of the Greater London Authority's (GLA) 'Olympic Legacy Supplementary Planning Guidance' (OLSPG), created in 2012, which covers both the site and the existing urban areas that surround it. (See Plate 4.1.)

Figure 4.2   The LCS neighbourhoods.
Source: The Legacy Communities Scheme (LCS) by the London Legacy Development Corporation.

Through its multiple and extensive drawings, specifications and schedules, the LCS establishes rules for future urban, form recognising the capacity to realise a vision of future place in a number of ways. It is defined as a designed 'framework' because these rules are based on assumptions about population growth, housing demand and local need, as well as of land and development economics that extend decades into the future. Strongly informed by the ideal of convergence, many of them represent ways of managing the realisation of local resources in terms of housing, employment, social infrastructure and amenities, whilst at the same time controlling the scale and layout of development across different planning delivery zones and development parcels. The whole reads as a subtle negotiation between development opportunity and spatialised standards of civic responsibility, and as a compromise between conceptions of the scope for private and public benefit.

Since the early iterations of the LMF, the 'deprivation gap' has been portrayed as a reflection of the Lea Valley's presence as a rift or 'tear' in the urban fabric of east London (LDA, 2009: 57). Echoing the SRF's emphasis on the role of disjointed development in producing the conditions of persistent deprivation, connectivity principles at both local and wider city scales are developed as vital tools for overcoming 'social exclusion'. Whilst the development of the Olympic Park and immediate post-Games transformation provided the opportunity to make connections across the valley's numerous waterways, the LMF and LCS show how delivery zones would be divided up to form a porous matrix of streets and public spaces that link across and into existing fringe neighbourhoods. In the OLSPG, enhanced connectivity is seen not only to afford more convenient and practical access to resources and amenities and to improve the viability of local businesses, but also to form the basis of a new neighbourliness, promoting 'community stability and cohesion' (GLA, 2012: 18). Implied in this is a strong faith in the capacity of a planned public realm to create the basis for social connectivity, which in somewhat idealistic terms downplays the potential politics of public space use and interaction, which researchers have highlighted as features of multicultural neighbourhoods (see, for example, Dines, Cattell, Wils and Curtis, 2006).

Connectivity forms one strand of the LCS's larger emphasis on the 'integration' and 'co-location' of different land uses, types of housing, services and social infrastructure. The result in terms of built form has, since the early days of the LMF, been a mixed-use neighbourhood approach that clusters future development into a series of focal development areas (Plate 4.2).

These are said to have the potential to acquire spatial as well as social 'characters' (LDA, 2009) that complement the diverse fabric of mixed-use neighbourhoods at their edges. Mixed-use development makes it possible, in a practical sense, for people to live and work in the same area, for land to be used efficiently, and for services, social infrastructure and commercial development to be collocated in innovative ways that reduce the demand for transport. These advantages are communicated in Plate 4.2, an 'illustrative zonal masterplan' for Planning Development Zone 8 in which the ingredients of a neighbourhood mix are portrayed. In the OLSPG it is asserted that mixed use development 'will help deliver convergence' (GLA, 2012: 23), not only by providing access to resources that the site has been seen to lack historically, but also by providing a spatial articulation of the ideal of harmonious diversity, which, as Lees (2003: 77–78) argues, was a feature of the discourse of the Urban Renaissance. The issue here is not with mixed-use development *per se*, but the reliance on urban form to realise convergence. In this, there appears to be all too little consideration of the potential for the most diverse and lively settings to be far from harmonious, as their life and offer is shaped by competition, market pressures and property values, as by social tensions, power imbalances and territoriality. It would be the careful management of these dynamics alongside urban form that would perhaps lead to convergence.

Echoing the quantitative analysis underpinning the convergence strategy, the emphasis of the LCS on closing deprivation gaps is also on quantitative data. In the lengthy 'Social Infrastructure and Housing' sections of the LCS, locations and levels of provision are described in square metres, percentages, distances, catchment areas, numbers of service providers or places, and are calculated to reflect the numbers and age profiles of anticipated future residents. The LCS makes clear that social infrastructure is designed to 'match the demands of future residents of the LCS' on the service catchments in which they will be located, but that existing residents will be encouraged to 'use and benefit from the services provided' (LLDC, 2011: 174).

The density of development clearly has a bearing on the amount of social infrastructure which is planned and provided. In this regard it is significant that the number of housing units originally planned to an upper limit of 10,000 has been reduced to 6,870. The emphasis has moved away since 2009 from one of maximising development potential to a more mixed scale, ranging from as low as 150 habitable rooms per hectare to up to 1,000 and averaging at between 450 and 600. The density targets provided for each neighbourhood in the LCS reflect the outcomes of negotiation between the potential for intensive development, the goal of realising convergence, and understandings of the relative economic viabilities of different densities and types of development. The LCS's 'Density Parameter Plans' represent the spatial outcomes of this negotiation. The density ranges offered to developers within these plans demonstrate recognition of the potential for the viability of the model to transform according to wider economic circumstances. However, the major danger would appear to be less that the conditions set for development through the LCS could make development unviable, than that what makes the convergence possible is the development of significant tracts of unaffordable housing. Since the fulfilment of convergence goals over 20 years through design and development of the Olympic site relies on the ability of the private sector to bear associated costs through various development phases, so it also depends on the ability of the London Legacy Development Corporation to manage the form and offer of development, but also and more crucially the politics of displacement by development within their piece of London.

Conclusions

When I began to write this chapter, I was struck anew by how deeply embedded the idea of convergence has become in Olympic legacy planning. On the one hand, it is hard to fault a long-term development process which sets out to address the historical problems of redevelopment by explicitly recognising local needs. And yet, as this chapter has shown, the task is far from straightforward. Convergence is predicated on an account of deprivation that pays relatively little heed to the real world in which it is both constituted and experienced through subtle and complex, not always negative relationships between people and places. This has a strong bearing on how the challenge of convergence is conceptualised in spatial planning and physical development terms. It may be too early to argue, following Young and Willmott, that planners 'have still put their faith in buildings, sometimes speaking as though all that was necessary for neighbourliness was a neighbourhood unit, for community spirit a community centre' (1962[1957]:198), but this is a real danger.

For all that this planning is seeking to carefully provide in terms of an adequate mix of place-based resources and amenities, the issue remains that as neighbourhoods in the Olympic site are not constructed around existing social groups, convergence can only really be realised on this spatial scale for imagined communities. Whilst this is not necessarily problematic in itself – after all development is realised in response to real growth and market pressures, and the careful control of provision is impressive – its difficulty lies in the promise of regeneration to existing local residents who remain vulnerable to displacement.

References

Beauregard, R. (2003). *Voices of Decline: The Postwar Date of U.S. Cities*. New York: Routledge.

Bernstock, P. (2014). *Olympic Housing: A Critical Review of London 2012's Legacy*. Farnham: Ashgate.

Campkin, B. (2013). *Remaking London: Decline and Regeneration in Urban Culture*. London: I.B. Tauris & Co. Ltd.

Carmona, M. (2009). 'The Isle of Dogs: Four development waves, five planning models, twelve plans, thirty five years, and a renaissance … of sorts'. *Progress in Planning*, 71, pp. 87–151.

Cattell, V. (1997). 'London's Other River: People, employment and poverty in the Lea Valley'. London: Middlesex University, Occasional Paper, School of Sociology and Social Policy.

Cattell, V. (2012). *Poverty, Community and Health: Co-operation and the Good Society*. Basingstoke: Palgrave Macmillan.

Davis, J. and A. Thornley. (2010). 'Urban regeneration for the London 2012 Olympics: Issues of land acquisition and legacy'. *City, Culture and Society*, 1(1), pp. 89–98.

Dines, N., V. Cattell et al. (2006). *Public Spaces, Social Relations and Well-being in East London*. Bristol: Policy Press for the Joseph Rowntree Foundation.

Fainstein, S. (1994). *The City Builders: Property, Politics and Planning in London and New York*. Oxford: Blackwell Publishers.

Greater London Authority (2012). *Olympic Park Supplementary Planning Guidance*. London: GLA.

The Five Host Boroughs (2009). *The Strategic Regeneration Framework*.

Hall, P. (1998). *Cities in Civilization: Culture, Innovation and Urban Order*. London: Weidenfield.

Hamnett, C. (2003). *Unequal City: London in the Global Arena*. London and New York: Routledge.

Hatcher, C. (2012). 'Forced evictions: Legacies of dislocation on the Clays Lane Estate', in H. Powell and I. Marerro-Guillamón (eds.) (2012). *The Art of Dissent: Adventures in London's Olympic State*. London: Marshgate Press.

Harvey, D. (1988). *The Condition of Postmodernity: An Enquiry into the Origins of Cultural Change*. Oxford: Basil Blackwell.

Harvey, D. (2008). 'The right to the city'. *New Left Review*, 53, pp. 23–40.

Harvey, D. (2003). *The New Imperialism*. New York: Oxford University Press.

Jacobs, J. (1993) [1961]. *The Death and Life of Great American Cities*. New York: The Modern Library.

Lees, L. (2003). 'Visions of "urban renaissance": The Task Force report and the Urban White Paper', in R. Imrie and M. Raco, (eds.) (2003). *Urban Renaissance? New Labour, Community and Urban Policy*. Bristol: The Policy Press.

London Development Agency (2009). People and Places, a Framework for Consultation: The Legacy Masterplan Framework.

London Legacy Development Corporation (2012). Local Development Scheme. http://www.londonlegacy. co.uk/media/LLDC-Local-Development-Scheme-FINAL-OCTOBER-2012.pdf

Marcuse, P. (1986). 'Abandonment, gentrification and displacement: The linkages in New York city', in N. Smith and P. Williams (eds.) (1986). *Gentrification of the City*. London: Unwin Hyman.

Malpass, P, and A. Murie (1999). *Housing Policy and Practice. 5th Edition*. Houndsmill and New York: Palgrave.

Olympic Park Legacy Company (2012). Legacy Communities Scheme EIA: Environmental Statement.

Oxford Economics. (2010). Six Host Boroughs Strategic Regeneration Framework – Economic Model.

Poynter, G and I. MacRury (2009). *Olympic Cities: 2012 and the Remaking of London*. Farnham: Ashgate.

Raco, M. and E. Tunney (2010). 'Visibilities and invisibilities in urban development: Small business communities and the London Olympics 2012'. *Urban Studies*, 2010, pp. 1–23.

The Six Host Boroughs and Mayor of London (2011). Convergence Framework: Annual Report 2009–2011.

The Six Host Boroughs and Mayor of London (2012). Convergence Framework: Annual Report 2011–2012.

The Six Host Boroughs and Mayor of London (2012). Convergence Framework and Action Plan 2011–2015.

Steadman-Jones, G. (1971) *Outcast London: A Study in the Relationship Between Classes in Victorian Society*. Oxford: Oxford University Press.

Supporting Healthier Lifestyles Strategic Regeneration Framework Steering Group (2011). Healthy Urban Planning in Practice for the Olympic Legacy Masterplan Framework.

Travers T., R. Tunstall and C. Whitehead, with S. Pruvot (2007). Population mobility and service provision: A report for London Councils. London: LSE London.

Watt, P. (2013). 'It's not for us'. *City: Analysis of urban trends, culture, theory, policy, action*, 17(1), pp. 99–118.

Whitehead, C. and T. Travers (2011). *The Case for Investing in London's Affordable Housing*. LSE London.

Yelling, J. (1986). *Slums and Slum Clearance in Victorian London*. London: Allen & Unwin.

Chapter 5
Legacy From the Inside

Ralph Ward

Introduction

London's claim to be the 'legacy games' is fair. The Mayor's decision to bid to host a Games based in Stratford, east London, was driven by the expectation that the investment it would require and the image it would generate, would boost the importance of Stratford as a location for new metropolitan development, commercial activity, and jobs (Institute of Government 2012: 13). This goal has been a core component of regeneration policy in east London for decades (Government Office for London 1996).

But the regeneration challenges of east London go some way beyond the development of Stratford. They centre on social issues linked to poverty: low incomes, poor educational attainment, low skills and unemployment. And the east London development deficit does not end at Stratford but exists across the sub-region as a whole. Large areas of underused and brownfield land still exist on a large scale, particularly in the old industrial areas of London's Docklands and the River Lea.

The regeneration rhetoric which accompanied the bid, and the Games, implied that these issues, too, were part of the urban agenda which the Games would address – or even resolve; the bid spoke, for example, of 'the transformation of the heart of east London for the benefit of all who live there' (London Candidate File 2003). The Olympic Park, it was widely said, would not be another Canary Wharf, which had become (unfairly) notorious as an island of international wealth creation detached from, and inaccessible to, the impoverished east London communities in which it squatted.[1]

Early, extensive and comparatively successful planning for the post-Olympic use of the Olympic Park and stadia has occurred in London. This seemingly simple outcome has proved to be beyond many past Olympic cities, and of itself offers impressive and important lessons for other bidding cities of which London is justifiably proud. Stratford is on track to become a metropolitan development location, and evidence suggests Olympic legacy is playing its part. Tumbleweed will – probably – not be seen blowing down empty streets of London E20.

Regional and social impacts, however, are less easy to locate. In practice, these bigger ambitions were always opaque and ill defined. Once a major part of the bid, they now seem largely eclipsed by the success of the event, and the progress of the immediate Stratford legacy. The Games did prompt a variety of new urban management structures and initiatives to attempt to address social problems, but these proved short-lived. Subsequent government cuts in local authority funding, together with its policy towards housing and benefits in particular, threaten to impact on east London communities in ways which may well undermine the positive outcomes of the Games. What, then, can London tell us about if, and how, a mega-event might deliver urban impacts that extend beyond the successful legacy development of the Park itself?

The Origins of London's Legacy

London's bid evolved through the coming together of two strands – the Mayor's desire to accelerate the 'regeneration' of east London and the desire amongst the sports community, nationally and locally, for the UK to mount a strong bid to host the event (Institute of Government 2013: 13–16).

---

1 Proximity does not of itself confer opportunity. The City of London has demonstrated for rather longer than Canary Wharf that London's global financial economy can be expected to bring few benefits to its immediate impoverished neighbourhood.

When London government was re-established in 2000 and a Mayor of London elected for the first time, candidates made no mention of a bid to host the Olympics. What *was* on the elected Mayor Ken Livingstone's agenda was to attempt to redirect investment and growth to east London to 'rebalance' the London economy (GLA 2004). East London's under-occupied and brownfield land offered a cheaper and more sustainable basis for growth than congested west and central London, and directing more income, wealth and development to east London offered to address social deficits which had existed in the area since its earliest days as London's industrial, working class zone (LDA 2007: 41–50).

This was not a time for Olympic adventures. London's Mayoral institutions, including Transport for London and the London Development Agency, were all newly formed and were finding their feet. London had long-standing problems which these new institutions were expected to address, rather than create new challenges for themselves. The limited budget of the Mayor and Greater London Authority (GLA) anyway left little extra for probable Olympic costs, and London was emerging bruised from its farcical attempt to deliver the 2005 World Athletic Championships.

At the same time, however, there *was* serious interest in a bid from a well organised sports lobby, London International Sport (LIS), which had support from the British Olympic Association (BOA). It was clear to them, following the failure of past Manchester and Birmingham bids, that the only UK candidate city of interest to the IOC was London. The Department of Culture Media and Sport had embarked on a national policy ('The Decade of Sport') to attract events to the UK and they watched with interest. They had sponsored the reconstruction of Wembley in 1996 as the national football and athletics stadium, with an eye to its potential use one day as an Olympic stadium, and although the athletics component proved problematic and was subsequently dropped from the project during construction, it was still seen as a contender.

LIS were well versed in London development policy and in 1999 had already orchestrated a review of potential locations that could practically form the basis of a bid. They had boiled the putative contenders down to a preferred short list of two: Wembley in north-west London, and Stratford in east London (LPAC 1999). So, when the Mayor took office there was a proposal on the table for him to consider. Not known as a sports enthusiast, what did appeal to the Mayor was the prospect of the substantial attention and funds the bid would bring to Stratford. He supported a bid on the condition that Stratford was the chosen location; a cost-benefit analysis appeared to confirm its viability (Arup 2002), the government, confident of a Paris victory, was roped in to guarantee the cost and a bid took shape.

Why Stratford?

Stratford had fallen on hard times since its Edwardian heyday as a national centre for rail manufacture and as the former 'shopping centre' serving a large Essex hinterland (Fraser 2012). But it retained excellent rail connections together with substantial vacant and underused land. It also had a potent strategic position. It was defined in the government's Thames Gateway Strategy as the Gateway's western hub (DOE 1996), and theoretically at least simultaneously provided a southern 'hub' for the emerging 'growth corridor' northwards from London to Stansted and Cambridge.

It also offered a combination of identity and accessibility that distinguished it from many of east London's more disconnected brownfield development locations in Docklands. Already its 'potential' had led the government to re-route the Channel Tunnel Rail Link through Stratford, and support the construction of an international and domestic Stratford high speed station (completed in 2007), which in turn had generated a plan to develop a giant new retail commercial and housing development, subsequently christened Stratford City, on land beside the station.[2] Other investment, in the form of the Docklands Light Railway and the Jubilee

---

2   Stratford City was a concept developed by London and Continental Railways, (LCR) successful bidders for the initial Channel Tunnel Rail Link (CTRL) concession from 1995 onwards. It utilises former rail land at Stratford transferred to the successful CTRL bidder. Land at Kings Cross was treated similarly.

Line, had been undertaken to link it to Canary Wharf. Longer term ideas around a possible Crossrail station were circulating.[3]

Stratford also, theoretically, offered several hundred acres of further riverside development land that was currently underutilised, in largely outdated low grade industrial use; theoretically, because redevelopment was hampered by intractable constraints, notably contamination, multiple (over 500) separate land ownerships, and extensive strategic utilities infrastructure, including a twin line of electricity pylons running across it. These constraints effectively ruled out, because of cost, any further growth of Stratford beyond 'Stratford City', despite the investment being placed there.

Construction of an Olympic Park offered a one-off opportunity to provide the money and muscle to tackle these constraints and liberate its development potential. It also offered to bring unparalleled national and international publicity, and, potentially, a new positive image that would serve to consolidate the Stratford City development, then still at the planning stage, and potentially attract a richer civic and cultural dimension to plans, which at the time were largely commercial and retail in character.

The Wider Expectations of an East London Bid

The Stratford Olympics also prompted more indirect and intangible expectations: that it might attract new investment and development more widely across east London and the Thames Gateway, and that it might generate social and community benefits that would address the social deprivation which characterises much of east London. These expectations derived more from the geographical fact of Stratford's location in the heart of London's most acute area of deprivation and its status as the 'western hub' of the Thames Gateway, than from any evidence based chain of logic that linked these outcomes to the presence of the mega-event. In practice the relationship, if any, between new development and the social welfare of the surrounding community is a matter of debate within the professional and academic world of planning and regeneration. It was, therefore, an argument that was handled with great caution by the boroughs,[4] who saw the value of the Games and its development legacy as the source of modest but useful numbers of construction and end user jobs, of affordable housing, and better social facilities – a new park, and retail, leisure and sports facilities – that would make the area a more satisfying place to live. Naïve expectations that this would impact on poverty on any scale were largely absent; on the contrary threats such as disruption, and higher land and housing costs, were more real (LB Tower Hamlets 2008).

Less guarded was expectation that a Stratford Olympics might attract new investment and development to other locations in east London. The Olympic site formed part of a wider development area known as the 'Arc of Opportunity', extending down the Lea to the Royal Docks, which had been the subject of a number of visioning initiatives over the years (LB Newham 1996). The putative housing and employment outcomes resulting from the development of the area as a whole tended to be conflated with those of the Olympic site itself, as potential Olympic outcomes.

The Transformation of East London: The Bid and Beyond

As Singapore approached, any hesitancy about the real regeneration value of the Games was not translated into the bid itself, or into the increasingly breathless rhetoric which accompanied it. Regeneration offered a novel argument and positive images which turned out to have enormous appeal to the authors of the bid and the bidding process, and arguably helped London clinch the decision in Singapore.[5] The London bid, it turned out, was not simply about having fun, or even showing off the city, it was about *doing good*. This played expertly into the IOC's growing concern not simply for an afterlife for park and sports facilities, but to see

---

3   Crossrail is the rapid east-west new generation underground link through central London, which, after many years' debate is now under construction and due to open in 2018.

4   The self-styled 'Host' east London boroughs of Greenwich, Hackney, Newham, Tower Hamlets and Waltham Forest contained most of the London 2012 Olympic venues.

5   Local children were famously paraded at the event itself.

legacies of broader value to the host city (Cashman 2002). In this London was well ahead of the IOC, and arguably has helped to define the IOC's approach to legacy.

More prosaically, but of importance to the Mayor and UK government, was the anticipated land value uplift consequent on Olympic investment. The financial receipts resulting from the translation of low value industrial land with high post-Games riverside development values offered to help mitigate the enormous cost of the Games and provide some foundation for longer term post Games legacy costs (Grant Thornton et al. 2007).

In terms of national domestic politics, 'regeneration' also helped to provide a rationale of sorts for spending so much money on a four week sports party, and for spending it, yet again, in London. Earlier decisions to select London locations for the new national football stadium and the Millennium Dome, rather than competing sites in other UK cities, had become a source of provincial grievance. But with east London by some way the statistically most deprived location in the country, it was unarguably the most deserving place to site an Olympic project with such profound regeneration outcomes.

Consequently when the London bid to host the Games was approved, uncritical expectations of the urban impact this might have were at their zenith. A striking statistic was deployed that continues to surface even today (DCMS 2012): that average life expectancy drops at each station as you travel east on the Jubilee Line. True, but exactly what this was intended to imply for the Olympic decision was not made explicit. Was 2012 actually going to help people live longer?

Planning for the Legacy of the Park

Planning responsibility for legacy, until the bid was won, initially lay with the London Development Agency (LDA). This put responsibility for both Games and legacy planning in the same hands, those of a body whose instincts lay with legacy outcomes rather than the event itself. From the outset, therefore, the use and form of the Park post-event was given detailed attention. A masterplan for the Olympic Park in both Games and legacy mode was commissioned by the LDA as part of the initial town planning consent required for the bid process, to demonstrate how the form chosen for the Games could be transformed into space with real legacy after the Games (LBs of Greenwich, Hackney, Newham, Tower Hamlets and Waltham Forest 2003). In the jargon of the time, it sought to demonstrate how the Games were simply a stage (albeit a grand stage) in the evolution of Stratford – a means to an end, rather than an end in itself.

The character and form of the post-Games park, in terms of urban land use and structure, was not firmly defined. The Mayor's first London Plan, then being drawn up, said little in hard terms about the legacy outcome it sought.[6] In addition to a legacy masterplan for the Park, a 'regeneration framework' was drawn up for the Lower Lea area, which sought to demonstrate how a bigger development legacy than simply the Park itself might be fashioned from the Olympic process (EDAW 2005).

Following the success of the bid and the award of the Games, the context changed. Delivery of the Olympic event, rather than legacy, became the (obvious) priority. The government, through the medium of a special purpose quango called the Olympic Delivery Authority (ODA), took a grip on the development of the Olympic Park, leaving the LDA responsible only for legacy. The principal legacy concern of the new Olympic Board, set up to oversee the Games and legacy, centred not unreasonably on the successful post-Games use of the Park and, in particular, the stadium and other arenas. As a result the concern for some of the less tangible aspects of Olympic legacy tended to become overshadowed. This was reflected in the 'Commitment to Sustainable Regeneration', jointly produced by the LDA and ODA in 2007 (LDA/ODA 2007). It provided a framework both for the immediate 'transformation works' that would be required immediately after the games, and for a longer term legacy costing plan. The budget had by now settled at around £7bn, focused almost entirely on the Games. Of this a mere £300m was allocated for what was called 'legacy transformation' i.e. the removal of some of the buildings and bridges and the remodelling of the Park and stadium into legacy

---

6　It simply spoke of the 'comprehensive development of Stratford as a new commercial, retail and residential area of London, making the most of its European links and its pivotal role connecting the London-Stansted-Cambridge corridor and Thames Gateway growth area'.

mode. Nothing further was allocated from the Olympic budget for post-Games development. A new budget would be required.

The LDA commissioned a Legacy Masterplan Framework (LMF) which followed fairly closely the same themes that characterised the earlier pre-bid legacy plans – largely a housing led environment, interspersed with the odd world class sports stadium, and parkland. An alternative point of view, that the Olympic site represented a commercial World City asset and that a transformative legacy should perhaps seek to encompass corresponding business, education, cultural and other civic activities, did not at the time receive significant attention. The Mayor's London Plan, which might have been the vehicle for exploring these issues, chose to say little further in terms of strategic priorities for the site. Possibly the commercial nature of Stratford City, whose Phase 1 retail mall was finally beginning to take shape, with a further approved 4 million square feet of offices in the pipeline, was thought to exhaust any credible commercial market in this part of London. In practice, Stratford City and the Olympic Park were (and still are) largely pursued as separate projects, beyond essential synchronisation of infrastructure works.

Planning for Wider Legacy

By 2007, planning for both Games and legacy had become centred on the Olympic Park. Where the development of the Park offered opportunities to bring wider infrastructural or employment benefits, these were exploited and exploited well (LDA 2012). The capacity of the combined heat and power system constructed for the Park extends well beyond its boundaries, and the LDA introduced skills, business and employment schemes to direct economic benefit towards the local area. At the same time, Olympic pressures inevitably led to decisions that created difficulties for the legacy planning and subsequent management; the construction of a largely temporary athletics stadium, for example, and the decision to relegate the one building with a supposed commercial legacy, the Media and Broadcasting Centre, to the least commercial location on the Park: its north-west extremity.

The broader impact if any of the Olympics on the welfare of the wider East End had become increasingly overlooked by the Olympic process. Even the physical development plan for the Park left untouched its peripheral, or fringe areas as they became called. Responsibility for creating new connections from the Park to its surrounding communities was redefined as a legacy rather than an Olympic problem. The term 'cliff-edges', used pejoratively to describe the literal and metaphysical separation of Canary Wharf from its hinterland, now re-emerged uncomfortably in an Olympic context. The Department of Communities and Local Government (DCLG), the Government department responsible for cities and regeneration, and increasingly regarded as the Government department accountable for the success of the east London legacy, began to see a potential 'reputational risk' emerging for the department and its ministers.

In the course of 2007, three new initiatives evolved to attempt to define a bigger east London legacy programme and establish the governance and implementation structures that might manage it. While the Olympic Board remained notionally 'in charge' of all things Olympic, including legacy, the planning and management of London legacy, at least, shifted towards those who had the main stake in its success and impact on London, and could bring an informed focus to bear.

The Olympic Park Regeneration Steering Group was established in May 2007 as a strategic governance body for the London legacy. Constitutionally it brought together all three levels of government: national, in the form of Secretary of State for the Department of Culture, Media and Sport (DCMS), and Secretary of State for the Department of Communities and Local Government (DCLG); regional, in the form of the Mayor; and local, in the form of the five leaders and mayors of the so-called Olympic 'Host Boroughs'. This was a unique structure and in principle offered extraordinary power, had the opportunity been grasped. Its name, however, largely gives it away. It tended to concern itself with plans for the Olympic Park in Games and legacy mode, and the resolution of potential differences between partners, rather than a more creative approach to legacy. Meetings were largely choreographed and ceremonial; on occasion its quarterly meeting was cancelled due extraordinarily to a perceived lack of significant business to consider. Part of the 'problem' lay in the lack of a clear and independent executive with the capacity to generate a legacy agenda, but principally it reflected a lack of will to explore a bigger legacy regeneration agenda than that of the Park.

The second initiative arose from the appointment of an LDA Director of Legacy. Hitherto, the LDA's legacy engagement was split between different teams – the property team buying the land, the planning team managing the SRF, the employment team managing skills and training initiatives, and so on. These teams performed very well in their particular fields, but the creation of a Director of Legacy finally brought focus to the process. The new Director, Tom Russell, was closely associated with the legacy programme built around the Manchester Commonwealth Games. Manchester had created a planning tool they called a 'Strategic Regeneration Framework' (SRF) which he sought to introduce into the London context

To consolidate this, the Department of Communities and Local Government used its influence to set in motion the 'East London Legacy Board', an assembly of the major public bodies and agencies whose responsibilities and programmes had an impact on the operation and regeneration of east London. It included government departments and agencies ranging from the Environment Agency, the Homes and Communities Agency, to Jobcentre Plus. The rationale of the Board was to develop a shared set of priorities for east London, to enable activities to operate in a more coordinated and effective way. A fairly classic corporate model, it met for over a year, but again lacked either the will or the executive capacity to develop an agenda or effective modus operandi.

Host Borough Unit and OPLC

In 2009 the structure of legacy management shifted again. A new body, the Olympic Park Legacy Company (or OPLC), was established to take over responsibility for the management of the Park legacy from the LDA (London Assembly 2010). It had been the intention of the LDA to set up a new organisation to take over the management and planning of post-Games legacy; the arrival of a new Mayor in 2008 added momentum to the process. The OPLC was a publicly owned company limited by guarantee, jointly owned by the government in the shape of the Department of Communities and Local Government, and the Mayor. The OPLC was defined wholly by the boundaries of the publicly owned Olympic Park and had no locus for wider legacy engagement; in some ways a mirror image, for legacy, of what the ODA was for the Games, though with a different set of powers and rather a different budget.

The OPLC was given clear responsibility to draw up a legacy development plan, investment programme and budget for the Olympic site, post-Games. The intention was for all land and sports venues, including the parkland, to be returned post-Games from the ODA to the OPLC, which would then devise suitable disposal and management arrangements for parkland and venues, and sell land for development. The main immediate task that the OPLC set itself was to review the Legacy Masterplan Framework it had inherited from the LDA. Once again, instead of taking the opportunity to explore some more transformative non-residential uses for the site, the OPLC endorsed the original housing led model, though this time seeking to reduce the scale and density to create an environment more reminiscent of established London residential streets (OPLC 2010).

Changes also took place in the leadership of the SRF, which in 2009 was effectively taken over from the LDA by the 'Host Boroughs'. The Host Borough Unit had existed since the award of the Games back in 2005, set up by the five east London local authorities in which Olympic venues were situated. Its stated aim was to be 'creative rather than defensive and to shape legacy strategy and outcomes' (Brown et al. 2012: 230) though in practice the Unit had little budget or capacity for the kind of analysis, research and policymaking that was required.

This changed in 2009, with the appointment of a new Chief Executive of the Unit, who saw the Olympics less as mechanism to deliver specific benefits, and more as an opportunity, or perhaps an obligation, for the city to finally address the social deficits which the disadvantaged local population of east London had been left to put up with for centuries and more. His vision of the SRF was a programme which in so far as it had connections with the Olympics at all, channelled the inspiration, goodwill and commitment generated by the Olympics into a set of far more fundamental changes to policy than the SRF had previously entertained. The target of the new SRF, christened the Convergence report, was over an approximate 20 year period, to reduce the myriad inequalities experienced by the deprived population of east London, such that east London's social conditions – employment, educational attainment, skills, health, and housing in particular – were improved to match the London average, rather than as today falling well below it (Host Borough Unit 2009).

The Convergence SRF was fundamentally a manifesto rather than a programme. Put together by a tiny team over a few months, it could not begin to develop a credible detailed programme for such a massive and fundamentally political, rather than technical, task. The problems of east London are lodged deep in its social history, and the social structure of London as a whole. What the Convergence Report did manage to establish, amongst a large and diverse political group ranging from the boroughs to the government, was agreement that something more needed to be done. Even David Cameron, in his inaugural speech as Prime Minister was moved to say, 'Let's make sure that the Olympic legacy lifts East London from being one of the poorest parts of the country to one that shares fully in the capital's growth and prosperity' (Winnett 2010). But little headway was made around what that might require, or how it might be achieved, particularly politically and organisationally. Shortly after the Prime Minister's speech, the government imposed the sharpest cuts in London on local government funding to several of the Host Boroughs. Unfortunately by placing the bar so high, the Strategic Regeneration Framework in its final form overlooked what issues, albeit perhaps modest, the Olympic investment might actually help to address in a more direct form. Ultimately, it has failed to provide a practical agenda around which to mobilise political and community energy in east London.

Current Legacy Structure

The management of Olympic legacy continues to mutate. The OPLC has been replaced by the London Legacy Development Corporation (LLDC), now wholly accountable to, and, in fact, notionally chaired by, the Mayor of London. It has significantly greater powers than the OPLC in that it has power to compulsorily purchase land, and is the local town planning authority for both strategy and planning decisions. It also covers a larger area, with boundaries extended into the fringe areas to the west and south of the Park. This is significant as it enables the legacy process to engage directly with physical development in the wider Lea Valley area. A further extension of the body down to the Thames is also under consideration.

The government, in the form of the Department of Communities and Local Government, has withdrawn from the London legacy agenda with a certain amount of relief. The Mayor's desire to take full responsibility for the legacy body chimed with the then Secretary of State's policy of 'Localism' – to relocate power and responsibility onto locally elected representatives. The government retains a small coordinating body in the Cabinet Office which oversees governmental inputs to the legacy process, but London legacy has now been largely delegated to the Mayor. Meanwhile the Host Borough Unit has been retitled the Growth Borough Unit, been shrunk further and is now engaged principally with monitoring the Convergence statistics. The future of the Unit is currently under review by the Host Boroughs.

Concluding Thoughts

Evaluating the impact of the Olympics on the regeneration of east London is difficult for several reasons. The timescales involved in urban change are generally rather longer than political context allows, and this is particularly the case where the hoped for urban impacts are so profound. This has not deterred the government's official evaluation process, or meta-evaluation, from concluding already that, '[w]hile the full legacy impacts on the regeneration of east London will not fully emerge for a number of years, what is already apparent is that the planning and regeneration for the Olympic games have made a significant contribution to the physical transformation of east London …' (DCMS 2013:163). Further difficulty is added by the lack of any clear definition of what this hoped for transformation of east London might look like, literally and metaphorically. It has a pleasing imprecision which appeals to politicians, bid writers and, one might suspect, the IOC, but it is a headache to the analyst. The absence of clear theories of change or chains of logic linking the Olympics to many of the expected outcomes, the social outcomes in particular, together with the wealth and range of other programmes operating in east London, and other investments such as Crossrail, makes the attribution of causality and the measurement of benefit into an art form.

Analysis of legacy often draws distinctions between 'types' of legacy – for example, between direct and indirect impact, between tangible and intangible legacies (see for instance Poynter, 2006). How these might be achieved is a complex and layered process, which can become lost in Olympic bids and subsequent programmes. Outcomes which are clearly beyond the realms of purely Olympic delivery and should be analysed and managed as such, nevertheless become reclassified as 'Olympic'. Mega-event bids – London is an excellent example – carelessly conflate them, and the Olympics somehow is expected to deliver something which decades of urban policy has failed to shift In practice, the more indirect and intangible the outcome, the less logically it can be achieved by the event itself, and the more it rests upon *what else* you do in parallel with the event – a consequence, in other words, of the city's urban, not Olympic, policy and programmes.

So, let us simplify this by removing the extraneous legacy noise generated by London's bid, and return to why we bid in the first place: to attract investment to Stratford that would consolidate its then current development plan and enhance the scale and range of its future development as a business centre for east London and the city as a whole, and to do this by holding an event that would create a development area of the highest quality and simultaneously generate a new image that would transcend radically the image that the area currently possessed.

On this analysis, London has done very well. The impact of the Olympics on Stratford's image, and on its civic cultural and commercial development prospects, has already been profound. Despite taking almost two years for the Park itself and major venues to be fully reopened to the public, the response has been positive, notwithstanding the fact that substantial areas remain development sites pending their disposal. The Stratford City shopping centre continues to trade very successfully, the former Olympic Village is now open for business and attracting town planning awards, and substantial unexpected new housing is under construction on land within the 'old' town centre which has long lain vacant.

Significantly, the Olympic site itself is attracting new commercial and cultural development linked to London's growth economy. Wistful pre-Games hopes that the Broadcasting and Media Centre might have a commercial digital media legacy have materialised. Less than two years after the Games, it already houses a data centre and operating TV studios, and space for digital education and training and small business incubation is in development, involving, amongst others, Loughborough University. Elsewhere on the Park, the Victoria and Albert Museum is negotiating to build new exhibition space devoted to contemporary design, and even though the initial plans for a new campus of University College London on the site of the Carpenters Estate were abolished, UCL is still in discussion over the development of a new satellite campus on a different site in east London.[7] Birkbeck, University of London has completed a new joint building with the University of East London within the existing Stratford town centre. If and when the International Rail Station might open for international services is not yet clear, but there is interest from international operators when Eurostar's monopoly of the route expires in 2018. Crossrail will open in 2019.

The Olympics will have the transformative effect on Stratford that lay at the heart of the bid. This demonstrates the overriding legacy importance of getting the Games 'right' and getting the legacy Park 'right'. Getting it right means:

- 'Putting it in the right place' – in other words, locating the event somewhere already defined as a development zone, where Olympic investment and image can consolidate established past (and future) policy, and in which expectations of future market interest are realistic.
- Holding a successful Games. This provides the feel good context around which a positive image can be generated.
- Undertaking authentic planning for legacy early on, with a focused client in place well before the event. London's arrangements were not perfect in this respect, but have been proved broadly effective.

Will this bring wider benefit to east London and, if so, in what form and how? This is not a matter for the Olympics. This is a matter for the city, and how it addresses its urban problems. Where the Olympics can potentially have an impact is to offer the city the rationale, motivation, and perhaps inspiration to review and improve how it organises itself to manage its urban renewal – in particular its engagement with its citizens.

The impact of the Olympics on London's political and professional capacity to address the regeneration of east London has been chequered. The hard edged legacy planning that has taken place and the machinery to achieve results have been focused almost wholly on the Olympic site. More strategic initiatives have been tested to develop

---

7   See the following UCL website for updates: http://www.ucl.ac.uk/olympic-park/faqs

a new understanding of, and approaches to, the future of the region, but none of these has become established, let alone succeeded in generating a clear set of strategic priorities around which to build a more comprehensive renewal programme. The implications of this 'failure' are more significant in the light of the Mayor's abolition of the LDA, which hitherto provided the capacity for London to think and intervene on a strategic basis.

Better orchestrated regeneration of east London is a step, therefore, that London has still to take. Given London's historic antipathy to strategic planning, this may take some time. The transformation underway in Stratford, authentically attributable to the Olympics, may give both the material and the motivation to embark seriously on the exercise.

References

ARUP (2002). *London Olympics 2012: Costs and Benefits.* London: ARUP/Insignia Ellis.

Brown R., G. Cox and M. Owens (2012). 'Bid, delivery, legacy – Creating the governance architecture of the London 2012 Olympic and Paralympic Games and Legacy'. *Australian Planner*, Volume 49, Issue 3, 2012, Special Issue: Planning For Olympic Games, pp. 226–238.

Cashman R. (2002). *Impact of the Games on Olympic Host Cities.* UAB Barcelona: Centre d'Estudis Olímpics (UAB).

DCMS (2012). *Beyond 2012. The Legacy Story.* London: DCMS.

DCMS (2013). *Meta Evaluation of the Impacts and Legacy of the 2012 Olympic Games.* London: DCMS.

DOE (1995). Thames Gateway Planning Framework RPG9a. London: DOE.

EDAW (2005). Lower Lea Regeneration Strategy. London: GLA/LDA.

Fraser N. (2012). *Over the Border: The Other East End.* London: Function Books.

GLA (2004). London Plan. London: Mayor of London/GLA.

Government Office for London (1996). Regional Planning Guidance for London RPG 3. London: GOL.

Grant Thornton et al. (2007). 'Towards an outline business plan for the Olympic Legacy Park Report'. London: LDA/ODA.

Host Borough Unit (2009). Strategic Regeneration Framework: An Olympic legacy for the Host Boroughs. London: Host Boroughs Unit.

Institute of Government (2012). Making the Games. London: IOG.

IOC (2009). Candidature Guidelines. Lausanne: IOC.

London Assembly (2010). Legacy limited: A review of the OPLC. London: London Assembly.

London Boroughs of Greenwich, Hackney, Newham, Tower Hamlets and Waltham Forest (2004). Olympic and Legacy Planning Applications, http://legacy.london.gov.uk/assembly/past_ctees/plansd/2004/plansdfeb24/plansdfeb24item08.pdf; accessed, May 10, 2014.

London Borough of Newham (1998). Arc of Opportunity Planning Framework. London: L.B. Newham.

London Borough of Tower Hamlets (2008). London 2012 Strategy and Programme. London: LB Tower Hamlets.

London Development Agency (LDA) (2008). Legacy Masterplan Framework. Mayor of London: LDA.

LDA (2012). Annual Report. Mayor of London: LDA.

LDA (2007). London Brownfield Sites Review: Report to London Assembly Environment Committee, http://legacy.london.gov.uk/assembly/reports/environment/lda-brownfields-review.pdf; accessed May 14, 2014.

London Organising Committee of the Olympic Games (LOCOG) (2012). Candidate File. London: LOCOG.

London Planning Advisory Committee (1999). Site options for an Olympic Village. London: LPAC.

ODA/LDA (2007). Commitment to Sustainable Regeneration. London: ODA.

OPLC (2009). Beyond the Games: The Future of the London 2012 Olympic Park, http://www.londonlegacy.co.uk/media/NM-Sustainability-Lecture-251110.pdf; accessed, May 18, 2014.

Poynter G. (2006). 'From Beijing to Bow Bells', London East Research Institute, Working Papers in Urban Studies. LERI: Mimeo.

Winnett R. (2010). 'David Cameron promises to bring growth to the regions', *Daily Telegraph*, 29th May, 2010. http://www.telegraph.co.uk/news/politics/conservative/7780503/David-Cameron-promises-to-bring-growth-to-the-regions.html; accessed May 10, 2014.

Chapter 6

This is East 20? Urban Fabrication and the Re-making of the Olympic Park: Some Research Issues

Phil Cohen

Introduction

The Olympic Park has been defined in many different ways: as a '*lieu de memoire*' for Olympophiles, a vision of the future city, a local public amenity for East Enders, a global visitor destination, a major venue for sports enthusiasts, and the location of a series of residential neighbourhoods eventually housing over 20,000 people. How can these different functions be combined, if at all? This chapter looks at the plans for the post-Olympic transformation of the site, and the translation of East 20 from the fictional address of the nation's favourite TV programme to the real *mise-en-scène* of a rather different kind of soap opera.

The chapter explores the complex process of urban fabrication: the creation of new infrastructure, its investment with meaning through official promotional discourses and locally situated narratives, and the evolving patterns of social navigation and use of the built and landscaped environment. It considers how the Park was 'imagineered' by the London Legacy Development Corporation (LLDC), how the tensions between 'vista' and 'enclosure' were negotiated in designing an organic landscape based on the principle of order-in-variety as part of a wider strategy to erase status distinctions between different housing tenure categories.

The chapter outlines a theoretical framework and methodology for analysing the strategies of inhabitation likely to be adopted by incoming residents to East Village, drawing on models of stake-holding, environmental perception and 'standpoint aesthetics'. In conclusion an argument is made for widening the interpretive community around issues of urban regeneration and Olympic legacy evaluation to include a broader range of voices than are usually heard in policy debates.

The Process of Urban Fabrication

East 20 represents an important part of the spectacular physical transformation of east London, a transformation that includes the rebranding of the area and its socio-economic regeneration. The post code itself may be 'borrowed' from *EastEnders*, but its 'imagineering' as part of the Olympic legacy breaks with the long standing tradition, epitomised by the TV soap, for representing east London as site of multiple deprivation. East Village itself has been described as a new piece of the city, one which brings together in a single place everything that is best about contemporary urban living. This includes a socially mixed community, lavish green spaces, nearby retail and leisure facilities, and generous transport connectivity. It is also an international visitor destination as an Olympic heritage site, and will be an important public amenity for East Enders. A whole new narrative landscape is thus in process of construction.

This process of urban fabrication has a dual aspect to it. It is about the creation of new infrastructure and about how those who come to inhabit it and make sense of it. The focus of the research that informs this chapter, is how East Village becomes a 'place', i.e. a meaningful location (Lewicka, 2011), but also potentially a 'community' with shared interests and activities (Keller, 2003). Residential place-making can often contain a strong social element of neighbouring (Keller, 2003; Young Foundation, 2010). At the same time, the relationship between neighbourhood, as a physical space, and community, as a set of shared interests and identities – can no longer be taken for granted in contexts where spatial mobility is an increasingly important aspect of certain urban lifestyles (Blokland, 2003; Savage et al., 2005; Watt and Smets, 2014).

For example, research on incomers to new private developments has highlighted somewhat contradictory responses. A form of pioneer community spirit has been identified whereby incomers bond on the basis of shared novelty and minor travails associated with moving in together (Lupi and Musterd, 2006; Watt, 2013). In contrast, research in the north of England (Savage et al., 2005) and Australia (Rosenblatt et al., 2009) has highlighted how incomers can develop a strong sense of identification with their new area *without* necessarily having any strong social links with neighbours. Such elective belonging is based instead upon an aesthetic appreciation of the quality of the built environment and physical landscape (Watt, 2013).

The on-going research underpinning this chapter investigates the configuration of cultural values, social attitudes and subject positions entailed in practices of place and community-making amongst incoming residents to East Village, focussing on different patterns of response to the new environment, the extent and type of neighbouring and community involvement, and orientations towards the LLDC vision.

There are three basic variables of place identity-making:

a. Material infrastructure – design of the built environment – physical geography;
b. Civic imagineering – brandscaping – narrative planning;
c. Patterns of inhabitation – locally situated meanings – cultural and social geography.

We are dealing with a process which is at once real (A), imaginary (B) and symbolic (C). Much official effort is expended by planning discourses in trying to weave these elements into a seamless web of representation, as if they could be neatly stitched together, but in fact for much of the time and in many places the relation between them is fluid, tense and even contradictory.

The Olympic Park: A Preliminary Overview

The *narrative* legacy of 2012 centres not only on the sporting exploits, and the media coverage of the event itself, nor even on the memories of athletes, spectators, volunteers and others who had a direct hand in its delivery, but on the way the Olympic Park, as the chief legacy site, is interpreted by those who live in and around it, who come to work or study there, who visit it as tourists or sports fans, or to picnic or walk the dog (Hopkins and Neal, 2012).

Our previous research into how community groups in east London have responded to the Olympics identified a number of key patterns of local stake-holding related to different kinds of social and cultural capital (Putnam, 2000; Cohen, 2013). There were those – the 'bridgers' –who had the confidence and resourcefulness to engage proactively with the regeneration process, to create partnerships or alliances with others in furtherance of their ends, and who adopted the Olympics as a platform for personal, professional or organisational advancement. In contrast, the 'bonders' were reactive and sought to maintain their own sense of internal cohesion, identity or integrity by creating little niches for themselves from which they could assert a proprietorial sense of the Games as 'their thing', by virtue of moral-cum-territorial claims staked over local amenities and resources. This model will be developed further in the following analysis and applied to the positions that incomers adopt in relation to the opportunities for civic participation afforded by residence in the Olympic Park.

Attachment to place is a multidimensional process, involving existential, moral, aesthetic, cultural and social choices, investments and evaluations, and these vary according to a range of biographical, spatial and structural factors (Savage et al., 2005; Duyvendak, 2011; Lewicka, 2011; Watt, 2006, 2009, 2013; Cohen, 2013). Just as there may be more to making yourself at home in a new place than the tenure of the house or flat which you occupy, so the sense of belonging may involve more than just buying into the developer's prospectus or getting on with the neighbours.

One critical aspect of both stake-holding and attachment to place is how the physical landscape and the built environment are variously perceived, narrated and invested with meaning as sites of boredom or excitement, beauty or ugliness, pleasurable amenity or hazardous traverse (Appleton, 1975, 1990). Our previous research with communities in Docklands has suggested that under some circumstances these constructions may become racialised (Rathzel and Cohen, 2006), but that they also bear on fundamental ways of navigating and dwelling in the world (Ingold, 2000). Some people feel at home in a landscape of exposure, in which wide open spaces and dramatic vistas symbolise for them the prospect of adventure and advancement, a launch pad for their ambitions, or a platform for performance; others prefer a landscape of seclusion, affording hideouts or other defensible spaces which screen them from unwanted intrusions. Such a landscape may represent a desire to find a refuge from the precariousness of everyday life, the instability of market forces, or personal trauma and is more characteristic of 'bonders' than 'bridgers'. The view to be tested here is that these preferred readings of the environment are likely to influence the way incomers go about the business of making themselves at home in the Olympic Park, shaping the kinds of arrival story they have to tell and the ways in which they negotiate the Olympic legacy.

In the mix: East Village as an Experiment in Housing Policy

An important aspect of East Village is that it is a mixed-tenure development. Such mixing of tenures, often owner occupiers with social rental tenants, forms an important strand within contemporary urban policy not only in the UK (Lupton and Fuller, 2009), but throughout Europe, North America and Australia (Arthurson, 2002; Bridge et al., 2012; Lelevier, 2013; Rose et al., 2013). Despite this, the evidence for the efficacy of mixed-tenure developments is somewhat inconclusive (Jupp, 1999; Atkinson and Kintrea, 2000; Bond et al., 2011). One important recent study in France (Lelevier, 2013) has noted the importance of the incomers' previous residential trajectories as well as the spatial layout of the development in influencing neighbourhood interactions; the latter has also been highlighted by Watt and Smets (2014). Our on-going research project is concerned to establish the housing histories of the incomers and their strategies of 'inhabitation' across all the tenure categories.

The East Village development offers a unique opportunity to engage with an ongoing public debate about the limits and conditions of social engineering through 'pepper-pot' housing schemes, precisely because it is *not* composed of the typical mix of owner occupiers and social rental tenants, but is instead formed from a range of tenants (market, intermediate and social) alongside shared owners (Chevin, 2012). This new mix is symptomatic of the fact that owner occupation is declining both nationally and in London as a proportion of the housing stock while private renting is growing in importance with the potential for increased levels of institutional investment (GLA, 2012; Centre for London, 2013; Sprigings, 2013).

A case study of the East Village is thus likely to provide important evidence regarding both the impact of the Olympic legacy on the regeneration of east London and about a process of urban fabrication designed to mitigate social distinctions and contribute to a convergence of life chances or lifestyles between different housing groups.

Some Key Concepts

From this preliminary analysis it is possible to briefly identify three aspects of the process of urban fabrication which bear on strategies of inhabitation: *stake-holding, environmental perception, and attachment to place* (Cohen, 2013).

Bonders and bridgers: Some Patterns of Stake-holding in the London Olympics

How do people come to invest – or disinvest – in a sense of place, emotionally, materially, and symbolically? What determines who adopts what position? Robert Putnam's distinction between what he calls 'bridgers' and 'bonders' is very pertinent to making sense of how people go about the business of owning – and sometimes disowning – the Olympics (Putnam, 2000).

In general terms, bridgers are individuals, groups or organisations that have the social capital, the confidence and resource, to engage proactively with the world in which they find themselves, and to create partnerships or alliances with others in furtherance of their ends. Bridgers would see the East End as offering a prospect on opportunities offered by the wider metropolitan economy, a place where people come to get a start in life. If they adopt a Londoner identity it will be as a means to widen the scope and scale of their activities.

Bridgers adopted the Olympics as a platform for personal, professional or organisational advancement, and behaved opportunistically to maximise their competitive advantage. They are strenuous networkers, and their involvement takes the form of rational calculating moves, which do not require any deep emotional investment or ideological commitment to the project, although they certainly do not pre-empt it. In fact, bridgers often form pressure groups to leverage resources from public bodies, and some organisations, like London Citizens, have been markedly successful in exacting concessions – in this case on minimum wage rates to be paid to the Olympic workforce – from the authorities. Bridgers are always on the look-out for new opportunities to further their cause, and they take risks, but if they are not reaping substantive rewards they tend to disinvest and move on to what they see as more interesting or beneficial projects. In other words, they behave most of the time according to a market economy of worth.

In contrast, 'bonders' are individuals, groups or organisations that have less social capital, but seek to maximise what they have by using it to maintain their own sense of internal cohesion, identity or integrity. Bonders are good at building niches for themselves in markets and institutions, but by the same token they tend to develop a silo mentality, are risk averse, and are always on the look-out for possible refuges from the winds of change. They are reactive rather than proactive, and disposed to feel anxious (and less aspirational) about regeneration projects; they are also more likely to experience and talk about loss of community and urban decline. For them the East End is an area that offers sanctuary and is greatly valued for that, but it is perceived as continually being threatened by invasion from outsiders. If they adopt a Londoner identity it will be as a platform to express these local concerns. Quite a number of the older residents came into this category.

If bonders adopted the Olympics they were likely to have a fierce, proprietorial sense of the Games as 'their thing', by virtue of moral-cum-territorial claims staked over local amenities and resources. They operate according to a moral economy of worth in which civic or bio-political values predominate. However, this intense and potentially long-term commitment was only likely to occur if they felt their claims and interests were being recognised and validated. If not, they quickly withdrew into disinterest or even outright opposition, and regrouped around their own local concerns. They could be reluctant to share 'their' Olympics or 'their' Stratford with other communities or organisations, whom they did not regard as legitimate stake-holders. Bonders were found equally within white and BME communities.

The distinction between bridgers and bonders is not primarily one of psychological disposition or socio-economic status, although it may have these as some of its correlates. Rather, it is related to the mode of emotional labour that is employed by potential stake-holders, and the type of social networks – concentrated or distributed – through which communities of interest or affiliation get mobilised around specific issues or stakes. The classic route from bonder to bridger was often through enrolment as a representative of some local interest group onto a public forum – and the Olympics provided a major conduit for such transitions. The positions adopted by bonders and bridgers not only relate to different stories about the East End and its immediate prospects, they are about different kinds of stakes that individuals, groups, or organisations may have in the Olympics. Here it is useful to distinguish between material and symbolic stakes; though these can be closely connected, they can also come into conflict.

Like all models, this one is simply a device to map a set of positions that are empirically found in a variety of strong and weak combinations. Agencies certainly shifted between these positions in the course of their involvement with the Olympics over time, and in response to the project's vicissitudes.

These positions could be summarised as Table 6.1 shows. It remains to be seen how far the incomers to East Village adopt or shift between these positions in relation to the opportunities for civic participation and community involvement that are open to them.

Landscapes of Exposure and Seclusion

The notion of investment always contains a libidinal as well as purely calculative component, although this may be disavowed. This usually takes the form of sentimental attachment to place. But what kind of sentiment?

In their book *Thrills and Regression*, Michael and Enid Balint characterise two kinds of emotional and spatial orientation to objects, linked to different ways of holding the mother's body unconsciously in mind (Balint, 1959). Philobats enjoy exploring the wide open spaces, are always on the look-out for new experiences and dares, like courting danger and the unknown, and see obstacles as challenges to their resourcefulness.

Table 6.1   **Typologies of 'bridgers' and 'bonders'**

	BRIDGERS	BONDERS
Economy of worth:	Market economy	Moral economy
Organisational stance:	Proactive	Reactive
Protagonist role:	Fixers and schmoozers	Peer' n Cheer leaders
Partnerships:	Tactical alliances	Closure strategy
Stake:	Platform of opportunity	'Our thing'
Heritage type:	Dividend/Payback	Heritage/Endowment
Typification:	Social entrepreneurs	'Hammers fans'

Source: Author's own work.

They travel hopefully because their psycho-geography consists of warm, friendly expanses which are felt to be safe and encompassing, a supportive stage for exciting performance; the infant has the whole wide world in its arms: the world is your oyster and you are its pearl! At the same time, this landscape is dotted with dangerous and unpredictable objects, threatening in their independence, thrilling in their challenge and representing hazards that have to be overcome. There is an underlying confidence that when things get risky or the going gets rough the wider world will click in and will provide resources to enable you to anticipate or head off potential disaster. The philobatic standpoint implies a position of epistemic trust, but also a penchant for masquerade, for taking risks with identity.

From an aesthetic point of view, philobats yield a primarily visual landscape centred on looking before and after oneself – with the self-serving as a vanishing point (Appleton, 1975, 1990). In terms of narrative genre, storylines are organized around omniscient, if not always reliable, first person narrators, and the story setting becomes a stage from which to show off performance skills. From this vantage point, even an economic crisis in which the world is turned upside down seems to yield exciting new possibilities.

In contrast, ocnephiles only feel safe when they stay close to home, when they are surrounded by familiar objects, signs and landmarks, where they feel literally in touch with their surroundings; they cannot bear the thought of exposing themselves to danger. It is the inn, not the road that attracts them, and they do not travel hopefully, if at all. They are always making little dens for themselves and looking for potential bolt holes in and against a wider world that is experienced as hostile or threatening. The ocnephilic universe thus consists of safe familiar objects separated by vast abysmal empty spaces. This is associated with a pervasive fear of being dropped, let down, losing or being torn away from objects. That is why there is so much clinging to the object, such intense attachment to place, in the belief that it will somehow shield you from external danger. Behind this lies the desire for a totally benign and protective environment, a world in which all risk and anxiety has been eliminated and one is held forever in the warm embrace of a protective family or state, guaranteeing permanent ontological security.

Aesthetically, this is a tactile landscape constructed through strategies for holding onto oneself when all is lost. Ocnephiles cling to a straight and narrow story line; they do not like embedded narratives, unreliable narrators, or convoluted plots that lead them off the beaten track. Any open vista becomes a source of danger and dread.

Balint sometimes writes about these figures as if they were real people, or at least personality types: philobats are extroverts, and potential claustrophobics, while ocnephiles are introverts who may become terrified of being out and about in public spaces. But he also indicates that these are object relations which exist in a variety of weak or strong combinations and may be distributed across many different kinds of *mise en scène.* They are key terms in what might be called a *standpoint aesthetics,* a comparative theory of aesthetic experience grounded in an analysis of its locally situated structures of perception. Table 6.2 shows a summary of the standpoint aesthetics in play here.

The different modes of attachment to place linked to these standpoints are represented schematically Figure 6.1.

Table 6.2   Key terms of standpoint aesthetics

THE PROSPECT	THE REFUGE
The commanding view	Hideouts and boltholes
Defensible space and enclosure	The cul de sac and den
Landscape of exposure and expansiveness	Landscape of seclusion or occlusion
The panorama and vista	The sheltered view
The uninterrupted gaze	The foreclosed gaze/scotomisation
Topophilia	Topophobia
Withdrawal into Olympian standpoint	Withdrawal into inner sanctum

Source: Author's own work.

THE PHILOBAT

- Warm, friendly expansive space with a few risky places of challenge or thrill
- Supportive platform for adventurous public performance
- Epistemic trust in map/territory correspondence – WYSIWYG
- Aesthetic quest for the sublime: making the landscape awesome
- Active navigation of the unknown
- Inhabitation through colonisation of public realm

THE OCNEPHILE

- Dangerous and unwelcoming world with a few safe and friendly places.
- Staying close to home, surrounded by familiar objects
- Tactile environment – keeping in touch with the self/clinging to others
- Epistemic distrust – sensitivity to map/territory disjunctures
- Aesthetic quest for the picturesque: making the landscape safe for viewing
- Inhabitation through niche building and furbishing of the private realm

Figure 6.1   Attachment to Place

Armed with these concepts, our current research aims to understand what is going on, on the ground, in East Village. How will different patterns of stake-holding, environmental perception and place attachment interact? Will there be a strong or weak correlation between seeing the Park as 'our thing', the quest for a picturesque landscape of seclusion, and staying close to home? Or between strenuous neighbouring and networking, an adventurous pioneering spirit, and appreciating the Park for its expansive vistas and panoramic views? The conclusion to this chapter sets out some of the political and circumstantial conditions that may provide answers over time to these questions.

Olympic Park: Towards a Contextual and Conjunctural Analysis

Imagining Community, Glossing Class

The American poet Wallace Stevens once famously said that people live not in places but in the description of places, and since its inception the LLDC has gone in for some strenuous re-description of the Olympic site, drawing on much the same Panglossian vocabulary as the original bid to promote their vision of the Park as offering the best of all possible urban worlds. The Olympification of host cities adds another dimension of civic imagineering (Rutheiser, 1996; Lukes, 2007; Hetherington, 2008; Muñoz, 2012). This is how the LLDC website advertises the Park to potential visitors:

> Imagine the best of London, all in one place. Tradition and innovation, side-by-side, in a landscape of quality family homes, waterways, parklands and open spaces – anchored by the London 2012 Olympic and Paralympic venues. The future Queen Elizabeth Olympic Park will offer all of this and more.
>
> It will take the best of 'old' London – such as terraced housing inspired by Georgian and Victorian architecture, set in crescents and squares, within easy walking distance of a variety of parks and open spaces.
>
> It will take the best of 'new' London – whether in terms of sport, sustainability or technology – to create a new destination for business, leisure and life. Above all, the Park will be inspired by London's long history of 'villages', quality public spaces, facilities and urban living, learning from the best of the past – to build successful communities for families of the future.
>
> Source: www.londonlegacy.co.uk

So it is a familiar story of something old, something new, something borrowed, something blue (waterfront development) – the tried and tested formula of what has been called 'recombinant' urbanism, which draws on traditional vernacular architectural idioms in conjunction with state-of-the-art construction and design technologies to produce a postmodern mix of built forms (Shane, 2005). The motif of the 'urban village' is central to this concept:

> Five new neighbourhoods will be established around the Park, each with its own distinct character. Some residents will live in modern squares and terraces, others will enjoy riverside living, with front doors and gardens opening on to water. With the right mix of apartments and houses, located close to the facilities communities need to develop and grow, the Park will have the foundations to become a prosperous, vibrant new piece of city.
>
> Source:www.londonlegacy.co.uk

The urban village is very much an invented metropolitan tradition and refers primarily to working class neighbourhoods in the inner city that either have become gentrified, or are where the 'gentry' have always lived – or at least since the 18th century (Butler and Robson, 2003; Butler and Hamnett, 2011). Jane Jacobs, the American urbanist who was an apostle of 'spontaneous un-slumming', saw the urban village as a model of piecemeal urban renewal in inner city areas threatened by 'slash and burn' redevelopment – an alternative regeneration strategy led by small businesses rather than large corporations (Jacobs, 1974; Alexiou, 2006). More recently, environmentalists have adopted the urban village as a symbol of historical individuality threatened by the culturally homogenising pressures of globalisation, as well as a model of local democracy and sustainable community development (Magnahi, 2005). Amidst cries of 'there goes the neighbourhood' as yet another Starbucks opens, the 'small is beautiful' school of urbanism has made significant inroads into both popular attitudes and professional planning practice over the last decade (Charlesworth, 2003).

Even though it is not in fact an appropriate model for the Olympic Park, given the very different circumstances of its conception, something of this philosophy has undoubtedly rubbed off on the Park designers. One of the features that gives the urban village its distinctive cosmopolitan atmosphere is the presence of artists. The Development Corporation has the ambition to make the Olympic Park into a new cultural quarter – 'a bit of Hoxton and a bit of the South Bank' was how one Oligarch described it – and as such a home and workplace for east London's growing creative class of artists, designers, and media folk. These are a relatively new phenomenon, not least in their mode of attachment to place (Florida, 1995; Landry, 1995).[1] For although they are global go-getters, constantly on the move, and definitely 'going places', they are as concerned with the cultural assets which make an area desirable as they are with the market value of their property; the aesthetics of a neighbourhood are as important to them as its material amenities, transport connectedness and social status, and their mobile privatism is tempered by their environmental concerns.

The fact that gentrification is very much the name of the Olympic Park game is underscored by its residential strategy. The legacy plan promises a 70/30 split between privately owned housing for affluent professionals and 'affordable housing' that in principle is available to lower-income groups. In fact, recent measures introduced by the government have stretched the concept of affordability upwards to include middle-income groups, whilst at the same time hiking rents in social housing up to 80 per cent of market rents, which will put them well beyond the pockets of the poor. And in some cases even the 70/30 cut is qualified by the cautionary 'if viable'. In East Village, the first of the new neighbourhoods, the housing association has promised that it will be 'nearly impossible' to tell the difference between privately rented homes and the social rents, and that its style of management will be 'tenure blind'. Unfortunately, the signs and symbols of social distinction are not confined to architecture, and can defy even the most egalitarian housing policy: no amount of 'pepper potting' can prevent the social radar of passers-by registering the tale that is told by door knockers, cars, prams, gardens, the presence or absence of curtains, and styles of external décor. Finally, the 8,000 – 10,000 jobs that it is claimed the Park will eventually create will be overwhelmingly concentrated in the knowledge economy, financial and professional services and the cultural industries, giving a further boost to the gentrification process, while a smaller number of people will be employed in the low-wage, low-skill sectors, primarily in the local hotel, catering and retail trades, or as office cleaners, site maintenance and security staff.

There is a crude enough spatial logic to this dual economy. The professional services class thrown up by the new economy needs another kind of service class to look after it; it needs people to wash, cook and clean for it, to mend its equipment, service its cars, mind its children and pets, minister to its recreational needs, staff its shops, wine bars and restaurants, improve its houses, fix its drains, and populate its neighbourhoods with a little local colour. This is precisely the role assigned to the post-industrial working class for whom the Olympic Park will provide some limited accommodation (Buck, 2002; Hutton, 2008).

The persistence of class distinctions is glossed over in the LLDC prospectus in a number of ways: firstly, by the reiteration that much of the housing will be for families and that 'family values' will prevail in the design of public amenities. In fact, in the context of the housing market a family home is simply a large house that has three or more bedrooms, and it can just as easily be occupied by a single childless but affluent owner. The possibility that many of the apartments will become company flats, as happened in the Barbican, another prestigious housing development linked to a cultural centre, or that the new housing will become a buy to rent investment opportunity, as has occurred in the Royal Docks, can certainly not be ruled out.

---

1   The definition of this 'class' has proved as elastic as that of 'creative industry'. In some usages it includes estate agents and hairdressers; in others it is confined to those working in traditionally defined areas of the arts.

Secondly, there is a great deal of talk about social inclusivity, but what this turns out to mean is that the site will have disability access and housing designed for life-time occupation, including special provision for senior citizens. While this is admirable, it rather dodges the fact that socio-economic status will continue to regulate and restrict access to these facilities; there is no sense in which this project could be regarded as redistributive in its effect on local housing classes. It is wheelchair access, not social access, that is the priority here. The outcome is more likely to be yet another example of 'splintering urbanism', offering a further prospect on global opportunity structures for those who are already fully paid-up members of the network society, while those who are dependent on the local, or informal economy, or the state, remain a marginal, even if not actively marginalised, presence (Graham and Marvin, 2001).

The strongest feature of the plan is its neighbourhood structure, which owes more than a little to Ebenezer Howard's vision of the garden city: housing, schools, shops, health and community centres and public space, including children's playgrounds – all are closely integrated into the urban fabric. There is no doubt that, taken as whole, it represents a significant advance on any previous post-Olympic site development. The only pity is that the main beneficiaries are likely to be wealthy investors and middle class gentrifiers, rather than local East Enders.

An Exercise in Imagineering

Much is made in the LLDC prospectus of the fact that the Park is an important public amenity for locals as well as a tourist destination for visitors to London. Here is how a day out in the Park is imagined:

> A day in the Park might start with a coffee and toast, soaking in the views of the Park and the striking 2012 Games venues. Your morning could feature a trip up the ArcelorMittal Orbit – to see the remarkable panorama across London – followed by some retail therapy at Westfield Stratford City. Lunchtime could include some exercise at one of the sports venues or some street art in the open spaces that will feature an exciting line-up of activities and performance. Your afternoon could be full of sport, whether trying your hand at BMX at the Velo park or watching world champions at the Aquatics Centre or the Stadium.

> And, to finish the day, you could enjoy dinner at one of the Park's restaurants – or head to Brick Lane, Green Street or other East London hotspots to enjoy local music and cuisine.

> Source: www.londonlegacy.co.uk

This 'visitor' is nothing if not an all-rounder, combining the tastes of *flâneur*, sightseer, sports enthusiast, shopaholic, fitness freak, gourmet and BMX biker all in one! But actually this little scenario is very revealing of what kind of public space is being envisaged. It is what Michel Foucault has called a 'heterotopia', an 'other' space, which juxtaposes in a single place a multiplicity of sites that are in themselves normally incompatible in scale and function and belong to quite different urban realms: the shop, the stadium; the garden, the observational tower, the terraced house, the pleasure ground (Rabinow, 2000; Duhaene, 2008). Heterotopias can be exciting and fun, but not everyone wants all the different elements of city life compressed – or jumbled up – in one space.

In any case, what most people enjoy doing in a park on a fine summer's day is nothing much: picnicking, sun-bathing, flirting, reading, listening to music, or just sitting around gossiping, while for those more actively inclined throwing a frisbee or kicking a ball about is the summit of their athletic ambition. There should be plenty of scope for all this relaxed (in)activity in the Olympic Park, especially in the ecological Northern Park which includes wetlands, woodlands and meadows, at least until required for commercial development. Still, it is a bit worrying – and symptomatic of the aspirational 'get-fit' Olympic agenda that the Park is supposed to embody – that none of the promotional videos or artists' impressions actually show people just lying around on the grass. They are either striding purposefully about, doing or watching sport, or jogging, no doubt egged on by Monica Monvicini's giant installation RUN.

There is another sense in which 'otherness' has been given a local resonance in the LLDC publicity. The frequent mention of 'East London hotspots' with their local music and cuisine, and similar references to events which will 'showcase local diversity and heritage', suggests that if *EastEnders* has a walk-on part in the post-Olympic spectacle it is to add a little local colour to the Park 'experience' by performing their cultures for the benefit of passing trade. It is the familiar 'order in variety' formula of British style multiculturalism: the civilising missionaries (here the Development Corporation) provides the order (in this case the planning framework), while the 'ethnics', the 'locals', the 'people', the 'others', furnish the variety in the urban setting (Cohen, 2012).

From a design standpoint, the layout of the Park, its configuration of venues, connecting paths and open spaces, draws explicitly on the tradition of English landscape gardening; but here order-in-variety is applied to the overall planning concept. The variety is provided by the sports venues, each of which has a distinctive architectural identity, and the order – or at least the harmonious confusion – is produced by the landscaping. The Park's chief design consultant, and now advisor to the LLDC, Ricky Burdett, is quite up-front about the fact that no 'one-size-fits-all' design brief was imposed on the architects and that they were encouraged to 'do their own thing'. He describes the result as 'fragmented but organic', an aptly postmodern image for the style of urbanism proposed (Shane, 2005; Sheard, 2005; Burdett, 2007). This architectural melange could all too easily result in what Rem Koolhas has called junkspace, 'a fuzzy empire of blur ... a seamless patchwork of the permanently disjointed'(Koolhaas 2001). This is a very different kind of spatiality from that order in disorder created when popular do-it-yourself urbanism emerges through grass roots community organisation (Sennett, 1973; Ward, 2002).

In fact the park has been subjected to a number of quite distinct landscaping strategies that co-exist in some degree of tension, some pulling it towards strong boundary maintenance and some towards a more open encounter with the wider environment (Hopkin and Neal, 2012). Some strategies are designed to make the park into a compact, defensible space: clamping strategies seek to enclose the site around itself to meet financing, security, maintenance and sustainability requirements; clustering strategies seek to give spatial articulation to a multiplicity of pathways, traffic flows and facilities, and a new nexus of productivity, investment and capital accumulation. On the other hand scoping – organising the landscape into a distinctive regime of envisagement and traverse – and scaling –positioning it within local/regional/national and global narratives – open it out to a wider set of interactions not so easily regulated.

This underlying spatial tension is camouflaged by the carefully modulated alternation of formal and informal design elements, knitted together to give the semblance of environmental 'organicity'. This manoeuvre takes a very concrete form. The biodiversity action plan for 45 hectares of new habitat, including wetlands, is designed to ensure that the fauna and flora disrupted by site remediation is replaced. Yet there will be no re-instatement of the 'alien species' that flourished amidst the richly polluted urban wilderness that was the pre-Olympic site, making it such an important place of pilgrimage for naturalists and ruinologists alike; no reprieve, then, for Japanese knotweed, Indian balsam and other exotic trespassers, not to mention the marijuana plants that flourished so wildly amidst the industrial ruins. The Park thus offers the visitor an artificial paradise of nature, an antiseptic isle from whose green and pleasant land all noxious weeds have been banished; it will be a beautifying lie about the cultural and environmental politics that have gone into its making.

Nevertheless, the early indications are that the Park will become a popular day out for Londoners, and for visitors to the city, especially in the summer with a lively programme of activities. Whether East Enders will take it to their hearts whether it will become the focus for territorial rivalries between youth gangs, already well entrenched in Hackney Wick and Stratford, and the extent to which it will become a story of two parks, it is still too early to say, and will require further research.[2]

Himalayan Balsam

Dwelling Places

The most ambitious and problematic aspect of the park legacy project is the attempt to integrate the remaining sports venues within an emergent urban fabric constructed around residential communities, and what are

---

2   The fieldwork is scheduled to commence in June 2014 and the first phase of the project is due to report in the Spring of 2015. For further information about this and related Olympic Park projects see www.livingmaps.org.uk

somewhat optimistically called 'employment hubs'. Sports stadia by their sheer physical size, and the fact that they remain empty for much of the time, and then briefly flood an area with a multitudinous and often vociferous population of spectators, exert an unsettling and even uncanny effect on their neighbourhoods (Bale and Moen, 1995). In the case of the Olympic Park the fate of the stadium, the biggest material asset (or in some views liability) as well as the symbolic flagship of the whole enterprise, came to focus public anxieties about the long-term viability of the 2012 project. The initial difficulty in finding a tenant and a sustainable post-Olympic use conjured up visions of other Olympic venues that have turned into ghost towns, haunted by their former glory, their only function to serve as cautionary monuments to the public folly or hubris that built them.

For better or worse there is nothing awesome or prophetic about the 2012 stadium. It is a good example of what has been called 'terminal architecture' – a huge oval shed for accommodating spectators and athletes with the maximum efficiency and minimum of fuss; as such it is indistinguishable from dozens of similar structures, in combining the envelope functionality of the aircraft hangar with the palatial uselessness of the architectural folly and the spiritual hydraulics of a cathedral where people come to worship sport (Pawley, 1998; King, 2004).[3] The sports stadium is a hybrid in another sense. It is part observatory, but without its panoptic vision, and part auditorium, but without its acoustics. It is a non-place, a transit zone, a space of flows (Augé, 1995). But the Olympic Stadium is a very special kind of non-place, because although it did not in itself generate any sense of local attachment, it was the epicentre of a global mediascape organised around the games, and in the post-Olympics it serves as focal point of public and private memory work as well as a place of pilgrimage for Olympophiles. It cannot but be regarded as a heritage site.

What was symptomatically missing from the LLDC's initial vision was any recognition that the urban fabric is made up of the stories woven into, around and about it by those who dwell there; yet it is through this protracted process of story-making that spaces become places with a specific local meaning and identity (Daniels, 2011).

The ways in which spaces become places also has a lot to do with different forms of stake-holding. For example there was an ambitious proposal to build a multi-faith centre, a classic exercise in bridging, but this had to be dropped because of vehement objections on the part of some local faith communities – bonders to a man (sic) to sharing ecumenical premises. For devout Muslims or Christians, their place of worship is a consecrated space which it would be sacrilege to share with non-believers. Football fans feel the same way about their home grounds, and routinely object to sharing them with rival teams. To someone who is not into football, a stadium may look like a giant shed, but to Chelsea supporters, 'The Shed' is their home from home, a focus of their attachment to the club, and a place of shared memories: it gives them a symbolic stake in the club's affairs – even and especially if, in material terms, Chelsea FC becomes the personal fiefdom of a wealthy Russian oligarch.

By no coincidence, this intimate connection between stake-holding and place-making has had a direct bearing on the fate of the Olympic Stadium. When the Olympic authorities entered into negotiations with West Ham United, their rival contender, North London's Tottenham Hotspur, cried foul to West Ham's original bid and threatened legal action. The second time round West Ham were successful in clinching the deal with the backing of London's Mayor and the LLDC.

The stadium saga illustrates the distinctions between different types of legacy-making. Located within a structure of conveyancing designed to maximise payback, the stadium can only be 'sold' as an investment opportunity and a source of future dividends from profit. Yet, it is also by definition a heritage site and potentially a popular *lieu de memoire*. Inscribed within the conveyance of a gift legacy it may be claimed as an heirloom or public endowment. This latter possibility is, of course, anathema to the champions of free market ideology. *The Daily Mail*, commenting on the decision to make West Ham the preferred bidder, ranted: 'Something for nothing. That is what is presumed West Ham are getting out of the Olympic stadium deal. A free ride. A gift from a grateful nation'.[4] In fact there are numerous strings attached to this 'gift', including a clawback on future profits that will ensure that West Ham's debt, already £80 million, will almost double and

---

3   Martin Pawley uses the concept to describe structures whose design is entirely determined by their function of containing large numbers of people or goods.

4   *The Daily Mail*, December 5, 2012.

whatever the return on their investment that enhanced status as a 'superclub' will bring, they will remain in the pockets of financiers for generations to come.

West Ham's bid in principle attempted to reconcile the two main economies of worth. Not only is the club aiming to ensure that the stadium has commercial viability and retains its value as a saleable asset, but it possesses a large, strategically located, supporter base that can adopt it as their home. Hammers' fans are drawn not only from east London but from the Cockney diaspora, so that every home game represents an ingathering of the tribe from Basildon and Brightlingsea to Billericay and beyond (Fawbert, 2005). When the club takes over the management of the stadium, this will not only give it a new lease of life but a distinctive new identity, no longer associated with the Olympics, of course, but as a focal point of east London's major sporting heritage. Yet West Ham's fans, bonders to a man, woman and child, are reluctant to make the move. As far as they are concerned, the Olympic Stadium is not an heirloom but a Trojan Horse; exchanging it for Upton Park, the club's ground for most of its history and hence a site of great sentimental attachment, would, in their eyes, be an act of *disinheritance,* not a dividend or the repayment of a symbolic debt.

The stadium example underlines the fact that it is not possible to simply read off stake-holder positions from socio-economic status. The cultural geography of east London, its patterns of affinity and affiliation, and its territorial claims and rivalries, many of them deeply embedded in the areas history and urban fabric, will play a decisive role in deciding the outcome of this gigantic piece of civic engineering. Ambitious claims have been made for the 2012 legacy, in particular that it will produce an effect of 'convergence', such that the structures of opportunity and life chances of children growing up in East 16 or East 20 will be the same as those of their peers in the most affluent parts of the metropolis. In many ways 'convergence' is an untenable – and untestable – proposition, given the intensified trend to ever greater structural inequality and the sheer scale of the contingencies that enter into long term regeneration effects (Piketty, 2014; Cohen, 2013). Stratford as an economic hub may be a bubble of commercial prosperity that will resist but its 'catalytic effect' shows no sign of extending into the hinterlands of outer east London, whose pockets of gentrification have quite different rationales (Bernstock, 2014). On the other hand, the population churn created by the advent of affluent globetrotters into East 20 is likely to militate against the formation of a stable, socially mixed neighbourhood which is the aim of the LLDC's community development programme. Nevertheless, as a rhetorical device 'convergence' speaks to a set of local aspirations for a better life which have long informed the political struggles of immigrant and working class communities in this part of the world. Our task as ethnographers will be to ensure that as far as possible those hopes, and the anxieties which inevitably subtend them, find an adequate space of representation.

Acknowledgement

I would like to thank Paul Watt for his input to the discussion of East Village and my colleagues at Living Maps for their general support.

References

Appleton, J. (1975). *The Experience of Landscape*. London: Wiley.

Appleton, J. (1990). *The Symbolism of Habitat*. Seattle: University of Washington Press.

Arthurson, K. (2002). 'Creating inclusive communities through balancing social mix: A critical relationship or tenuous link?' *Urban Policy and Research* 20(3), pp. 245–261.

Atkinson, R. and K. Kintrea (2000). 'Owner-occupation, social mix and neighbourhood impacts', *Policy & Politics* 28(1), pp. 93–108.

Augé, M. (1995). *Non-Places: Introduction to an Anthropology of Supermodernity*. London: Verso.

Balint, M. (1984). *Thrills and Regression*. London: Free Associations Press.

Bernstock, P. (2103). *Olympic Housing*. Farnham: Ashgate.

Blokland, T. (2003). *Urban Bonds*. Cambridge: Polity Press.

Bond, L., E. Sautkina, and A. Kearns (2011). 'Mixed messages about mixed tenure: Do reviews tell the real story?' *Housing Studies* 26(1), pp. 69–94.

Bramham, P. and J. Caudwell (eds.) (2005). *Sport, Active Leisure and Youth Cultures*. Eastbourne: Leisure Studies Association.

Bridge, G., T. Butler and L. Lees (2012). *Mixed Communities: Gentrification by Stealth?* Bristol: Policy Press.

Buck, N. et al. (2002). *Working Capital: Life and Labour in Contemporary London*. London: Routledge.

Butler, T. and C. Hamnett (2011). *Ethnicity, Class and Aspiration: The New East End*. Cambridge: Polity Press.

Butler, T. and G. Robson (2003). *London Calling: The Middle Classes and the Remaking of Inner London*. Oxford: Berg.

Centre for London (2013). Institutional Investment in London's Market Rented Sector. London: City of London.

Charlesworth, E. (ed.) (2005). *City Edge: Case Studies in Contemporary Urbanism*. Oxford: Architectural Press.

Chevin, D. (2012). New Urban Living for London: The making of East Village. Commissioned by the Smith Institute in partnership with Qatari Diar Delancey (QDD) and Triathlon Homes, available online: http://www.geography. org.uk/download/GA_conf13TheOlympicLegacyNewUrbanLivingforLondonthemakingofEastVillage. pdf (accessed August 2014)

Cohen, P. (2012). 'A Beautifying Lie? On Kitsch and Culture @ the Olympics'. *Soundings*, Vol. 50, 2012.

Cohen, P. (2013). *On the Wrong Side of the Track? East London and the Post Olympics*. London: Lawrence & Wishart.

Daniels, S. et al. (ed.) (2011). *Envisioning Landscape, Making Worlds: Geography and the Humanities*. London: Routledge.

Davis, J. (2012). Urbanising the Event: How past processes, present politics and future plans shape the London Olympic Legacy. PhD Thesis London School of Economics

Duhaene, M. (2008). *Heterotopia and the City: Public Space in a Postcivil Society*. London: Routledge.

Duyvendak, J.W. (2011). *The Politics of Home: Belonging and Nostalgia in Western Europe and the United States*. Basingstoke: Palgrave Macmillan.

Fawbert, J. (2005). 'Football fandom, West Ham United and the "Cockney diaspora": From working-class community to youth post-tribe?' in Bramham P. J. Caudwell J. (eds.) (2005). *Sport, Active Leisure and Youth Cultures*. Eastbourne: Leisure Studies Association.

Fink, J. (2011). 'Walking the neighbourhood, seeing the small details of community life: Reflections from a photography walking tour', *Critical Social Policy,* 32(1), pp. 31–50.

Florida, R. (1995). *Cities and the Creative Class*. London: Routledge.

Foucault, M. (1986). 'Of Other spaces: Utopias and heterotopias'. *Diacritics,* 16/1 (Spring 1986), pp. 22–27, translated from the French by Jay Miskowiec (first published as "Des Espace Autres", *Architecture/ Mouvement/Continuité*, 5 (October 1984), pp. 46–49).

Greater London Authority (2012). The Mayor's Housing Covenant: Making the private rented sector work for Londoners. London: GLA.

Hetherington, K. (2008). *Capitalism's Eye: Cultural Space and the Commodity*. London: Routledge.

Hopkins, J. and P. Neal (2012). *The Making of the Queen Elizabeth Olympic Park*. London: Wiley.

Humphry, D. (2013). 'Inside out: A visual investigation of belonging in a London neighborhood', in M. Kusenbach and K. E. Paulen (eds.) (2013). *Home: International Perspectives on Culture, Identity, and Belonging*. Frankfurt am Main, Bern, Bruxelles, New York, Oxford, Warszawa, Wien: Peter Lang, pp. 249–268

Hutton, T. A, (2008). *The New Economy of the Inner City*. London: Routledge.

Ingold, T. (2000). *Perception of the Environment: Essays on Livelihood, Dwelling and Skill*. London: Routledge.

Jupp, B. (1999). *Living Together: Community Life on Mixed-Tenure Estates*. London: DEMOS.

Keller, S. (2003). *Community: Pursuing the Dream, Living the Reality*. Princeton: Princeton University Press.

King, A. (2004). *Spaces of Global Cultures: Architecture, Urbanism, Identity*. London: Routledge.

Koolhaas, R. (2001). 'Junkspace', in: Chung, C. J., J. Inaba, R. Koolhaas, S.T. Leong (eds.) (2001). *The Project on the City 2: Harvard Design School Guide to Shopping*. Cologne: Taschen: 408–421.

Kusenbach, M. and K. E. Paulsen (eds.) (2013). *Home: International Perspectives on Culture, Identity and Belonging*. Frankfurt a.m.: Peter Lang.

Landry, C. (1995). *The Creative City*. London: DEMOS.

Lelévrier, C. (2013). 'Social mix neighbourhood policies and social interaction: The experience of newcomers in three new renewal developments in France'. *Cities*, Volume 35, pp. 409–416.

Lewicka, M. (2011). 'Place attachment: How far have we come in the last 40 years?' *Journal of Environmental Psychology* 31, pp. 207–230.

Lukas, S. (2007). *The Themed Space: Locating Culture, Nation, Self*. Plymouth: Lexington Books.

Lupi, T. and S. Musterd (2006). 'The suburban community question', *Urban Studies* 43(4), pp. 801–817.

Lupton, R. and C. Fuller (2009). 'Mixed communities: A new approach to spatially concentrated poverty in England', *International Journal of Urban and Regional Research* 33(4), pp. 1014–1028.

Magnaghi, A. (2005). *The Urban Village: A Charter for Local Democracy and Sustainable Development*. London: Zed Books.

Muñoz, F. (2006). 'Olympic urbanism and Olympic villages', *Sociological Review*, Volume 54, Supp 2.

Pawley, M. (1998). *Terminal Architecture*. London: Reaktion Books.

Piketty, T. (2014). *Capital in the Twenty First Century*. Boston: Harvard University Press.

Putnam, R. D. (2000). *Bowling Alone: The Collapse and Revival of American Community*. New York: Simon and Schuster.

Rabinow, P. (ed.) (2000). *The Essential Works of Michel Foucault Vol II: Aesthetics*. London: Penguin.

Rathzel, N. and P. Cohen (2006). *Finding the Way Home: Young People's Perceptions of Place in London and Hamburg*. Göttingen: Vandenhoeck & Ruprecht Press.

Rose, D., A. Germain, M.H. Bacque, G. Bridge, Y. Fijaklow and T. Slater (2013). '"Social mix" and neighbourhood revitalization in a transatlantic perspective: Comparing local policy discourses and expectations in Paris (France), Bristol (UK) and Montréal (Canada)'. *International Journal of Urban and Regional Research*, Volume 37, Issue 2, pp. 430–450.

Rosenblatt, T., L. Cheshire and G. Lawrence (2009). 'Social interaction and sense of community in a master planned community', *Housing, Theory and Society*, 26(2), pp. 122–142.

Rutheiser, C. (1996). *Imagineering Atlanta: The Politics of Space in the City of Dreams*. London: Verso.

Savage, M., G. Bagnall and B. Longhurst (2005). *Globalization and Belonging*. London: Sage Publications.

Shane, D. (2005). *Recombinant Urbanism: Conceptual Models in Architecture, Urban Design and City Theory*. Chichester: Wiley Academic.

Sheard, R. (2005). *The Stadium: Architecture for the New Global Culture*. Singapore: Periplus.

Sprigings, N. (2013). 'The end of majority home-ownership: The logic of continuing decline in a post-crash economy', *People, Place & Policy*, Online 7(1), pp. 14–29.

Trumpbour, R. (2007). *The New Cathedrals: Politics and Media in the History of Stadium Construction*. Syracuse University Press, New York.

Vigor, A., M. Mean and C. Tims (2004). *After the Gold Rush: A Sustainable Olympics for London*. London: DEMOS & IPPR.

Watt, P. (2006). 'Respectability, roughness and "race": Neighbourhood place images and the making of working-class social distinctions in London', *International Journal of Urban and Regional Research*, 30(4), pp. 776–797.

Watt, P. (2009). 'Living in an oasis: Middle-class disaffiliation and selective belonging in an English suburb', *Environment & Planning A*, 41(12), pp. 2874–2892.

Watt, P. (2013). 'Community and belonging in a London suburb: A study of incomers', in M. Kusenbach and K.E. Paulsen (eds.) (2013). *International Perspectives on Culture, Identity and Belonging*. Frankfurt a.m.: Peter Lang.

Watt, P. and Smets, P. (2014). *Mobilities and Neighbourhood: Belonging in Cities and Suburbs*. Basingstoke: Palgrave Macmillan.

Young Foundation (2010). *Neighbouring in Contemporary Britain*. London: Young Foundation.

Chapter 7

Legacies of Turin 2006 Eight Years On: Theories on Territorialization in the Aftermath of the Olympic Games

Egidio Dansero, Alfredo Mela, Cristiana Rossignolo

Introduction

Over the last twenty years the critical debate on mega-events has intensified significantly (Girginov, 2013; Gold and Gold, 2011; Lenskyj and Wagg, 2012). Mega-events have become a popular object of research in urban and regional studies. However, not all research reflects adequately on the theoretical. The growing interest is linked to the acknowledgement, often uncritically, of a positive role of mega-events in urban and territorial policies, as catalysts and accelerators of urban change and renewal (Essex and Chalkley, 1998). This positive expectation stems from the presentation of success stories often celebrated at international level, to the extent that they have imposed themselves as "good practices". In fact, event organizers regularly attempt to overstate the positive impact and underrate the negative effects (Sandy et al, 2004), even if there is a large series of failures and behind the "lights" of the success stories more than a few shadows are hidden, related to gentrification, social exclusion and displacement, environmental destruction, social conflicts (Cashman, 2010; Dansero et al., 2011; Essex and Chalkley, 2004; Hayes and Karamichas, 2011; Hiller, 2000; Lenskyj, 2002; Spilling, 1998) and 'huge sunk costs' (Davidson and McNeill, 2012: 1626).

Among many issues addressed, this chapter reflects upon the territorialization of the Olympic Games; as a moment of the outstanding production of territory, à la Raffestin, at different scales and in symbolic, physical and organizational terms. This perspective allows, in our opinion, a more complex view of the mega-event with regard to the dynamics and the policies of the host cities and regions, supporting critical reflection upon the idea of "planning legacy". In assessing the Olympic legacy, if the immediate effects represent a field already known and studied, then it becomes crucial to take into account those variables, in terms of territorialization, that might take place over subsequent years.

When, in 2006, the XX Olympic Winter Games came to Turin the event was used to give a further and decisive thrust to the city's Post-Fordist transition. Turin, eight years on from its Olympics, is an interesting field for understanding the long-term impacts (legacies) on local territories and environments, with special attention given to the different scales involved: Turin and its metropolitan area and the Alps. From this perspective it can be seen that the beginning of the global economic crisis for Turin represented, and is still representing, a break in its urban development thereby suggesting that local/global relationships need to be reconsidered within this longer term timeframe.

This chapter is organized as follows: the first section presents some reflections on the mega-event as a process of territorialization, seen as the "production of territory". The following four sections focus on the case study – the Turin Olympic Winter Games – through the examination of some specific issues: the *new* geography between the city and the mountains, the territorialization process and the legacy for the city, the change of the city image and the legacy following the global economic crisis. The chapter concludes with further reflections on the relationship between the Olympic territorialisation and the new geopolitical trend in favour of hosting mega-events.

The Territorialisation of the Mega-event

A key element of the mega-event is its "extraordinary" nature that fits inside, and often over, the relatively "ordinary" territorial transformations. From this perspective the reflection on mega-events raises the issue between the nature of ordinariness and the extraordinary: between a "before", an "after" and "during" in which there are specific plans to manage the temporary system linked to the event. The mega-event can be interpreted as the construction and the consequent activation of an Olympic territory which in reality is a temporary spatial system, intended to last for the duration of the event, which rests and is superimposed on the hosting territory. If mega-events are by nature only transitory, it is, therefore, essential to carefully plan to produce truly lasting legacies (Chalkley and Essex, 1999; Moragas et al., 2003).

A mega-event, and the Olympics in particular, can be seen as a process of territorialization, or in other words, as a production of territory, which is in turn a space produced by the action of a player who carries out a programme: a space to which human energy and work has been applied, with anthropological value (Raffestin, 1980; Raffestin and Butler, 2012). Olympic territorialization takes place on different scales, from the process that leads to the selection of the host site, to the latter's transformation to make it suitable for hosting the event, the period of de-territorialization that often follows the event and coincides with the dismantling, and sometimes the abandonment, of some of the infrastructures associated with it, and to the re-territorialization which may occur when the territory that hosts the event is able to appropriate its legacy in full (Dansero and Mela, 2007).

The Olympic mega-event structures space, differentiating it by selecting certain localities and discarding others. It requires that space be transformed in order to adapt it to its needs and in so doing it acts as a standardizing impulse. However, the relationship is ambivalent, as the encounter between the Olympic world and the locality chosen in the "common place" of the mega-event is primarily a relationship of force between the plurality of players that see this "common place" as their best bet for implementing their strategies. The Olympic mega-event both seeks and consumes spatial differences, but it can also end up by producing them (Raffestin, 1980; Klauser, 2012). This depends on the uncertain outcome of the "negotiation" between the standardizing tendencies of a supra-local player – the IOC, the sponsors – which try to impose its restrictive view of a territorial complexity that it is not always able to metabolize and make part of its own perspective, and local strategies and resistance on the other hand, which are the outcome of a conflict between different visions of the territory and its potential for change. This territorialization can be seen as an encounter – and clash – between different territorializing acts, and takes place on several levels.

The T-D-R (territorialization-de-territorialization-re-territorialization) cycle is specifically produced by the mega-event and can thus be interpreted as the production of a "project territory" modelled on the mega-event's needs. Olympic territorialization, moreover, inasmuch as it is the production of new territory, is interwoven with the "normal" dynamics of change that are already operating in the "context territory" through a combination of T-D-R cycles that are independent of the mega-event (Figure 7.1).

Regarding the territory's symbolic transformation, this is expressed through names (the stadium, the boulevard, the Olympic villages) that can often last well beyond the event, but can also be an opportunity for building strategies that identify the name for a variety of symbolic and material purposes (marketing the event, justifying the projects involved, or creating a "territorial quality stamp").

There can be no doubt that reification, or in other words the material transformation of the territory, is the most obvious aspect of Olympic territorialization, and the one that tends to last longest through the construction of infrastructures that are directly connected with the event (the sports and tourist facilities) or support it (the road systems connecting the venues), as well as all the other material changes that surround it. Finally, the Olympic territory is shaped by distributing functions, activities and people in a space so that it can be effectively managed. This spatial organization can be extremely complex and highly articulated.

The risks, now a familiar topic in the debate that surrounds the Olympic legacy, but which still lurk as threats when preparing bid files, when carrying out the Olympic program, and above all when managing the aftermath of the event, consist in producing an excess of territorialization, which rather than eliminating earlier territorial shortcomings (in public services, mobility, etc.), can lead to heavy debts for the future in terms of reusing buildings. Conversely, "around the world, most Olympic cities have seen many of their facilities demolished or else left underused or in disrepair" (Davidson and McNeill, 2012:1628).

The focal point of our argument concerns the territorial appropriation of the mega-event; that is the ways in which a host territory can settle the dialectic between an extraordinary event and the ordinariness of the society-environment-land relationship. These reflections find contact points with what Hiller (Hiller, 2003) theorized when he identified the Olympics as a phenomenon of interest in the spatial sciences because they

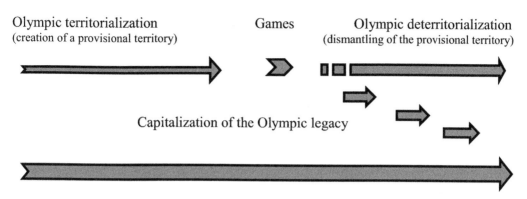

Figure 7.1   The T-D-R cycle
Source: Authors' own concept and design, 2014.

affect the normal process of decision taking in urban planning. There is a contrast between the logic and objectives of the IOC and the host city: while the mega-event planners are interested in the short term, the host cities put emphasis on the long-term and post-event.

Turin 2006: A New Geography Between the City and the Mountains

Referring to the theoretical frameworks described above, concerning the processes of territorialization and using a wide series of studies related to the analysis of Games spatial impacts, before the event and the evaluation of actual impacts after it (Cashman and Horne, 2013; Bondonio, Guala, Mela, 2008; Crivello, Dansero, Mela, 2006; Dansero and Mela, 2012; Guala and Crivello, 2006), the following section will now analyse the tracks imprinted on the territory by the legacy of the Turin Olympic Winter Games. A first consideration concerns the spatial dimension of the Olympic territory. Winter Olympics can involve territories of different scale: some Games were heavily concentrated in space, while in other cases they involved a broader and complex spatial system. In general, while the organization of the Summer Games is strongly correlated to the host country and often to its capital, the Winter Games, on the contrary, usually affect just a region (Chappelet, 2010).

In fact, "since 1964 [the Winter Olympics] have not been awarded to small towns in the mountains but to cities with several thousand inhabitants, sometimes at a fair distance from the ski runs: Innsbruck, Grenoble, Sapporo, Sarajevo, Calgary, Nagano, Salt Lake City and Turin" (Chappelet, 2010).Vancouver and Sochi can now be added to this list. And it also applies to the case of Turin, where the organization of the Games led to a close relationship between the city and two main Alpine valleys. It is, however, worth noting that the main distinctive features of Torino 2006 can be seen in spatial terms (Dansero and Mela, 2012), with the explicit construction of the hosting territory through the integration of different areas and networks that had never been thought of together before. The "Olympic region" (see Plate 7.1), in the strict sense, was a significant portion of the Province of Turin comprising of Bardonecchia to the West, Torre Pellice to the South and Turin to the East (Dansero and Puttilli, 2012).

As Essex and Chalkley (2011) observed, the Olympics was an instrument of wider regional integration: the Games involved the construction of a heterogeneous and, in many respects new territory, which coincided neither with the administrative subdivisions nor with homogeneous areas in terms of socio-economic development.

In 1998, Turin saw its candidature as a sterling opportunity to step up the pace of post-Fordist re-territorialization, easing the transition away from the old "one-company town" to a model based on a plurality of different roles. The Olympics seemed to be able to speed up the Master Plan projects (1995) along the rail line that cuts across the city ("the backbone"), a new north-south avenue with a mixture of urban restructuring and event facility projects arranged strategically along it. Most of the event facilities were located in the southern sections of the rail line (Lingotto and Piazza d'Armi). On the other hand, the Alpine towns saw the mega-event as a chance to boost their competitiveness in winter tourism through image-building efforts and by extending and improving their infrastructures and accommodation facilities. It follows that Olympic territorialization was conceived to be built on the long-standing economic base and place-specific resources of these areas, with their concentration on snow sports, renewing their infrastructures and the attractions they can offer to tourists.

The "regionalization" of the Olympics has been one of the short-term results in terms of the central role that the Games assigned to an urban space located at a significant distance (as much as 90 kilometres) from the mountains, and in relation to the rediscovery of a historic relationship between Turin and the Alps (Bontempi, 2006). However, the spatial dimension that formed during the organization of the Olympics and which favoured a close relationship between the city of Turin and the mountain regions quickly dissolved in the following period. In particular, the activity of territorial governance, imposed by the organizational needs during the seven years prior to the Games, immediately stopped when those needs ceased and the strategic interests of the city and the Alpine areas returned to being distinct if not opposed in many ways. Many factors played a role in this development. Some of them were connected with the Winter Olympics: in particular, the widespread perception among the Alpine population of an imbalance between the city and the mountains in the media representation of the Games. Turin has been accused by some Alpine stakeholders of having spread the image of a purely urban event, obscuring the role played by the Alpine valleys and thereby decreasing the possibility of obtaining the benefits of creating a new positive image for these areas. Other causes of divergence between the city and the mountains concerned problems not associated with the Olympic event, such as, for instance, the conflicts related to the construction of a high-speed railway between Turin and Lyon (part of the Mediterranean Corridor). In this case, a large part of the Alpine population accused the city of favouring the construction of an infrastructure that will not have a positive impact on the areas crossed by the railway, but would cause economic and environmental damage.

Therefore, referring to the T-D-R cycle (Figure 7.1) and evaluating the more recent processes the de-territorialization effects have been short-lived and the "ordinary" territorialization dynamics have quickly reassumed a prevalent role. A similar consideration could be made by focusing the attention only on the Alpine territory: although the organization of the Games has allowed the renovation of the ski resorts, the tourism model of the valleys has remained essentially unchanged, continuing to rely almost exclusively on a tourism related to the winter season and the snow sports. So, the dismantling of the Olympic territorialization until now has not been followed by a process of re-territorialization. In particular, the problems related to the reuse of the two large facilities – such as the ski jump ramp *Stadio del Trampolino* in Pragelato and the bobsled track in Cesana – are unsolved. These facilities, in fact, had no future from the outset and are, as expected, largely redundant in terms of both use and exchange value (Legambiente, 2007).

Turin 2006: The Territorialization Process and the Legacy for the City

Conversely, if we limit the evaluation of the territorialization processes only to the city of Turin (see Plate 7.1), the judgement concerning the Olympic legacy must be more nuanced. In fact, looking at the long-term trends, it can be recognized that the Olympics have contributed significantly to the advancement of territorial transformations that the city administration intended to promote. This applies, in particular, to the recovery of former industrial areas located along the "central backbone" of the city and the reorganization of the system of sports facilities. However, the Olympics also played an important role as a "catalyst" of processes that originated from political decisions not related to the event. The Olympic legacy, therefore, is visible in all of the changes that were brought about and have been accelerated by the needs related to the Games' organization, maybe even more so than in the areas that were actually built as Olympic venues. According, therefore, to processes identified in Figure 7.1, Olympic territorialization has contributed to an overall re-territorialization of the urban system, interacting in a complementary way with "ordinary" dynamics. Post-Games, each of the areas directly affected by the Olympic works, has followed specific paths, determined by the characteristics of those works or by other dynamics, derived from different projects.

The development of the Turin Olympic Village, in the south of the city, which hosted 2,500 people during the Olympics, is of particular interest. The affected area was more than 100,000 square meters and at the centre of it was the structure of the former General Fruit Market, whose origins date back to 1934. A pedestrian bridge, built in shapes so as to make it a symbol of the event, linked it to the *Lingotto*, which housed the headquarters of Turin 2006. Soon it became evident that the buildings that had housed the athletes were in a state of decay, which made it difficult to give them a new designation. More recently, these homes have been occupied by refugees, predominantly refugees escaping outbreaks of violence in Libya in 2011. Currently, there are about 400 inhabitants. Public offices are present in other buildings; one of them was recovered thanks to a project, which led to the creation of 42 new low-cost apartments that host families, singles, college students and people with temporary housing problems.

The case of the Media Village, in the northern part of the city, is quite different: the buildings that had hosted nearly 1,500 journalists have been converted into 426 apartments of public property, destined to house a population of low social status. This complex of buildings is part of a new neighbourhood, "Spina 3", an old district of steel production, an area of more than 1 million square meters. Although the district as a whole is characterized by a certain degree of social heterogeneity, the buildings show a remarkable concentration of social problems (Bianchetti and Todros, 2009; Olagnero and Ballor, 2010). Nevertheless, the introduction of policies promoting active citizen participation and conflict mediation has led to an acceptable degree of social integration in the neighbourhood (Conforti et al., 2012).

Therefore, the two abovementioned cases have led to different outcomes: in the first case the de-territorialization of Olympic space has left a noticeable void in the area and the processes of re-territorialization have encountered acute difficulties. In the second case, the process was more gradual, because the "ordinary" dynamics of recovery of an industrial area have resulted in new and substantial investments and have stimulated social integration policies. This is also the case for the other Media Village in the north-east part of the city. This Village was designed and engineered primarily based on post-Olympic user requirements, mainly including students and academics. Four buildings were realized with a total of 280 rooms and 330 beds, used as university residences from 2007 as part of the project for the new university campus in the former Italgas industrial area.

Turin 2006: The Image Change

What has been said so far concerns the material dimension of the transformation of the Olympic territory, namely that of "reification". The symbolic dimension did not necessarily follow a parallel path; however, it is necessary to distinguish between changes that affect the city as a whole and those that refer to specific locations involved in the Olympic event.

From an overall point of view, there is no doubt that Turin has profoundly changed its image since 2006; the Olympics, therefore, have emerged as a turning point for the city's image among its inhabitants and the general public. Although the data related to tourism have shown a significant, but not striking growth in this sector, the surveys carried out a few years after the Games have revealed that Turin is widely regarded as a city of culture, art, food and a good quality of life (Bondonio and Guala, 2012). Also regarding these changes it may be observed that the Olympics have acted as a catalyst within a process of cultural change to which various factors contributed in a synergistic way. However, from the perspective of the wider population, the Olympics often appear as a decisive factor, the Games being seen as synthesizing all other processes of urban change.

Nevertheless, the symbolic traces left by the Olympics on places of particular relevance for the event have often proved somewhat fleeting. Only a few places carry the memory of the Games in their names; among these is the stadium that hosted the opening and closing ceremonies, which is now designated as the "Olympic" stadium and hosts the matches of the Torino F.C. team. The city's other football club, Juventus F.C, now has its own stadium, where the stadium built for the World Cup in 1990 previously stood. The different symbolic connotations of both stadiums are now widely due to the rivalry between the two clubs, much more than to the symbolic legacy of the Games. Among other Olympic symbols, the area designated for the athletes' awards, the "Medals Plaza", was a temporary installation within one of the most important historical sites of Turin, *Piazza Castello*, and, therefore, with its dismantling the consolidated symbolism of the place has immediately resumed its dominant role. The same process of symbolic de-territorialization of the Olympic space has occurred with other elements of the urban furniture, for instance images of the mascots; there are now few traces of them, mostly in peripheral areas.

In short, we can say that the symbolic changes to the city owe much to the Turin 2006 event, but at the same time few of them retained a direct trace of the Olympic symbolism, with its specific content related to winter sports, international competition, passion and so on. This symbolism was to become just one part of a wider place branding strategy that was to recreate a creative, vibrant, cosmopolitan image of the city, one that displaced its reputation as a city with a heavy industrial crisis related to its reliance upon Fiat (Vanolo, 2008).

Turin 2006: The Legacy Following the Global Economic Crisis

In assessing the Olympic legacy, the global economic crisis – which began a couple of years after the event – has represented and is still representing a new break in the process of urban development and also affects the reflection on Turin's event and the evaluation of the long-term effects of the Games.

In public debate the Olympics are seen as a crucial event. It is recognized that the Games helped to present Turin as a city that had made the transition into the post-industrial world, a city that was able to achieve independence from the exclusive link with the automotive industry, while maintaining a vocation as a technological city capable of great organizational efforts. However, the financial crisis of 2008 also gave strength to the arguments of those who oppose the hosting and organization of major events, especially when resources for public investment are rapidly declining and the possible economic effects of mega-events are always uncertain and may occur only after several years. Thus, many argue that, in this phase of recession affecting many western economies, mainly countries with emerging economies are interested in organizing mega-events, for reasons of national prestige. It is, hence, not a coincidence that in 2012 Italy decided not to put Rome forward as a candidate for hosting the 2020 Olympic Games.

Although these arguments mainly relate to future events, to some extent they also retrospectively change the image of the Winter Olympics in Turin 2006.Public opinion of the Winter Games in Turin is, for instance, tainted by the fact that the debt per capita of the city of Turin is the largest among the Italian cities (about 3.5 billion Euros at the end of 2010), although, in reality, direct investment for the Olympics consisted only 7 per cent of the total debt (Bondonio and Guala, 2012). Above all, there is, however, a widespread impression that the beginning of the crisis only two years after the Olympics reduced the potential positive effects of the event on the city and that, in any case, this kind of experience, with similar investments, cannot be repeated for a long period. Thus, the Olympics of 2006 tend to be seen as the culmination of a long and fruitful phase of transformation of the city, followed by a new period of uncertainty, whose outcome is not easy to predict.

Conclusions

The lessons for urban planning to be learnt from the Turin Olympic Winter Games refer to the theme of the cycle of T-D-R: consideration of the post-event utilization of facilities must not be left as an after-thought, but should instead be a prime consideration in the infrastructure planning. In fact, both tangible and intangible effects do not occur automatically, but they have to be planned for and must be integrated into long-term development strategies. To secure effective processes of re-territorialization local/global relationships also need to be considered. If the recent global crisis requires the rethinking of the territorialization of the Olympics, it is therefore right to ask what happens now in terms of legacy.

Until 1992, the number of bidding cities remained generally low, but over the past two decades, a growing interest at national and international levels has produced new forms of competition for hosting mega-events – be they in the field of sport, culture or politics. For instance, the list of cities competing for the 2012 Summer Olympics comprised three cities at the top of the international urban hierarchies – London, Paris, New York – plus some second tier world cities – Moscow and Madrid. The growth of interest shows that mega-events are a great opportunity for a host city or country to exhibit their specialist know-how and capacity for innovation. Yet, it is interesting to note that until ten years ago host cities were predominantly located in Europe, and to a lesser extent, in North America and Asia, hence in regions of advanced economic development (Essex and Chakley, 1998).

In the competition for hosting mega-events a new geo-economic and geo-political element has emerged in the last few years. After a "geographical turnaround" – from Europe and North America to Asia, Africa and the Arab world – the BRICS now dominate the scene, reflecting their increasing economic hegemony and their willingness to project their powerful identity (Müller, 2011, 2012; Müller and Steyaert, 2013). In this context, 'with global events impinging on local decision-making, the Olympic Games and other mega-events turn urban politics into urban *geo*politics' (Müller and Steyaert, 2013:146). In fact, as Caffrey (2008: 808) observes: 'Games allow each state's proxies to compete without killing each other'. The past experiences of mega-events of the old continent were more closely linked to the image of the host city, such as for example, models of big urban transformation of Barcelona in 1992 and London in 2012. Now things are different, as Müller explains:

"Among bidders, the strong growth of emerging markets has created the necessary capital and infrastructure base as well as technical know-how to put together and finance sophisticated applications that meet and exceed the requirements of governing bodies such as the IOC and FIFA. At the same time, due to the sovereign debt crisis, Western states have become less willing to foot the substantial public bill of mega-events. [...] On the selection side, awarding mega-events to emerging countries contributes to opening up new markets with considerable growth potential, which is of particular interest for the corporate sponsors that fund the lion's share of these events. What is more, host cities in these economies often have less financial and planning constraints in hosting mega-events". (Müller 2012: 695)

Moreover, there is a noticeable shift towards hosting mega-events outside Western Europe and North America. This trend should be seen as today's new forms of affirmation of emerging and upcoming States as illustrated by the Olympic Games in Beijing (2008) and in Rio de Janeiro (2016), the Olympic Winter Games in Sochi (2014) and in Pyeongyang (2018), the FIFA World Cup in South Africa (2010), in Brazil (2014), in Russia (2018) and Qatar (2022). Cornelissen (2010) and Müller (2012) maintain that this new trend demonstrates that the emerging states seek spaces to show signal achievements and diplomatic stature and that hosting cities are no longer simply looking for jobs and investment.

The 2008 Olympics in Beijing seems to be emblematic: in fact, it was recognized as an event of great relevance not only in sports, but also on the geopolitical and geo-economic levels, both global and local. Adopting the slogan "One Games, One World", China attempted to show its integration into the international community, having completed its programme of modernization.

In such countries as China, the creation and deployment of a positive image internationally seems to be ranked higher than the more material impacts on growth or infrastructure in these settings (Berkowitz et al., 2007; Müller, 2011). We can consider for example the case of the Olympic Winter Games in Sochi in 2014 that 'serve as an instrument to show to the world that Russia, besides being an energy superpower that likes to flex its military muscle, should also be taken seriously as a global player in the game of leisure and tourism' (Müller, 2012: 698).

Some experts, however, have pointed out that this new geopolitical trend in the emerging markets could bring some risk linked to the prevalence of the political purposes that "often feature a more hierarchical planning culture, less pressure on financial resources, less concern for environmental issues and more profound urban transformations in a push for modernisation than cities in the West (e.g., Abramson, 2007; Stanilov, 2007)" (Müller, 2012: 694).

If this analysis is valid, it could imply a possible future divergence between emerging and mature economies on the issue of Olympic territorialization. The former countries could see the organization of a mega-event above all as a geopolitical choice and, at the same time, as an opportunity to attract foreign capital to their main urban centres. Thus Olympic territorialization would play the role of a media showcase that imposes itself on the standard spatial dynamics in a rapid and sometimes forceful way. On the other hand, older developed countries might see this type of territorialization more as a risk than as an opportunity; as a consequence they may forego bidding or try to define – even by negotiating with the IOC – a "softer" and more sustainable model of event, not only with regard to its impact on the environment, but also in relation to its economic dimensions and the spatial needs of the host city.

References

Abramson, D. B. (2007). 'The dialectics of urban planning in China', in F. Wu (ed.) (2007), *China's Emerging Cities: The making of new urbanism*. London: Routledge, pp. 66–86.

Berkowitz, P., G. Gjermano, L. Gomez and G. Schafer (2007). 'Brand China: Using the 2008 Olympic Games to enhance China's image'. *Place Branding and Public Diplomacy*, Vol. 3, pp. 164–178.

Bianchetti, C. and A. Todros (2009). 'Abitare Spina 3'. *ASUR*, 94, pp. 63–72.

Bondonio, P. and Guala C. (eds.) (2012). *Gran Torino. Eventi, turismo, cultura, economia.* Roma: Carocci.

Bondonio, P., C. Guala and A. Mela (2008). 'Turin 2006 OWG: Any legacies for IOC and Olympic territories?', in R. K. Barney et al. (eds.) (2008). *Pathways: Critiques and discourse in Olympic research.* Ninth International Symposium for Olympic Research, ICOS. London, Ontario (CA), pp. 151–65.

Bontempi, R. (2006). *Torino. Città delle Alpi. City of the Alps.* Torino: TOROC.

Caffrey, K. (2008). 'Olympian politics in Beijing: Games but not just Games'. *The International Journal of the History of Sports*, 25(7), pp. 807–825.

Cashman, R. (2010). 'Impact of the Games on Olympic host cities: University lecture on the Olympics' [online article]. Barcelona: Centre d'Estudis Olímpics (UAB), International Chair in Olympism (IOC-UAB) http://olympicstudies.uab.es/2010/docs/cashman_eng.pdf (accessed 25.02.2014).

Cashman, R. and Horne J. (2013). 'Managing legacy', in Frawley, S. and D. Adair (eds.) *Managing the Olympics*. Palgrave London: Macmillan, pp. 50–65.

Chalkley, B. and S. Essex (1999). 'Urban development through hosting international events: A history of the Olympic Games'. *Planning Perspectives*, 14 (4), pp. 369–394.

Chappelet, J. L. (2010). 'A short overview of the Winter Olympic Games: University lecture on the Olympics' [online article], Barcelona: Centre d'Estudis Olímpics (UAB), International Chair in Olympism (IOC-UAB), http://olympicstudies.uab.es/2010/docs/chappelet_eng.pdf (accessed 26.02.2014).

Conforti, L., C. A. Dondona and G. Perino (eds.) (2012). Metamorfosi della città. Torino e la Spina 3, IRES Piemonte, Torino: Regione Piemonte. http://www.ires.piemonte.it/convegni/Spina3_pubblicazioneires-regionepiemonte14–12–12.pdf (accessed 14.01.3014).

Cornelissen, S. (2010). 'The Geopolitics of Global Aspiration: Sport Mega-events and Emerging Powers', *International Journal of the History of Sport*, 27, pp. 3008–3025.

Crivello, S., E. Dansero and A. Mela (2006). 'Torino, the Valleys and the Olympic Legacy: Exploring the Scenarios', in N. Müller, M. Messing and H. Preuss (eds.) (2010), *From Chamonix to Turin. The Winter Games in the Scope of Olympic Research*, Kassel: Agon, pp. 377–393.

Dansero, E., B. Del Corpo, A. Mela and I. Ropolo (2011). 'Olympic Games, conflicts and social movements: The case of Turin 2006', in G. Hayes and J. Karamichas (eds.) (2011). *Olympic Games, Mega-Events and Civil Societies*. Basingstoke: Palgrave Macmillan, pp. 195–218.

Dansero, E. and A. Mela (2007). 'Olympic territorialization'. *Revue de Géographie Alpine | Journal of Alpine Research*, 95 (3), pp.16–26.

Dansero, E. and A. Mela (2008). 'Per una teoria del ruolo dei grandi eventi nei processi di territorializzazione', in N. Bellini and A. Calafati (eds.) (2008). *Internazionalizzazione e sviluppo regionale*. Milano: Franco Angeli, pp. 461–488.

Dansero, E. and A. Mela (2012). 'Bringing the mountains into the city: Legacy of the Winter Olympics, Turin 2006', in H. Lenskyj and S. Wagg (eds.) (2012). *The Palgrave Handbook of Olympic Studies*. Basingstock: Palgrave Macmillan, pp. 178–194.

Dansero, E. and M. Puttilli (2012). 'From Ford to the Olympics: The development of an emblematic ski resort – Sestriere'. *Journal of Alpine Research*, Revue de Géographie Alpine, 100(4), pp. 1–13.

Davidson, M. and D. McNeill (2012). 'The redevelopment of Olympic sites: Examining the legacy of Sydney Olympic Park'. *Urban Studies*, 49(8), pp. 1625–41.

Essex, S. and B. Chalkley (1998). 'Olympic Games: Catalyst of urban change'. *Leisure Studies*, 17 (3), pp. 187–206.

Essex, S. and B. Chalkley (2004). 'Mega-sporting events in urban and regional policy: A history of the Winter Olympics'. *Planning Perspectives*, 19 (2), pp. 201–232.

Essex, S. and B. Chalkley (2011). 'Driving urban change: The impact of the Winter Olympics', in J. Gold and M. Gold (eds.) (2011). *Olympic Cities.* London: Routledge.

Girginov, V. (ed.) (2013). *Handbook of the London 2012 Olympic and Paralympic Games.* London: Routledge.

Gold, J. R. and M. M. Gold (eds.) (2011). *Olympic Cities. City Agendas, Planning, and the World's Games, 1896–2016, Second Edition.* New York: Routledge.

Guala, C. and S. Crivello (2006). 'Mega-events and urban regeneration. The background and numbers behind Turin 2006', in N. Müller, M. Messing and H. Preuss (eds.) (2006). Fro*m Chamonix to Turin. The Winter Games in the Scope of Olympic Research.* Kassel: Agon Sportverlag, pp. 323–42.

Haugen, H. O. (2005). 'Time and space in Beijing's Olympic bid'. *Norwegian Journal of Geograph*y, 59, pp. 221–227.

Hayes, G. and J. Jaramichas (eds.) (2011). *Olympic Games, Mega-Events and Civil Societies. Globalization, Environment, Resistance.* Basingstoke: Palgrave Macmillan.

Hiller, H. (2000). 'Mega-events, urban boosterism and growth strategies: An analysis of the objectives and legitimations of the Cape Town 2004 Olympic bid'. *International Journal of Urban and Regional Research,* 24, pp. 439–458.

Hiller, H. (2003). 'Toward a science of Olympic outcomes: The urban legacy', in de Moragas M., C. Kennett and N. Puig (eds.) (2003). *The Legacy of the Olympic Games, 1984–2000,* Olympic Museum, 14–16 November 2002. Documents of the Museum, IOC, Lausanne, pp. 102–109.

Klauser, F. R. (ed.) (2012). 'Theme issue: Claude Raffestin'. *Environment and Planning D,* 30(1), pp. 106–190.

Legambiente (2007). Dossier. L'eredità olimpica di Torino 2006, http://www.legambientepiemonte.it/doc/ Olimpiadi-pdf/08–02–07%20Dossier_eredita%20olimpica%20Torino2006.pdf (accessed 23.01.2014).

Lenskyj, H. J. (2002). *The Best Olympics Ever? Social Impacts of Sydney 2000.* Albany: State University of New York.

Lenskyj, H. and S. Wagg (eds.) (2012). *The Palgrave Handbook of Olympic Studies.* London: Palgrave Macmillan.

Moragas, M. de, C. Kennett and N. Puig (eds.) (2003). The Legacy of the Olympic Games 1984–2002: International Symposium, Lausanne, 14th–16th November 2002. Lausanne, International Olympic Committee.

Müller, M. (2011). 'State dirigisme in megaprojects: Governing the 2014 Winter Olympics in Sochi'. *Environment and Planning A,* 43(9), pp. 2091–2108.

Müller, M. (2012). 'Popular perception of urban transformation through mega-events: Understanding support for the 2014 Winter Olympics in Sochi'. *Environment and Planning C: Government and Policy,* 30(4), pp. 693–711.

Müller, M. and C. Steyaert (2013). 'The geopolitics of organizing mega-events', in J. M. S. Munoz (ed.) (2013) *Handbook on the Geopolitics of Business.* Cheltenham: Edward Elgar, pp. 139–150.

Olagnero, M. and F. Ballor (2010). Convivenza e cooperazione sociale in contesti urbani: il caso degli ex Villaggi Olimpici a Torino, Rassegna Italiana di Sociologia, 51, 3, pp. 429–458.

Raffestin, C. (1980). *Pour une géographie du pouvoir.* Paris: Librairies techniques.

Raffestin, C. and S. A. Butler (2012). 'Space, territory and territoriality'. *Environment and Planning D: Society and Space,* 30(1), pp. 121–141.

Roche, M. (2000). *Mega Events and Modernity: Olympics and Expos in the Growth of Global Culture.* London: Routledge.

Sandy, R., P. J. Sloane and M. S. Rosentraub (2004). *The Economics of Sport. An International Perspective.* New York: Palgrave Macmillan.

Spilling, O.R. (1998). 'Beyond Intermezzo? On the long-term industrial impacts of nega-events: The case of Lillehammer 1994'. *Festival Management and Event Tourism,* 5(3), pp. 101–122.

Stanilov, K. (2007). 'Urban planning and the challenges of post-socialist transformation', in K. Stanilov (ed.) (2007). *The Post-socialist City: Urban Form and Space Transformations in Central and Eastern Europe after Socialism.* Dordrecht: Springer, pp. 413–425.

Vanolo, A. (2008). 'The image of the creative city: Some reflections on urban branding in Turin'. *Cities,* 25(6), pp. 370–382.

Chapter 8

The Politics of Mega-event Planning in Rio de Janeiro: Contesting the Olympic City of Exception

Anne-Marie Broudehoux and Fernanda Sánchez

Introduction

As host of the 2014 finals of the Férération Internationale de Football Association (FIFA) World Cup and of the 2016 Summer Olympic Games, Rio de Janeiro is one of the rare cities in recent history to receive the world's two top sporting mega-events, within a two-year span. The impacts of this unprecedented "double whammy" in terms of infrastructure projects and the transformations of the city's urban landscape are tremendous, and have important social, political and economic ramifications.

This chapter examines the role of sporting mega-events in the transformation of the urban landscape, to understand some of their impacts upon local population groups. It investigates the production of the "Olympic City" as a complex image construction process, which mobilizes multiple agents and requires important social, spatial and political reconfigurations (Sánchez 2010). Based on the accelerated implementation of large urban projects and on the projection of an illusory image of successful urbanity, these image construction initiatives often result in radical territorial transformations and in socio-spatial exclusions that have lasting impacts upon vulnerable populations directly affected by these interventions (Broudehoux 2007).

This work rests upon the hypothesis that mega-events are increasingly being instrumentalized by local political and economic elites to remake the city in their own image. Through an analysis of many of Rio de Janeiro's event-related projects, the chapter examines the effects of mega-event planning upon the projection of an exclusive vision of urbanity. It suggests that this planning vision can open the way for the state-assisted privatization and commodification of the urban realm, and serve the needs of capital while exacerbating socio-spatial segregation, inequality and social conflicts.[1]

Urban Governance in Rio de Janeiro: From Strategic Planning to Event-City

Rio de Janeiro is a city long concerned with the production and dissemination of a positive urban image. As Brazil's national capital for over two hundred years, and the nation's prime tourist destination, Rio's symbolic profile has always influenced its development. Rio de Janeiro's incursion into the realm of mega-events began in the early 1990s, with the hosting of the 1992 World Summit on the environment. Over the next decade, the city would more firmly embrace mega-events as a core promotional strategy to attract global capital, hosting, among others, the 2007 Pan American Games, the 2010 World Urban Forum, the 2011 Military World Games, the 2013 World Youth Day, the 2014 FIFA World Cup, and the 2016 Olympic Games. The city pursued a strategic policy of hosting of mega-events as a means of positioning itself among great world cities and as a unique opportunity to attract global interest, to stimulate inward investment, and to improve the city's "primitive accumulation of symbolic capital" (Torres Ribeiro 2006).

---

1   Much of this analysis rests upon the collective work of a series of scholars involved in two integrated Brazilian research laboratories, the Laboratório Globalização e Metrópole (Globalization and Metropolis) of the Fluminense Federal University (UFF) in Niterói, and of the Laboratório Estado, Trabalho, Território e Natureza (ETTERN) (State, Work, Territory and Nature) at the Federal University of Rio de Janeiro (UFRJ). Researchers from these groups have since the early 1990s explored emerging planning models (Sánchez 2010; Bienenstein 2001), analysed large urban projects and studied mega-events (Vainer 2009; Lima Junior 2010; Oliveira et al. 2007). Over the years, international researchers have also collaborated with these groups and helped define some of the issues related to the relationship between mega-events and urban development in countries of the global South (Freeman 2012; Gaffney 2012; Broudehoux 2013; Stavrides 2010).

The coming of Rio's moment in the global spotlight (*O Momento Rio*) was facilitated by several factors, including the adoption, in the 1990s, of "strategic planning", a neo-liberal mode of governance. In Brazil, this managerial approach was perceived as the only viable option to engage in the "mythical" inter-urban competition for an increasingly mobile global finance capital and to face the new conditions imposed by globalization (Vainer 2009). Planning strategies focused on improving the city's image, repackaging its assets and marketing its competitive advantage in order to attract foreign investors, wealthy residents, international tourists and agents of the new creative class.

In the 1990s, Rio de Janeiro's urban politics underwent important transformations, which deeply impacted the city's institutional structure and its approach to management. The period was marked by a new rapprochement between the city's public and private sectors, which revived an old alliance between real estate capital and the city's executive power (Bienenstein 2001; Oliveira 2013; Lima Jr 2010). Planning activities were limited to licensing, inspection, and the promotion of adaptive strategies serving the real estate market and the privatization of public services. This political re-orientation greatly impeded the implementation of policies that could have helped reduce social inequality (Oliveira 2013).This demonstrates that the issue of social inequality is not central to the new urban agenda and not a determining factor in the current framing of urban interventions.

By the mid-1990s, Rio de Janeiro had adopted a mode of governance modeled after the strategic planning approach pioneered in Barcelona, whose acclaimed Olympic revitalization was widely emulated in Latin America. Barcelona's success in mobilizing private sector resources to revitalize its urban infrastructure, and its use of culture and sporting events to rejuvenate a depressed urban image, attract external capital, and position itself in the global economy, greatly influenced Rio de Janeiro's new urban vision (Ferreira 2010). In 1995, the Strategic Plan of the City of Rio de Janeiro *(Plano Estratégico da Cidade de Rio de Janeiro),* Latin America's first strategic plan, was adopted. It vowed to restore tourism as the city's "natural" vocation and to insert Rio in the mega-events circuit as a viable way to enhance the city's global image and stimulate inward investment (Acioly 2001).

Rio's embracing of sporting mega-events as a development strategy was facilitated by a rare political alignment at the municipal, provincial and federal levels, with a strong political alliance between President Lula's Worker's Party, Rio de Janeiro's state governor, Sergio Cabral, and Rio's mayor, Eduardo Paes. This unified political front was behind Rio's successful Olympic bid, after two failed attempts to get the Olympics (2004, 2012), and the hosting of the 2007 Pan American Games.

Mega-events and Urban Redevelopment

In recent decades, many cities around the world have taken an entrepreneurial approach to urban territorial management in order to efficiently compete on the global market (Harvey 1989). In the context of global economic restructuring and interurban competition, the staging of high-profile events such as world exhibitions, international conferences or sporting events has come to be perceived as a choice development strategy and a potent vehicle for post-industrial adjustment (Hiller 2012; Short 2008; Greene 2003). Hosting mega-events is seen as a rare opportunity for place promotion, helping enhance global visibility through media coverage and advertising. It is also perceived as a panacea for economic regeneration, stimulating domestic consumer markets while capturing mobile forces of capital (Gold and Gold 2011; Hiller 2000).

In recent years, mega-events have become more than mere catalysts for urban development, they are now powerful engines in the neoliberal reconfiguration of the city, promoting the privatization and commodification of urban space and the implementation of market-oriented economic policies (Vanwynsberghe et al. 2012; Hayes and Horne 2011; Peck and Tickell 2002). They represent important tools to leverage urban interventions, allowing local governments to reprioritize the urban agenda while facilitating the accelerated implementation of key projects and attracting investors to finance them (Hiller 2006; Smith 2005; Chalkey and Essex 1999). For some critics, the urban impact of mega-events goes even deeper, affecting the reconfiguration of power structures at the local and national level and contributing to the imposition of a new neoliberal order on the city, marked by a high level of exceptionalism (Vainer 2011).

For Brazilian sociologist Carlos Vainer (2009), Brazil's neo-liberal mode of governance is characterized by an exercise of power that is at once authoritarian, based on selective participation, and marked by the growing role of the private sector in urban management. This form of neoliberal authoritarianism, described by Zizek (1999) as "post-political", also rests upon the construction and consolidation of consensus, where open debate around urban issues is discouraged, and dissent is replaced by compromise and pacification. This characterization echoes Swyngedouw's (2010) description of the competitive city as dependent upon a reconfiguration of the political order, with the growing participation of private actors and other unelected agents in the act of governing. According to Vainer (2009), Brazil's own brand of strategic planning radicalizes the Catalan model by reinterpreting the integration of public and private sectors advocated by Castells and Borja (1996) into a submission of the common good to private interests. By relying upon institutional flexibility, strategic planning allows local authorities to reformulate planning regulations to better serve investor interests, by adapting zoning and land use plans and granting tax exemptions and other derogations (Lima Junior 2010).

Brazilian scholars claim that it is by generating an artificial crisis, prompted by a discourse of fear, violence and economic decline, that neoliberal leaders have managed to generate popular consensus about the need for major urban interventions (Vainer 2009; Arantes 2009). For local authorities, the need to respond to such crisis in a rapid and appropriate fashion justifies bypassing lengthy political discussions and legitimates resorting to authoritarianism (Gusmão 2013) [2]. In this context, mega-events' tight schedule and fixed deadlines, help generate a sentiment of urgency – what Stavrides (2010) calls an "Olympic state of emergency" – thereby creating unique, exceptional conditions that facilitate and accelerate the realization of large-scale urban projects.

For Vainer (Vainer, 2011), this state of emergency has a profound impact upon local planning conditions, turning the host city into a "city of exception", where exceptions literally become the rule. This characterization draws upon Agamben's (2005) "state of exception", a theory that describes the suspension of laws in times of crisis and emergency in order to face an unexpected 'necessity'. Mega-events would thus represent essential tools to help by-pass the democratic political process in the implementation of mega-projects. They would legitimize the adoption of an exceptional politico-institutional framework that authorizes the relaxation of certain rules and obligations in the implementation of urban interventions that will benefit the event. The event-city would thus be characterized by the disruption of accepted legal and social norms; the suspension of established procedures, restrictions and controls; the reformulation of planning regulations; the circumvention of existing laws; the lifting of safety standards; and the introduction of highly restrictive regulatory instruments to ensure compliance with the stipulations of local and global organizers and better serve investor interests (Hayes and Horne 2011; Lima Junior 2010).

Furthermore this state of exception would facilitate the imposition of extra-legal forms of governance, allowing non-elected agents, including beneficiaries of international capital sponsorship like the IOC and FIFA, to play a key role in local decision-making (Gusmão 2011). As a result, public-private coalitions thus enjoy extraordinary powers to carry out massive urban transformations without any form of accountability. The scope of event-related projects to be undertaken and the magnitude of demands made by international organizations like the FIFA and the IOC have given these coalitions the license to take exceptional measures in order to reshape the city for the needs of the event, its sponsors and local partners.

Mega-events also result in the creation of self-governing territorial enclaves, constituted as special autonomous zones or states within the state, where sovereign law is suspended and political and ethical responsibilities are blurred. Gusmão (2013) describes the political and juridical autonomy that is granted to mega-events franchise holders like the FIFA and the IOC, as leading to the emergence of a parallel form of government and a parallel form of justice. These organizations are thus able to remake the city, in both its

---

2 Nelma Gusmão de Oliveira (2013) sees a strong convergence between the authoritarian character of neoliberal planning practices and the production of mega-events, marked by the direct inference of executive powers in the act of legislation. For her, one of the great dangers of strategic planning, especially in the context of the Olympic state of emergency, lies in its depoliticizing power. In her view, strategic planning rests upon the construction and consolidation of consensus, which leads to the gradual disappearance of the political in favour of the consensual. Vainer (2009), has similarly defined entrepreneurial planning as leading to a radical negation of political space, not solely in the disparition of the rigid separation between public and private sectors, but in the submission of public interest to private ones.

temporal and spatial dimensions, by imposing their own time frame upon urban development, and creating archipelagos of extraterritoriality, which often become exclusive commercial territories for their sponsors and commercial allies (Broudehoux forthcoming).

Rio de Janeiro's Mega-Events in Historical Perspective

The 2007 Pan American Games represented Rio's first incursion into the realm of sporting mega-events. Many projects initiated for this event would later be realized for the 2014 FIFA World Cup and the 2016 Olympics (Oliveira 2009). Project implementation patterns that were developed for the Pan American Games would also be adopted in the realization of the other two mega-events. For example, despite official claims that event-related interventions would be evenly distributed among four urban areas, investments were concentrated in upscale sectors of the city, especially in elitist Barra da Tijuca (Mascarenhas et al. 2011). The rationale for locating activities in this wealthy suburb was to protect the athlete's safety and well-being, while ensuring that the city's world image would not be tainted by glimpses of violence, disorder and poverty. It was also motivated by a desire to consolidate this area's emergence as a secondary urban center by stimulating real estate development.

Investments for the Pan American Games were marked by their ad-hoc nature and market-friendly character, a total disregard for long-term development and a lack of concern for their local social and spatial context. For example, the João Havelange Olympic Stadium benefited from many exceptional conditions, including the relaxation of local zoning regulations, which allowed the construction of this twelve-story structure in a low-rise residential sector (Sánchez and Bienenstein 2009). Although officials had promised that the stadium would help revitalize this working class neighborhood, little was done to improve local conditions while traffic congestion was greatly enhanced (Martins da Cruz 2010). In April 2013, the seven-year-old stadium was closed for an undetermined period, after inspectors found major structural defects in its roof. Its very short lifespan reopened public debate about the waste of public funds on sporting infrastructure, and allegations resurfaced concerning overbilling during construction.

Another example is Vila Panamericana, the athlete's village, whose construction on reclaimed marshland in Barra da Tijuca, was clearly intended to stimulate the local real estate sector and to accelerate the local urbanization process. Although publicly funded,[3] the village was later ceded to a private real estate company to be commercially exploited as a residential condominium, prompting critics to denounce the transaction as a direct transfer of public funds to the private sector (Mascarenhas et al. 2011).

One important legacy of Rio's Pan American Games was the creation of several resistance movements to protest against evictions and denounce injustice, resulting in a few key victories. An example of such organized resistance is Vila Autódromo, a well-established low-income neighbourhood near the future Olympic park in Barra da Tijuca, whose existence was threatened by repeated eviction orders. Its persistence in demonstrating its rights and claiming its legitimacy warranted its survival, although it would come under threat once more with the realization of the Olympic Games. Many of the resistance groups born during the Pan American Games, such as the Comitê Social do Pan *(Social Committee for the Panamerican Games),* survived the event and have morphed into organizations fighting the Olympics and the World Cup.

The Pan American Games would thus set the tone for the planning of future mega-events. Olympic investments would similarly be concentrated in Barra da Tijuca, where the dream of building a post-card perfect vision of modern Rio is still very much alive. Olympic projects continue to be planned as instruments of real estate development, with the implementation of modern infrastructure to help promote land valuation and boost the local tourism industry. However, the most lasting impact of this initial venture into event-led redevelopment was the powerful consensual rhetoric that would paint mega-events as a panacea for the ongoing urban crisis, and a quick fix solution for urban regeneration.

---

3 Its construction was funded by the Worker's Support Fund (FAT), which is managed by the Caixa Economica Federal (CEF), a Brazilian public bank.

Rio de Janeiro's Mega-Events in Political Perspective

Projects planned for the 2014 World Cup and 2016 Olympics confirmed the market-friendly orientation that prevailed in Rio, at once elitist, segregationist and exclusive. Rio's mega-events were put forward by policies that promoted the concentration of power and capital and the privatization of public space and services. Local economic agents took advantage of a favorable social, economic, and political conjuncture to push the adoption of many opportunistic projects. The municipal government similarly used public projects as self-serving marketing initiatives.

More than US$20 billion of public funds would be spent in Rio de Janeiro on Olympic-related investments. These investments have strengthened the role of these private companies in the transformation of the city's urban landscape, especially in transportation management. Some of the greatest beneficiaries of these projects are large private engineering and construction firms like Odebrecht and OAS Ltd, who have expanded their influence in urban affairs, winning numerous bids for major public works and event-related projects (Oliveira 2013). A handful of firms have thus seized most major urban projects, including the construction of the new Bus Rapid Transit corridors, Rio's subway expansion, the Maracanã Stadium makeover and Porto Maravilha, the city's vast port revitalization project.

Olympic preparations also confirmed the centrality of Barra da Tijuca as a priority development area, despite being already blessed by generous public investment. Mass transportation projects are clearly favoring Barra over the working class northwestern periphery, which is ten times more populous but do not enjoy adequate public transportation (Oliveira 2013). Thanks to the Olympics, more than US$4 billion of public funds are being spent on a sixteen kilometers subway expansion towards Barra da Tijuca, where many Olympic venues, including the Olympic Village, will be concentrated. City Hall with federal funds are also spending US$1.3 billion to build exclusive lanes for the new Bus Rapid Transit system connecting Barra da Tijuca to the city center. Covering a total distance of 150 km, these BRT lines represent the largest one-off transportation investment in Rio de Janeiro's history and the centerpiece of Rio 2016's transportation plan. Their construction will result in the dislocation of tens of thousands of residents and will have permanent environmental impacts on local wetlands, now subjected to intense real-estate development and accrued vehicular traffic (Gaffney et al. 2012).

An emblematic example of the exceptional urbanism that has marked Rio's mega-events preparation is the polemical renovation of the Maracanã stadium, site of the World Cup's closing match and the Olympics' opening and closing ceremonies. This mythical stadium, long considered a symbol of Brazilian identity and of Rio's popular culture, had already undergone a costly (US$250 million) renovation before the 2007 Pan American Games. However, in preparation for the World Cup, FIFA requested an in-depth reform that would deeply alter the Maracanã's architecture, spatial organization, and social accessibility. Part of the renovation involves the creation of a brand exclusion zone around the stadium, thanks to an exceptional legislation adopted for the FIFA World Cup and the Olympics and backed by special federal decree. It allowed the establishment of an exclusive, monopolistic commercial territory, banning the sale of products and the placement of advertizing from non-sponsor companies for the duration of the events. While Brazilian law prohibits the sale of alcohol within a vast perimeter around sporting stadia, this rule was lifted during the World Cup, for the exclusive benefit of an American beer company. At the same time, traditional food-sellers were banned from the area, where they had long held the right to sell pre-game churrasco (barbecue) to generations of fans (Gusmão 2013).

These restrictions and transformations were the object of vehement contestation by local population groups. Fans associations and members of the general public accused FIFA of denaturing the Maracanã by dramatically reducing the amount of affordable seating and forbidding a host of popular uses that had marked its history and identity (Mascarenhas et al. 2011; Bienenstein et al. 2014). People took to the streets to protest the adulteration of this beloved public institution, travestied into a multi-use arena for concerts and spectacles. A series of judicial actions were taken against the state's plan to privatize the Maracanã, after spending more than half a billion dollars of tax payers money bringing it up to FIFA standards.

Multiple evictions in the stadium's vicinity to establish a safety perimeter and create open-air parking lots also sparked a host of protest movements. In March 2013, Rio's riot police brutally evicted dozens of indigenous squatters from an abandoned museum near the Maracanã. Since 2006, native people from across

Brazil had used the building, known as *Aldeia Maracanã* (Maracanã Village) as a safe haven when visiting Rio to study, sell crafts, or receive medical attention. In a spectacular display of force that was broadcast around the world, shock troops used pepper spray and tear gas to disperse more than one hundred unarmed protesters (Sánchez 2013). Maracanã was also at the heart of the June 2013 protest movement that rocked all of urban Brazil. During the 2013 Confederations Cup, an event that serves as a testing ground for the FIFA World Cup, several actions denouncing the tremendous social and economic cost of hosting mega-events took place near the stadium. By the time the Cup was held in 2014, more evictions and demolitions had been carried out, but none of the planned parking lot and facilities had been built.

Rio's Port Revitalization as an "Exceptional" Olympic Project

Porto Maravilha, Rio's port revitalization project is unprecedented in the city's history, both in scope and in cost. The project is affecting five inner-city districts, spending US$4 billion to turn five square kilometers of devalued housing and industrial buildings into upscale office and residential towers. First-rate science and art museums, shopping malls and a new cruise-ship terminal anchor the project, while docks and warehouses are converted into shopping and entertainment venues. The project plans to quadruple the area's present population of 25,000 and to increase the port capacity from 50,000 to two million passengers a year (Porto Maravilha 2011). The project embodies the complex dynamics of mega-event planning and exemplifies many aspects of the "city of exception" theory (Vainer 2011).

Although the port revitalization project was not part of the original Olympic bid, local authorities aggressively lobbied the IOC to transfer Olympic functions to the port area once the city's candidacy was secured (Bentes 2011). By establishing Porto Maravilha as one of the great Olympic legacies to the city, authorities managed to capitalize upon its association with the Olympic brand to help market the real estate project (Ferreira 2010). The 2016 target date also fostered a sense of imminence and helped justify the haste with which many planning decisions were taken. While efforts to locate major facilities, including the Olympic Village, in the district failed, the IOC reluctantly allowed a few lesser projects to be built in the port, including the technical operations center and the media village (Cidade Olímpica 2011). But these decisions were short lived. By the end of 2013, these few Olympic-related projects were transferred back to Curicica, near Barra da Tijuca, a move justified by the City government as being linked to cost reduction and economies of scale.

As an urban project, Porto Maravilha is exceptional in many regards, but especially in its reliance upon the private sector. Not only does it represent the largest public-private-partnership (PPP) in Brazilian history, but its entire construction and management have been placed into the hands of private interests. A private corporation, the Port Urban Development Company (CDURP) was set up in 2009 to coordinate the project and court international investors, while the provision of all public infrastructure and services for the coming 15 year were contracted out to a private consortium, the Concessionária Porto Novo, made up of three of Brazil's largest engineering and construction firms – Norberto Odebrecht Brasil, Carioca Christiani-Nielsen Engenharia and OAS Ltd,

Researchers like Gusmão (2011) have found evidence suggesting possible cases of collusion between the state and private enterprises in the reconfiguration of the legal framework that facilitated Porto Maravilha's realization. She identifies entire sections of the municipal decrees that established the PPP's structure, determined its territorial limits, and suggested modifications to the master plan, to be identical and appear to have been lifted from a private sector proposal for the port's revitalization, submitted in 2009 (Gusmão 2011). The three enterprises behind this proposal were Odebrecht, Carioca and OAS Ltd, the same companies later selected as the unique eligible contenders for the realization of Porto Maravilha (Gusmão 2011). These companies already benefitted from the economic activity generated by the two mega-events, especially Odebrecht which was responsible for the construction of four of the twelve FIFA stadia in Brazil, for the refurbishing of the Maracanã, as well as the construction of Rio's new Bus Rapid Transit system.

This consortium will be responsible for managing the project, clearing the land and upgrading urban infrastructure, as well as providing basic services such as street lighting, drainage and garbage collection in the district until 2024 (Porto Maravilha 2011).

Also exceptional was the port's redevelopment process, which was facilitated by extraordinary political interventions, financial innovations and legal decrees passed in exceptional circumstances. For example, the municipal decree that authorized a public-private partnership with the Porto Novo consortium was passed within weeks of Rio winning its Olympic bid in 2009 and was adopted in an emergency fashion that failed to allow sufficient time for public scrutiny. This in spite of the fact that there was no precedent in Brazil for granting responsibility for the realization of public works and the maintenance and provision of public services for an entire urban district to a single contractor (Gusmão 2011). Other municipal laws were adopted in a similarly urgent manner, including decrees giving tax benefits to CDURP and other businesses wishing to settle in the port area or to participate in its redevelopment.

Yet another example of the exceptionalism that characterizes Porto Maravilha's realization has to do with the innovative financial instrument that was adopted to help finance its second phase, and which have been denounced as a form of financialization of real estate speculation (Rolnik 2011). Additional Construction Potential Certificates or CEPACs are purchasable rights to build beyond the established limit in a specific area of the city. These titles, which can be traded on the stock exchange and subjected to speculation, allow private developers to maximize land use and increase their profit margin (Faulhaber 2013). In Porto Maravilha, CEPACs represent the right to build above the six stories legal limit, to up to a height of fifty stories.

According to an analysis carried out by Jorgensen (2011), in order for investors to recoup the value of the CEPACs, residential or office space in Porto Maravilha will be the most expensive in Rio de Janeiro, making it unlikely for affordable housing to be built in the area. CEPACs have thus been described as a prime instrument of both privatization and gentrification of Rio's port area, as many poor families will have to be relocated from the port area to the far suburbs (Freeman 2012). Since the success of the project relies upon the potential for this devalued part of town to become prime real estate, which is contingent on a drastic alteration of the area's socio-economic makeup, in Porto Maravilha, gentrification and social exclusion will not be the accidental by-products of market-led urban renewal, but the necessary conditions for the project's successful realization (Broudehoux 2013). In August 2013, Mayor Eduardo Paes admitted that the CEPACs would privilege commercial use over residential use. Favela housing removal, the scarcity of new housing projects and lack of investment in rehabilitating the existing housing stock, coupled with an absence of means to capture the added value of real estate speculation to finance low cost housing projects, confirm the planned gentrification of the area. Already, at the end of 2014, 995 families had been removed (Comitê popular 2014).

Resisting Porto Maravilha: The Case of Morro da Providência

A particularly sensitive area within Porto Maravilha is Morro da Providência, a hillside favela of 5000 located at the heart of the vast real estate project. As Rio's first informal settlement and the oldest in Brazil, this 115 year-old favela has housed generations of port workers and refugees from the slave trade. Long ostracized for reasons of racial prejudice, the settlement was further marginalized by the arrival of violent drug gangs in the 1980s. Although the community's wholesale relocation is politically unfeasible, Providência's proximity to Porto Maravilha is clearly problematic for the real estate valuation of the project. But local authorities have found another way to limit its negative impact on the glamorous project: Providência will be trimmed, tamed and turned into a tourist attraction. Under the cover of the state-sponsored favela upgrading program *Morar Carioca,* various infrastructure projects will reduce the size of the settlement, while multiple cosmetic interventions, and the establishment of a permanent police pacification unit (UPP) will neutralize its threatening image (Freeman 2012).

At the heart of Providência's touristic transformation is the construction of a cable car system, completed in 2013, which connects residents to public transportation networks while giving tourists access to magnificent views of the harbor. The cable car connects Providência to Rio's main railway station and to Samba City, a popular tourist attraction and entertainment venue. The construction of this cable car required the relocation of dozens of families, while the construction of a funicular leading to the uppermost portion of the community will cause more demolitions. In all, nearly one-third of the community is threatened by port-related interventions. Since the Federal programme, *Minha Casa Minha Vida,* responsible for rehousing evictees, plans to build no more than two per cent of relocation homes near Providência, it is unlikely that displaced residents will be rehoused close to their original home (Galiza 2013).

Like elsewhere in Rio, where event-related interventions threaten local communities, Providência residents complain about the absence of information, dialogue, and transparency concerning the project, and what they see as a blatant disrespect for their community. Local activists criticize the construction of expensive, image-driven cultural projects as part of Porto Maravilha, while more urgent needs should be prioritized, especially in terms of education, health, and the fight against poverty. Many have mobilized to fight evictions and housing rights violations, but they know they are fighting an uphill battle in taking on Olympic-related projects.

Throughout Rio, resistance to the neoliberal reconfiguration of the city and to the violent dislocation of the poor has been weakened by the consensual power of mega-events, their strong symbolic appeal and their status as spectacular global media magnets (Broudehoux 2013). Another major impediment to the development of efficient resistance movements is the opaque and highly bureaucratic housing relocation process. By keeping residents in the dark as to project specifics, individualizing the negotiation process, and scattering displaced residents across the territory, authorities have managed to limit collective action and to accelerate project implementation (Broudehoux 2013).

To increase their visibility and leverage, Citizens have joined forces with other community groups, including the *Fórum Comunitário do Porto* (Port Community Forum), an alliance of residents, scholars, activists and community leaders created in January 2011 to fight for the rights of port area residents. Others are members of the Comitê Popular da Copa e das Olimpíadas *(People's Committee for the Cup and the Olympics)*, a vast coalition created before the Panamerican Games to limit the negative impacts of mega-events. In Providência, a series of well-organized protests succeeded in attracting media attention and convinced authorities to postpone some interventions and to make concessions regarding their implementation. For fear of more controversy, it took the city almost two years to inaugurate Providência's completed cable car, in July 2014, during the World Cup festivities. Several other infrastructure projects were left unfinished, including the funicular project.

Conclusion

As great moments of collective euphoria and generators of a state of emergency, mega-events allow the creation of exceptional moments in a city's history. Market agents and their political allies take the opportunity of this moment of exception to remake the city according to their own device, by implementing measures that resonate with their political and economic interests. The chapter sought to demonstrate many ways in which mega-events are instrumentalized to justify the adoption of neoliberal urban policies, helping them appear at once urgent, necessary, and unavoidable. These spectacular events have a powerful boosting effect, helping create the perfect conditions for an authoritarian transformation of the city, according to the dictates of corporate agents, while exacerbating some of the most exclusionary aspects of neoliberal urban policies.

The chapter also underlined how the realization of the Olympic city relies upon an aggressive, state-sponsored form of gentrification, as government incentives make it both safe and attractive for speculators to appropriate urban territories, and benefit from unlocked land values. It demonstrated how event-related projects are at the source of much economic dispossession, with the re-directing of funds that had been allocated to poverty alleviation, towards the business sector. In Rio de Janeiro, many gentrifying projects, which benefit the real estate and engineering industries, have been financed by funds dedicated to social projects or low-income housing, such as the cable cars, funded by the federal program for accelerated development or (PAC), intended for sanitation and housing provision, or the *Minha Casa Minha Vida* federal low cost housing program used to finance the relocation of those displaced by event-related projects. Funds from the workers' retirement fund (FGTS) have also been used to finance the controversial Porto Maravilha project, source of some of the greatest population displacement in the city (Sanchez and Broudehoux 2013).

Mega-events are not just unwilling facilitators in the neo-liberalization of the city, they also play an active role, benefiting directly from the marketization and privatization of urban land and assets. With their uncompromising focus on image, and on securing the benefits promised to sponsors, broadcasters and franchise holders, they have allowed market actors and a market logic to increasingly control urban governance without any regard for transparent democratic accountability. Mega-events thus both accelerate and consolidate neoliberal hegemony and help secure the advances of private power holders over common interest.

But recent events have proven that the poor and the excluded are not duped by the mega-events spectacle and will no longer tolerate seeing their cities being sold to private corporations. In June 2013, over a two-week period that coincided with FIFA's Confederations Cup, Brazil was shaken by social movements on a scale rarely seen in the last decades. Demands focused on public transportation, education, health, and housing but also called for a radical transformation of Brazilian society and a deep reform of the exercise of political power (Badaró 2013). In Rio de Janeiro, where mobilization was the strongest, the issue of public spending on sporting events figured prominently among recriminations. Demonstrators denounced the arrogance and brutality of ruling coalitions, especially those with vested interests in mega-events, including media agencies, large national corporations, real estate speculators and a host of international businesses with close links to the FIFA and the IOC. According to Vainer, (2013), it was their blindness, pretension and violence that brought together in collective action, hundreds of thousands of hitherto un-politicized youth.

The work accomplished by the Comitê Popular da Copa e das Olimpíadas played a significant part in raising public awareness to the public policy problems posed by mega-events. This organization has condemned shady deals in the construction of World Cup stadia, unjustified evictions and controversial demolitions, especially near the Maracanã, all of which entered the public debate and were widely featured in slogans and on posters during the protests. What came out of the crisis was a rich and eloquent form of symbolic resistance that strategically used emblematic stadia for the expression of discontent. Slogans creatively merged social demands, mega-event criticism and complaints about Rio's city branding initiatives, symbolically challenging the consensus that had hitherto prevailed.

As the realization of mega-events proceeds, public criticism and denunciation are multiplying. The multitude on the streets is beginning to reveal the fragility of a system that may have appeared all-powerful, but which is now swaying in the face of contestation and in light of multiple revelations about its abusive practices. While the outcome of these social movements remains uncertain, it appears that the political landscape of mega-event planning may already have been irreversibly altered.

References

Acioly, C. (2001). 'Reviewing urban revitalisation strategies in Rio de Janeiro: From urban project to urban management approaches'. *Geoforum,* 32(4), pp. 501–530.

Agamben, G. (2005). *State of Exception.* Chicago: University of Chicago Press.

Alegi, P. (2008). 'A nation to be reckoned with: The politics of World Cup Stadium construction in Cape Town and Durban, South Africa'. *African Studies,* 67(3), pp. 397–422.

Arantes, O. (2009). 'Uma estratégia fatal. A cultural nas novas gestoes urbanas', in Arantes, O., C. Vainer and E. Maricato. *A Cidade do Pensamento Unico: Desmanchando consensos.* Petrópolis: Vozes, pp. 11–74.

Badaró, M. B. (2013). A Multidão nas Ruas: construir a saída para a crise política. (The multitude on the streets: Building an exit for a political crisis) Unpublished text, History Department, Federal Fluminense University.

Bentes, J. et al. (2011). Perspectivas de Transformação da Região Portuária do Rio de Janeiro e a Habitação de Interesse Social. Paper presented at: 14th annual meeting of the National Association of Researchers in Urban and Regional Planning (ANPUR); Rio de Janeiro.

Bienenstein, G. (2001). Globalização e Metrópole: a relação entre as escalas global e local: o Rio de Janeiro. Proceedings of the 9th annual meeting of the National Association of Researchers in Urban and Regional Planning (ANPUR); Rio de Janeiro, Brazil.

Bienenstein, G. L. Mesentier, B. Guterman and V. Teixeira (2014). 'A Batalha pela preservação da alma do Maracanã', in Sánchez, F., Bienenstein, G., Oliveira, F. and P. Novais, (eds) (2014). *A Copa do Mundo e as Cidades. Políticas, Projetos e Resistências.* Niterói: EDUFF, pp. 175–204.

Broudehoux, A. M. (forthcoming). 'Mega-events, socio-spatial fragmentation and extraterritoriality in the city of exception: The Case of pre-Olympic Rio de Janeiro'. *Urban Geography.*

Broudehoux, A. M. (2013). 'Sporting mega-events and urban regeneration: Planning in a state of emergency', in Leary, M. E. and J. McCarthy (eds.) (2013). *The Routledge Companion to Urban Regeneration*. London: Routledge, pp. 558–568.

Broudehoux, A. M. (2007). 'Spectacular Beijing: The Conspicuous construction of an Olympic metropolis'. *Journal of Urban Affairs*, 29(4), pp. 383–399.

Castells, M. and J. Borja (1996). 'As cidades como atores políticos'. *Novos Estudos*, 45: pp. 152–166.

Chalkey, B.S. and S.J. Essex (1999). 'Urban Development through hosting international events: A history of the Olympic Games'. *Planning Perspectives*, 14(4), pp. 369–394.

Cidade Olímpica (2011). Porto Olímpico, futura referência em habitação de qualidade. [accessed 21 March 2014]. Available from: http://www.cidadeolimpica.com/porto-olimpico-sera- referencia-em-habitacao-de-qualidade

Comitê Popular da Copa e Olimpíadas (2014). Megaeventos e Violações dos Direitos Humanos no Rio de Janeiro. Public report. Rio de Janeiro.

Faulhaber, L. (2013). Rio Maravilha: práticas, projetos e intervenções no território. Niterói: EAU, Federal Fluminense University.

Ferreira, A. (2010). 'O Projeto "Porto Maravilha" No Rio De Janeiro: Inspiração ee Barcelona e Produção a Serviço do Capital?' *Revista Bibliográfica de Geografía Y Ciencias Sociales*, 15, pp. 895: 20.

Freeman, J. (2012). 'Neoliberal accumulation strategies and the visible hand of police pacification in Rio de Janeiro'. *Revista de Estudos Universitários* 38(1), pp. 95–126.

Gaffney, C., F. Sanchez, G. Bienenstein and T. Gomes (2012). The River in Transition: transportation discourse and impacts in Rio de Janeiro. Paper presented at International Congress. Latin American Studies Association; San Francisco.

Galiza, H. (2013). Mega-events, housing rights, and social inequality: The case of Porto Maravilha, Rio de Janeiro. Paper presented at the American Association of Geographers Meeting, Los Angeles, April 2013.

Greene, S. J. (2003). 'Staged cities: Mega-events, slum clearance and global capital'. *Yale Human Rights and Development Law Journal*, 6, pp. 161–187.

Gusmão de Oliveira, N. (2011). Força-de-lei: rupturas e realinhamentos institucionais na busca do "sonho olímpico" carioca. Paper presented at: 14th annual meeting of the National Association of Researchers in Urban and Regional Planning (ANPUR); Rio de Janeiro.

Gusmão de Oliveira, N. (2013). O Poder Dos Jogos e Os Jogos De Poder: Os Interesses Em Campo Na Produção De Uma Cidade Para O Espetáculo Esportivo [dissertation]. Rio de Janeiro: IPPUR, Federal University of Rio de Janeiro.

Harvey, D. (1989). 'From managerialism to entrepreneurialism: The transformation in urban governance in late capitalism'. *Geografiska annaler*, 71, pp. 3–17.

Hayes, G. and J. Horne (2011). 'Sustainable development: Shock and awe? London 2012 and civil society'. *Sociology*, 45(5), pp. 749–764.

Hiller, H. H. (2012). *Host Cities and the Olympics: An Interactionist Approach*. London: Routledge.

Hiller, H. H. (2006). 'Post-event outcomes and the Post-modern Turn: The Olympics and urban transformations'. *European Sport Management Quarterly*, 6(4), pp. 317–332.

Jorgensen, P. (2011). Tentando entender a Operação Urbana Porto do Rio' Online posting. [accessed 19 June 2014] Available from: http://abeiradourbanismo.blogspot.com/2011/10/tentando-entender-operacao-urbana-porto.html

Lima Jr, P. N. (2010). Uma estratégia chamada "planejamento estratégico". Rio de Janeiro: 7 Letras.

Martins da Cruz, M. C. (2010). Do Mississipi Carioca ao Estádio Voador. Forjando espaços de legitimação na indiferença [thesis]. Niterói, Brazil: Federal Fluminense University.

Mascarenhas, G., G. Bienenstein and F.Sánchez (2011). O Jogo Continua. Megaeventos e Cidades. Rio de Janeiro: Press of the State University of Rio de Janeiro (EDUERJ).

Oliveira, A. (2009). O Emprego, a Economia e a Transparência nos Grandes Projetos Urbanos. Paper presented at: Globalização, Políticas territoriais, Meio Ambiente e Conflitos sociais [Globalization, Territorial Policies, the Environment, and Social Conflicts]. Second Meeting of the ETTERN-IPPUR-UFRJ on; Vassouras, Brazil.

Oliveira, F. L. (2013). Urban Politics and Olympic Games in Rio. Paper presented at: Annual Meeting of the American Association of Geographers; Los Angeles.

Peck, J. and A. Tickell (2002). 'Neoliberalizing space'. *Antipodes,* 34(3), pp. 380–404.

Porto Maravilha (2011). Official website. [accessed 19 June 2014] Available from: http://www.portomaravilha. com

Sánchez, F. and A. M. Broudehoux (2013). 'Mega-events and urban regeneration in Rio de Janeiro: Planning in a state of emergency'. *International Journal of Urban Sustainable Development*, 5(2), pp. 132–153.

Sánchez, F. and G. Bienenstein (2009). Jogos Pan-Americanos Rio 2007: Um Balanço Multidimensional. Paper presented at: International Congress of the Latin American Studies Association; Rio de Janeiro.

Sánchez, F. (2010). A Reinvenção das Cidades para um Mercado Mundial. Chapecó (Brazil): Argos, UNOChapecó, 2nd ed.

Sánchez, F. (2013). Aldeia Maracanã: é assim que se faz uma Copa? Brasil de Fato. [accessed 22 July 2014] Available from: http://www.brasildefato.com.br

Short, J. (2008). 'Globalization, cities and the Summer Olympics'. *City,* 12(3), pp. 321–340.

Smith, A. (2005). 'Reimaging the City: The value of sport initiatives'. *Annals of Tourism Research,* 32(1), pp. 217–236.

Stavrides, S. (2010). The Athens 2004 Olympics: Modernization as a State of Emergency. Paper presented at: International conference on Mega-events and the city. Federal Fluminense University; Niterói, Brazil.

Swyngedouw, E. (2010). Post-Democratic Cities For Whom and for What? Paper presented at: Regional Studies Association Annual Conference, Budapest.

Torres Ribeiro, A. C. (2006). 'A Acumulação Primitiva do Capital Simbólico', in Jeudi, H. P. and P. J. Berenstein (eds.) (2006). *Corpos e Cenários Urbanos: Territórios Urbanos e Políticas Culturais.* Salvador (Brazil): Press of the Federal University of Bahia (EDUFBA).

Vainer, C. B. (2011). Megaeventos e a Cidade de Exceção. Paper presented at: 14th annual meeting of the National Association of Researchers in Urban and Regional Planning (ANPUR); Rio de Janeiro.

Vainer, C. B. (2009). 'Pátria, empresa e mercadoria: notas sobre a estratégia discursiva do Planejamento Estratégico Urbano', in O. Arantes, C. Vainer and E. Maricato (eds.) (2009). A Cidade do Pensamento Unico: Desmanchando Consensos. Petrópolis: Vozes, pp. 75–103.

Vainer, C. B. (2013). Megaeventos, meganegócios, megaprotestos. [Accessed 22 July 2014] Available from: http://www.ettern.ippur.ufrj.br/ultimas-noticias/196/mega-eventos-mega-negocios-mega-protestos

Vanwynsberghe, R. et al. (2012). 'When the Games come to town: Neoliberalism, mega-events and social inclusion in the Vancouver 2010 Winter Olympic Games'. *International Journal of Urban and Regional Research*, 36(2), pp. 3–23.

Žižek, S. (1999). *The Ticklish Subject: The Absent Centre of Political Ontology.* London: Verso.

PART III
Intangible Legacies

Part III – Introduction
Intangible Legacies

Gavin Poynter and Valerie Viehoff

'First of all, there are tangible and intangible legacies. In the same vein, certain authors speak of hard and soft legacies, or of physical and spiritual legacies,… A new conference facility built for a mega event, such as the one hosting the International Broadcast Centre in Vancouver during the 2010 Olympic Winter Games, is a tangible legacy. By hosting the 1994 Commonwealth Games in facilities that were for the most part temporary – of which no trace remains – the city of Victoria (Canada) demonstrated that these games could be held in medium-sized cities, which thus constitutes an intangible legacy for the Commonwealth Games Federation (CGF)'.

Chappelet J. P. (2012) 'Mega Sporting event Legacies: A multifaceted concept' *Papeles de Europa* 25 (2012): 76–86

As Jean-Luc Chappelet suggests in his considered analysis of the concept, legacy has two main dimensions, the tangible and the intangible. In his example, derived from the Victoria Winter Olympics held in 1994, he suggests that the temporary facilities provided for the events held in Victoria left no lasting tangible legacy for the host city. However, its hosting of the event enabled the Commonwealth Games Federation (CGF) to conclude that a medium-sized city could be host to future Games. For the CGF, this was a very useful intangible 'legacy', what could be called a knowledge enhancing legacy. The Victoria Winter Olympics illustrates another dimension of the 'multifaceted' nature of legacy (tangible or intangible), the recipients of a 'positive' legacy may be one or more of a diverse range of stakeholders such as international sports federations, national or local political elites or local enterprises and communities. Indeed, between these groups, a positive legacy for one may be received as a negative legacy by another when, for example, the urban infrastructure improvements associated with the event creates rising land values that tend over the longer term for lower income communities to be displaced by higher income residents. The character and complexities of intangible legacies are the themes of Part 3 of this text.

Identifying and Analysing Intangibles

An aspect of legacy that often marginalizes discussion of the 'intangible' is, paradoxically, its adoption as a narrative by organisers for legitimizing (usually public) expenditure on the mega-event. From preparation to post-event phases, legacy outcomes become subject to accountability[1]. The definition and assurance of 'legacy', alongside cost, is typically at the heart of stakeholders' agendas. The indices of legacy take on a prominence in the planning, management and governance of the event. The depth or superficiality of legacy measures, and the seriousness with which the measurement processes and the dissemination of findings is taken, becomes itself a further index of the credibility of the 'good' event. While macro and micro economies

---

1   Some authors have offered analytical approaches to measuring the 'intangible' by using, for example, the contingent valuation method (CVM) but these tend to be presented within the framework of cost/benefit analysis and their focus is more, perhaps, on 'impact' rather than the conception of legacy. See, for example, 'The value of Olympic success and the intangible effects of sport events – A contingent valuation approach in Germany', by Pamela Wickera, Kirstin Hallmannb, Christoph Breuerb and Svenja Feilerb in *European Sport Management Quarterly*, Volume 12, Issue 4, 2012 Special Issue: Managing the Olympic Experience: Challenges and Responses, pp. 337–355.

can be tracked, housing, construction projects and employment rates monitored, and approval ratings collated and compared, there are certain other intangible factors that may be ignored. In particular, these may include the immaterial or 'cultural' factors such as city image or the assertion or renewal of national identity arising from hosting a global event such as the Olympics (MacRury and Poynter 2010) as well as the less readily auditable changes in morale, including 'soft' legacies such as civic pride or individuals' aspirations. At the supra-individual level, it is possible that the mega-event may generate 'can-do' attitudes within host cities; attitudes that inform subsequent or future major urban re-development or regeneration projects (MacRury and Poynter 2009). It is precisely this multifaceted nature of legacy, particularly in its intangible expressions, that may be excluded from the study of mega-events because of the complexities involved in their measurement.

Sociologists have sought to develop frameworks for understanding the longitudinal impacts of mega-events especially on the cities and urban spaces in which they take place. Hiller, for example provides a linkages model to conceptualise the mega-event in the context of urban processes; a model that seeks to distinguish elements of the event as dependent and independent variables. Forward linkages seek to identify how the event itself is the cause of effects, an example might be community pride. Backward linkages refer to the stated policy objectives sought by holding the event, such as enhancing the attractiveness of a location to encourage inward investment, and parallel linkages are residual to the event itself, since they inter-act with many other factors, and may lead to unexpected outcomes such as changes in the image of a particular area of a city located near to the mega-event site or changes in the travel patterns of urban commuters. Such linkages are not influenced directly by the event organisers, hence their unexpected character (Hiller 2000: 181–205).

Hiller's model, in moving on from simple cause/effect approaches, provides the scope for the inclusion of the 'intangible' and offers but one example of how over the last decade or more social theorists have generated an extensive literature on the often hidden or intangible dimensions arising in urban areas that host mega-events. Such studies have examined changes in citizens' perceptions of public places and spaces including the architectures of surveillance and the privatization of public spaces and, more positively, the galvanizing effects on local communities arising from the volunteering programmes associated with sports mega-events (Fussey et al 2011; Minton 2012; Sadd 2012).

In this Section the intangible is considered in some, though not in all, of its complex dimensions. Other Sections address aspects of the intangible, such as city image, branding and place making. Here, the focus is on what Chappelet has called 'top down' and 'bottom-up' legacies (Chappelet 2012: 82). Top-down legacies refer to those sought by the event organisers and the political and business elites whose policies and financial support typically provide the framework for the mega-event and 'bottom-up' are those which are not planned by elites or organisers but often arise from the actions, attitudes and values of citizens located within the host city/nation or visitors from without. Top-down legacies may be informed by the knowledge and experience shared within and between cities that host mega-events. In this sense legacy relates to the intangible in the form of the sharing and management of knowledge and expertise (Chapters 4 and 6). 'Bottom-up' may arise from the activities of, in particular, civic or educational agencies and institutions for whom a mega-event may provide opportunities to progress their own aims and purposes (Chapter 5).

Unsurprisingly, the chapters draw especially upon the experiences of cities that have hosted the Olympic Games. It is, perhaps, the Olympics that have been most exhaustively researched in relation to the intangible and their implications for urban processes. The IOC's adoption of the legacy narrative and its encouragement of host and potential host cities to accumulate, evaluate and distribute knowledge between them is but one reason for such a focus on the Olympic Games as Berta Cerezuela and Chris Kennett's chapter reveals.

References

Fussey, P., J. Coaffee, G. Armstrong and D. Hobbs (2011). *Securing and Sustaining the Olympic City*. Farnham: Ashgate.

Hiller, H. (2000). 'Toward an Urban Sociology of Mega-events'. *Research in Urban Sociology*, Volume 5, pp. 181–205.

MacRury, I. and G. Poynter (2010). 'Team GB' and London 2012: The Paradox of National and Global Identities'. *International Journal of the History of Sport*, 27(16–18), pp. 2958–2975.

MacRury, I. and G. Poynter (2009). London's Olympic Legacy. A think piece report prepared for the OECD and Department of Communities and Local Government, http://www.uel.ac.uk/londoneast/publications/; accessed August 1, 2014.

Minton, A. (2012). *Ground Control*. London: Penguin.

Sadd, D. (2012). Mega-events, community stakeholders and legacy: London 2012. http://eprints.bournemouth.ac.uk/20305/; accessed May 19, 2014.

Plate 4.1   The LCS development in the context of London's urban villages
Source: London Legacy Development Corporation.

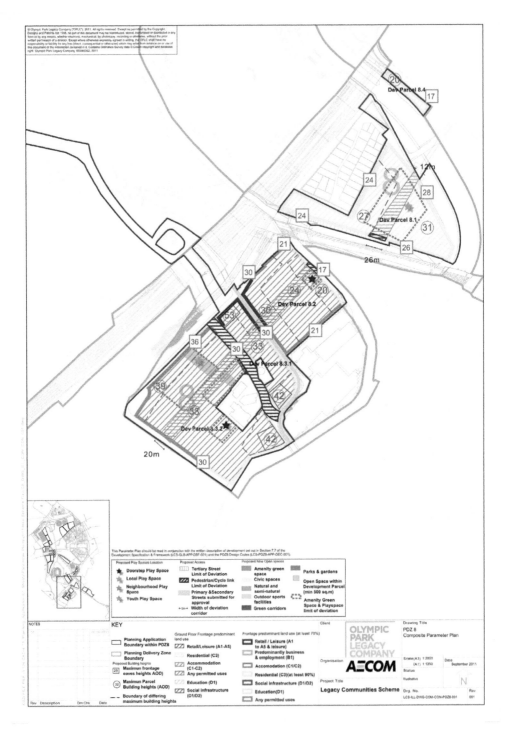

Plate 4.2   PDZ 8 Composite parameter plan
Source: London Legacy Development Corporation.

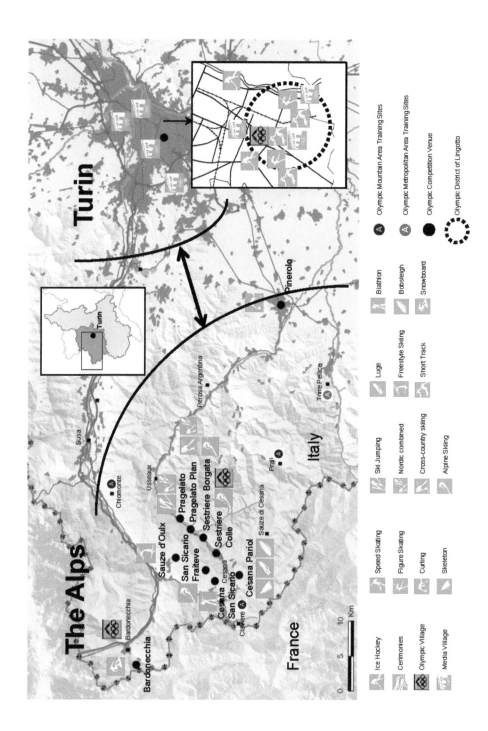

Plate 7.1   Turin 2006: The Olympic Region
Source: Politecnico e Università di Torino

Plate 12.1   Computer generated view of Queen Elizabeth Olympic Park with its future neighbourhoods

Source: London Legacy Development Corporation.

Plate 13.1   Sites of Olympic sporting venues
Source: Puplic Properties Company, Athens.

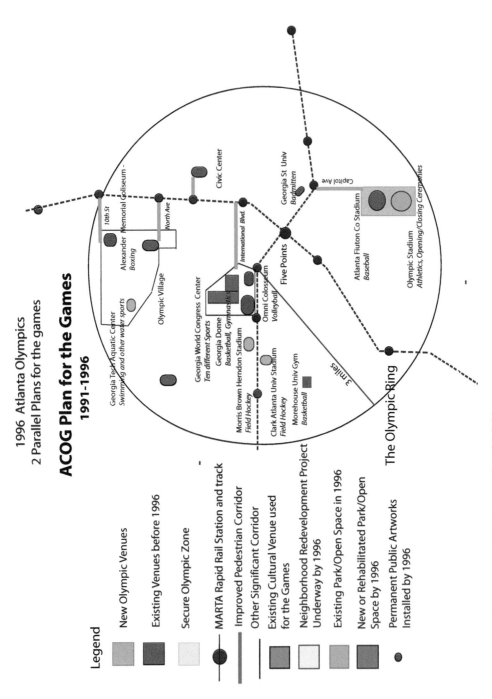

Plate 14.1   The "Twin Peaks of Mt. Olympus": ACOG Plan
Source: Randal Roark.

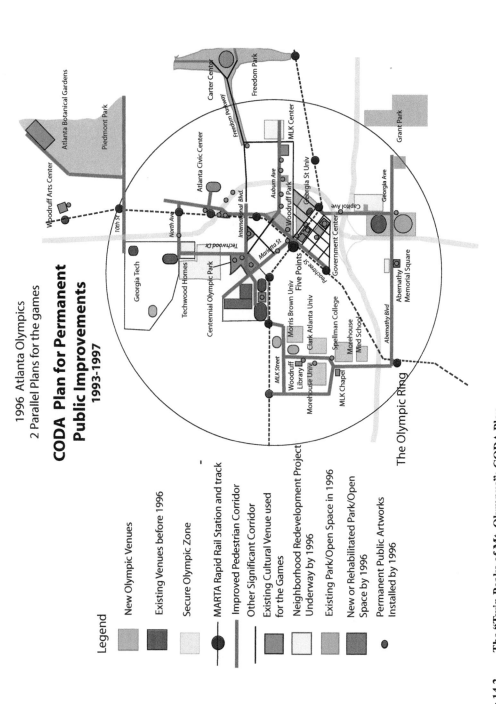

Plate 14.2   The "Twin Peaks of Mt. Olympus": CODA Plan
Source: Randal Roark.

Plate 15.1   Moses Mabhida and ABSA stadiums

Source: Roger and Pat de la Harpe Photography.

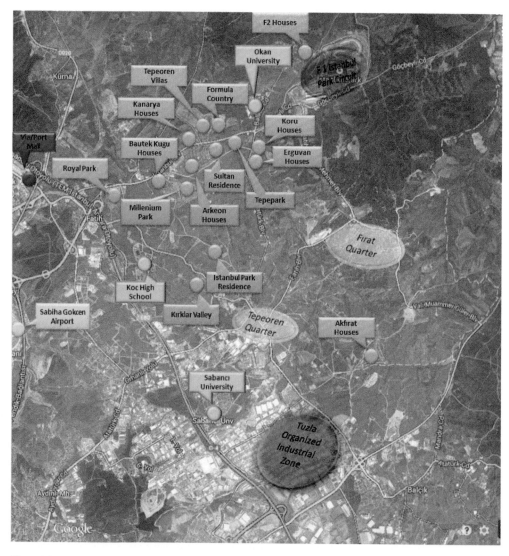

Plate 16.1   Spatial distribution of investment and developments around the Istanbul Park
Source: Eryılmaz, 2012.

Chapter 9

Intangible Learning Legacies of the Olympic Games: Opportunities for Host Cities

Berta Cerezuela and Chris Kennett

Academic discourse on event legacy (Andranovich et al., 2001; Cashman, 2003, 2006; Getz, 2007; Leopkey, 2009; Hall, 1997; Masterman, 2009; Moragas, Kennett and Puig, 2003; Preuss, 2007) has demonstrated how the success of event legacy relies on a combination of tangible and intangible elements. Hosting the world's biggest event provides a unique learning opportunity for the multiple stakeholders involved, and host cities have a unique learning opportunity to maximise. This exploratory chapter focuses on how learning can take place as part of the process of staging an event; how current learning methods and mechanisms are applied in the case of the Olympic Games; and identifies how host public sector actors are missing a learning opportunity as major stakeholders. A call is made for increased awareness and strategic planning for Olympic knowledge legacy within host cities.

Olympic Legacies

'Legacy' is a relatively new concept in Olympic discourse but since Sydney 2000 has become prevalent in the organisation of recent Olympic Games, and in particular at the London 2012 Games. Although Leopkey (Leopkey, 2009) found the concept of legacy was used as far back as the official reports of Melbourne 1956, its contemporary use is more closely linked to the emergence and consolidation of the Olympic Games (in particular the Summer Olympic Games) as a modern day mega-event (Roche, 2000). The increasing scale and scope of the Games at the very end of the 20th century saw them expand into a multi-billion dollar commercial venture, a global brand, the most watched sports event in media history, and importantly for this chapter, events of a magnitude capable of transforming host cities.

The Barcelona 1992 Olympic Games were a watershed moment for the concept of Olympic legacy. This legacy was identified retrospectively through the research of Moragas and Botella (Moragas and Botella, 2002) in one of the only consolidated attempts by researchers, until that moment, to capture the multidimensional longer-term impacts of an Olympic Games ten years after their celebration. These Games also marked a turning point in the celebration of modern Olympic Games in terms of their scale and explicit use as a catalyst for economic, urban and social change in the city, and creating what came to be known as the 'Barcelona model' for the organisation of the Olympic Games (Kennett and Moragas, 2005).

Post 1992 saw Atlanta (1996) host a largely commercial Games with a more short-term approach, likened by Cashman to a firework that burned brightly and then disappeared (Cashman, 2003). Sydney, Salt Lake City, Athens, Torino, Beijing, Vancouver, London and more recently Rio de Janeiro, all followed to differing degrees the transformational model that was established in Barcelona and were Games that strategically aimed to bring about some form of lasting, deeper change in the host cities: legacy. Their propensity to leave a deeper, more complex, long-term imprint on a city, required research that went beyond short-term economic impact studies and worries over 'white elephant' venues after the Games had left town.

While there is no agreed definition of Olympic legacy, some attempts have been made. Cashman warned that 'legacy is an elusive, problematic and even dangerous word' (Cashman, 2003: 33). Its ambiguity means it is subject to multiple interpretations and potential manipulation, particularly when used as a vehicle to achieve wider, often political objectives by stakeholders. Despite this, general agreement exists among researchers that Olympic legacy involves multi-dimensional, long-term impacts including economic, political, cultural, infrastructural, environmental and social legacies. These legacies can be tangible or intangible, as defined and classified by Leopkey (Leopkey, 2009).

However, as Cashman recognised, legacy is often associated with positive results, whilst ignoring negative outcomes of hosting the Games (Cashman, 2006). The Montreal 1976 and Athens 2004 Olympic Games are clear examples where Olympics left some negative long-term legacies in the form of debt or unused sports facilities. Thus, the planning of positive legacies in bidding processes can result in unplanned outcomes of a potentially negative nature, reminding Olympic organisers that the Games are possibly the most high-risk event in contemporary society (Flyvbjerg and Stewart, 2012).

Due to the significant investment necessary to host and organise events such as the Olympic Games, legacy is often used as a tool for justification, transparency and accountability by both event organisers and government agencies in the host city. Legacy is now a key asset included in bid projects and has become an integrating part of the event concept that requests from strategies and policies to plan, monitor and evaluate the outcomes of the event.

There is growing evidence of the importance of intangible legacies that have until now been overshadowed by the focus by Olympic Games organizers and researchers alike on the economic and financial factors; measuring short and medium-term impacts such as direct and indirect investment; infrastructure development; employment; housing; urban change; or tourism-related revenues (Sherwood, Jago and Deery, 2005). Much less research has been undertaken into the intangible impacts of hosting an Olympic Games, such as people-related impacts (skills development, education, volunteering, knowledge generation); image projection and city marketing; collective emotion and memory (feel good factors, civic pride, shared values); network effects (physical and/or virtual); or culture (identity, cultural expression). These more transient, elusive impacts can act as driving forces for the development of a longer-term legacy after the celebration of an Olympic Games and could be considered as a key part of the desired 'return on investment' to the host city and its citizens.

Amongst intangible legacy, intellectual legacy appears as one of the key assets that can be left by an event such as the Olympic Games to the host community as a result of intense learning demands during the planning, staging and delivering process. Intellectual legacy, as defined by the IOC, refers to 'all information and knowledge assets (both tangible and intangible, […] [that] can be both short and long term and can take the form of new skills, know how, expertise, content, methods, tools and other capabilities that have been produced or acquired for the purposes of staging an Olympic Games'. (IOC: 2012a).

Indeed, as the Olympic Games have grown in scope and scale, the organisational complexity of delivering the Games on time, on budget and meeting the increasingly ambitious objectives established by Organizing Committees of Olympic Games (OCOGs), the role of Knowledge Management in the event process has become increasingly important in order to ensure the generation, capture and sharing of intellectual capital generated from the Games. In addition, the prominence of legacy in the current Olympic model poses the need for methodologies to measure and understand the local impact of the Games in order to maximise benefits. The IOC has introduced the Transfer of Knowledge (TOK) programme to ensure knowledge transfer between host cities, as well as the Olympic Games Impact (OGI) system to measure the impacts of the Games in the host city.

However, the focus of current knowledge management processes during the organization of the Games take place within and between Olympic actors and focuses on securing a successful hand-over between host cities that minimises risk. This is not primarily aimed at satisfying the learning needs from other major stakeholders, such as public administration actors in host cities and countries. An important opportunity exists to learn through the Games and to capitalise on new knowledge as a return on public investment, maximising opportunities to develop and ensure an intellectual legacy from the event. In addition, the methodological difficulties to measure intangible legacies, as well as the time-scoped limitations of impact studies to monitor long term impacts,

Learning Opportunities From Organising and Hosting Events

All events, regardless of their size, typology or complexity share common characteristics such as their one-off nature, perishability, uniqueness, intangibility or fixed time scale (Shone and Parry, 2001); their flexible structure and openness to innovate and adapt to a complex and changing environment (Hanlon and Jago, 2000); or their pulsating organisational nature (Toffler, 1990); that grow and then shrink quickly after the event. While some of these characteristics might pose challenges for developing learning processes, they also provide great opportunities for developing new skills, know-how or expertise, and improving performance and capabilities through evaluation and knowledge management processes.

Event evaluation is a continuous process along the event life cycle aimed to accomplish three main functions: 1) assess the management process and identify improvements; 2) create an event profile as a communication and marketing tool; and 3) measure the level of success in achieving set objectives and outcomes by using key performance indicators (Bodwin et al., 2011). Therefore, event evaluation is a tool for analysis and improvement to 'critically observe, measure and monitor the implementation of an event in order to assess its outcomes accurately' (Bodwin et al., 2011: 630). Event evaluation involves pre-event evaluation or feasibility studies; the monitoring and control processes during the planning and implementation phases; and the post event evaluation that measures the event outcomes and opportunities for future improvement (Bodwin et al., 2011).

As shown in Figure 9.1, the event evaluation process is therefore the key component to measure the event impact and legacy. It has traditionally focused on economic evaluation, in particular the cost-benefit analysis, linked to the increasing public investment in events and the consequent need for accountability to taxpayers. However, there has been an increasing attention, both from event stakeholders and academia to evaluate events from a triple-bottom-line approach (Sherwood, Jago and Deery, 2005). However, evaluation should not only be seen as a process in the closure stage of the event and should be planned from the early stages to ensure that evidence-based decisions can be made on the value and significance of the event, amongst others (Mallen and Adams, 2013).

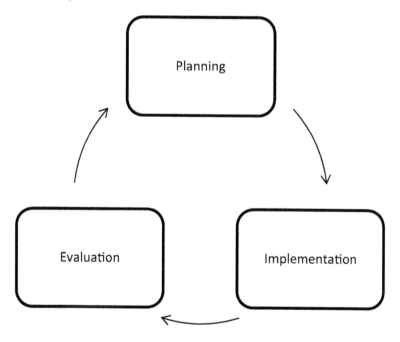

Figure 9.1   Event evaluation graph
Source: Adapted from Bodwin et al., 2011.

Event evaluation relates closely to knowledge management process, aiming to improve the performance of an organisation by sharing organisational knowledge – i.e. 'information combined with experience, context, interpretation, and reflection. It is a high-value form of information that is ready to apply to decisions and actions' (Davenport and Prusak, 1998: 43) – to enhance organisational improvement. Organisations adopt Knowledge Management projects to achieve different aims, including the creation of knowledge repositories, improving knowledge access, enhancing the knowledge environment and managing knowledge as an asset (Davenport and Prusak, 1998).

Therefore, managing knowledge is a key factor for developing a competitive advantage for all organisations as it supports individual, team and organisational learning by making a better use of an organisation's intellectual capital and increasing and enriching traditional ways of accessing expertise (Davenport, De Long and Beers, 1998; Lahti and Beyerlein, 2000; Jarrar, 2002).

In the case of the Olympic Games, this approach of combining event evaluation and knowledge management process and integrating them as part of the Games Management Model that has been adopted by the IOC in the establishment of the Transfer of Olympic Knowledge and Olympic Games Impact programmes.

The demands of staging such an event require an intense and constant learning process for all those involved in the process of staging the event, from the bidding phase to the post-Games period.

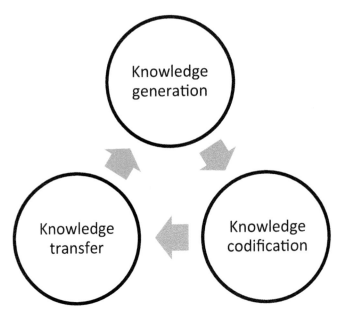

Figure 9.2   The Knowledge Management process
Source: Adapted from Davenport and Prusak, 1998.

As part of the whole management process, from bid to closure, organisers define a strategy to foresee the knowledge and information needs of internal and stakeholders for each phase. They are intensive-knowledge organisations. From the bidding phase to the post-event phase, they are immersed in a constant and diverse learning process. They simultaneously learn from the experiences of previous host cities, as well as their own unique experiences. They may also learn about many issues such as transport systems, sport facilities requirements, ticketing or volunteer management. Information and Knowledge Management can assist to achieve operational excellence in complex environments, by improving operational efficiency; reducing reinvention and duplication; and supporting effective decision-taking processes (Halbwith and Toohey, 2005).

The learning opportunities and need for Knowledge Management at an Olympic Games start at the very beginning of the event management process when cities decide they will attempt to make an official bid to be hosts. Cities wanting to host the Olympic Games have to be endorsed by their respective NOC to submit their application to the IOC. During this process, a bid committee is established, formed by representatives of members of the public administration and other actors. Once the application is accepted by the IOC, cities become Candidate Cities and are required to submit a candidature file to the IOC. The IOC Evaluation

Commission evaluates the bid to assess the city's potential to host the event based on technical criteria. The Commission produces an evaluation report and a list of Candidate Cities to be drawn by the IOC Executive Committees to be submitted to the IOC Session for election (IOC, 2013: Rule 34).

The preparation and presentation of the bid involves a period of reflection for the city and of self-analysis. Not only is the city's capacity to host the Games in infrastructure, organizational ability, and in financial terms questioned, important strategic issues are raised as to why (if at all) a city wants to host the Games, the impacts this will have and the legacies that may result. Bid cities enter into a fierce competition to be chosen as hosts and hold themselves up to intense international as well as local scrutiny and are judged on their potential to join (or consolidate their status in) an exclusive club of 'Olympic cities' deemed capable of putting on the biggest show on Earth.

Once the Games are awarded to a city, a Host City Contract is signed to 'set out the legal, commercial and financial rights and obligations of the IOC, the Host City and the NOC of the host country in relation to their specific Olympic Games' (IOC, 2009: 7–100). The Organising Committee for the Olympic Games is established and the process of planning and organising the event starts. During this process, the IOC supports the organising committee and many stakeholders are involved in the organisation of the event, which the OCOG needs to work with.

In order to understand the scale of the Games, London 2012 involved 10,500 athletes from 204 countries; 2,961 technical officials and 5,770 team officials; an approximate workforce of 200,000 people, including more than 6,000 staffs, 70,000 volunteers and 100,000 contractors. 8.8 million tickets were available for spectators that made 20 million city journeys during the Games. Over 21,000 accredited media communicated the Games to a potential 4 billion worldwide audience. The Games sporting programme included 39 disciplines in 26 sports that took place in 34 sporting venues and over one million pieces of sport equipment were needed.

Olympic Learning Systems in Action

Since 2000 the International Olympic Committee has managed TOK to capture knowledge from the organising committee in a structured way so that it can be made available for the IOC itself and for other bidding and host cities (Felli, 2003). The programme's origins are in Recommendation 16 of the 2000 Commission, which identified limitations in access to information and resources for monitoring and controlling preparations for the Games. The aim of this programme is to 'provide assistance to future Organising Committees and offer them general information and examples of the practices required to be able to develop and implement the best solution for its environment' (Felli, 2003: 125).

The IOC Olympic Games Knowledge Management Programme is an integrating element of the Olympic Legacy Management approach, which also involves as key elements the Olympic Charter, the 21 Agenda, the bid dossier, the host city contract and the Olympic Games Impact Study (OGI) and other impact studies (IOC, 2012b).

The Olympic Games Knowledge Management Programme aims to create, consolidate and disseminate information as well as to facilitate learning through experience. It provides a knowledge management model for event owners, which is integrated into the Games management model. This programme runs throughout the Games' organisation cycle and has a clear focus on minimising risk. The IOC adopts the role of coordinator and facilitator and stimulates a culture of knowledge within the organising committee (IOC, 2012a).

The first set of actions includes initiatives to ensure the proper management of the information and documentation generated during the different phases of the event and their preservation for the future. Actions focus on gathering explicit knowledge and include a series of guidelines and tools for document management that organising committees can adapt (Bianchi, 2003). It also establishes an obligation for host cities to maintain and preserve corporate records once the Games are over. Since the Beijing Olympic Games in 2008 the host city's contract includes clause 25, which establishes that 'the OGOC must guarantee that, both in the period prior to and the period after the Games, archives related with the Games must be securely managed and maintained, and the IOC must have free access to these archives' (Bianchi, 2003: 162). In addition to this clause, appendix L of the contract states that Organising Committees must provide the IOC with a minimum of documents (which are defined in the TOK Action List) for inclusion in its documentary archive at the

Olympic Museum Lausanne and its Olympic Studies Centre, whose mission is to preserve the memory of the Olympic Movement.

The Olympic Games Knowledge Management Programme does not only refer to information and documentation. It also deals with the tacit knowledge generated during the organisation and celebration of the event and how this knowledge should be organised and managed to contribute to the consolidation of a knowledge base for post hoc use in the organisation of future Olympic Games. The TOK programme includes different sources for obtaining this knowledge through elicitation and collaboration processes: documentation, personal experiences and the offer of services (IOC, 2012a).

TOK documentation mainly consists of guidelines and manuals that are periodically revised and explain the ways in which different Organising Committees have applied the requirements established by the IOC to different functional areas, as shown in Table 9.1. These guidelines are complemented with more detailed information and documentation on some of the areas, such as VTOK (Visual Transfer of Knowledge), an image database. Once the Games are over, the Organising Committee is obliged to present the IOC with an official report that in the most recent editions of the Games, follows a pre-established structure based on four key areas related with the candidature, organization, celebration and impact of the event. These guidelines also respond to the demand from academics for sufficient information to be able to make comparative studies (Felli, 2003). The information that is collected and produced by TOK is managed by the IOC via an extranet that is accessible by candidate cities and Organising Committees.

Table 9.1   Technical Manuals and Guides

• Accreditation and Entries	• City Activities
• Design Standards for Competition Venues	• Finance
• Sport	• Food and Beverage Services
• Olympic Village	• Information and Knowledge Management
• Accommodation	• NOC Service
• Transport	• Olympic Games Impact (OGI)
• Media (Broadcasting and Press)	• Olympic Torch Relay
• Ticketing	• Signage
• Brand Protection	• Venues
• Marketing Partner Services	• OCOG Marketing
• Protocol and IOC Protocol Guide	• Olympic Hospitality Centre
• Workforce	• Brand, Identity and Look of the Games
• Medical Services	• Digital Media
• Ceremonies	• Legacy (guide)
• Communications	• Environmental management (guide)
• Games Management	• Cultural Olympiad (guide)
• Paralympic Games	• IPC Accessibility (guide)
• IOC Session and Related Meetings	• Spectator Experience (guide)
• Arrivals and Departures	

Organising Committees are provided with an international network of experts on the different functional areas of the organisation of the Games to which they may refer should they need to, as well as a series of seminars and workshops. Likewise, they also participate in the Observer Programme that is organised during the Games, which enables them to obtain first-hand information about the preparations and operations during the celebration of the event. Members of National Olympic Committees, international federations and experts that revise TOK guides also participate in this programme. Finally, one of the key items of the programme is the Games debriefing, a three or four day session that is organised during the months after the event has ended and at which the Organising Committee makes its evaluation and establishes recommendations for the Organising Committees of subsequent Games.

The second main part of the Olympic learning system is the Olympic Games Impact (OGI), which provides a consistent methodology to assess the impact of the Olympic Games in the host city, its environment and citizens, and aims to offer a 'means of understanding the effects of certain actions undertaken and to make adjustments if necessary' (IOC, 2009). Therefore, OGI is not aimed to be predictive, but to identify tendencies and results.

OGI analysis goes beyond traditional economic impact analysis to cover triple bottom line economic, social and environmental areas of sustainable development. OGI includes 120 of indicators with both context and event focus that can be tailored to the characteristics and needs of each host city and their stakeholders. In addition to these mandatory indicators, each host city can select the optional indicators to cover or suggest additional ones. OGI covers a 12-year period starting 2 years prior to being elected as a host city until 3 years after the event is held.

By allowing tailoring, OGI has the potential to become a learning tool on how impact can be maximised from investments made in the planning and staging process of events. However, OGI has important limitations when identifying and measuring the longer-term impacts and wider legacy of the Games, in particular in terms of intangible legacies such as intellectual legacy.

In order to assist the TOK and OGI programme, the IOC Games Management Model has developed information and knowledge management procedures as part of the different phases of the event management process. During the bid, initiation, planning, operational-readiness and event time phases, information and knowledge management focuses on risk limitation, maximising efficiency and ensuring quality standards. This is achieved by assisting in planning and decision-taking processes, ensuring the flow of information amongst stakeholders and providing information and content management services.

The post-event phase is mainly focus on intellectual capital and involves gathering information and knowledge to support the evaluation and closure processes (IOC, 2012a). During this phase, it focuses on supporting the OCOG reporting processes, the OGI study as well as managing the legacy collections (i.e. corporate archives) that would remain in the host community as a key element of the information and knowledge legacy of the Games, and therefore, its intellectual legacy. The post-Games period is, therefore, another key moment for local public administrations to maximise the capture information and knowledge resulting from staging the event to then be used and share as a key asset for competitive advantage for the city and its citizens.

Figure 9.3   Games Management Phases
Source: Adapted from IOC, 2012a.

Olympic Learning Opportunities Post-Games: A Missed Opportunity for Host Cities

Public sector organisations as one of the main stakeholders in event management processes in Olympic cities are presented with an important opportunity to develop learning through the Games by capitalising on new knowledge as a return on investment. This legacy of an Olympic Games is not only connected to public accountability for the economic investment needed to host the Games; but can also be a tool for continued public policy development.

Public sector organisations play a lead role in developing event strategies to guide their involvement, priorities and decision-making, and use events as a tool for development. As stated by Bowdin et al. (Bowdin et al., 2011) local government involvement in events can vary from owning and managing venues; acting as consent authority and regulatory body, coordinating public services such as transport, police or health during the event; a funding body economically supporting the event; an event marketer or the event organiser or co-organiser. Their learning interests from the event will vary depending on their role in the event.

Indeed, knowledge management is not a new practice in the public sector. Socially-constructed or private-generated knowledge has been used extensively and managed by government agencies in the process of defining, planning and implementing public policies (Riege and Lindsay, 2006). Knowledge management practices in public governance have also become increasingly important due to recent public sector cutbacks and the need to deliver with higher level of effectiveness, efficiency accountability and transparency. Therefore, delivering better and more cost-effective services by enhancing successful partnerships and providing a high return to tax payers is one of the critical challenges faced by governments.

Literature suggests that knowledge management strategies by public sector should not only focus on capturing, management and exploiting knowledge, but to transfer certain knowledge back to stakeholders, creating a knowledge sharing environment while demonstrating public accountability (Barnes et al., 2003, Riege and Lindsay, 2006). Two-way knowledge partnerships are gaining increasing importance and become 'not only a success factor but a differentiating competitive factor' for good policy development (Riege and Lindsay, 2006). However, knowledge shared by government partnerships is mainly of tacit nature (Wiig, 2002) and presents more challenges to be managed. Literature review on knowledge legacy and knowledge management strategy for recent editions of the Olympic Games confirms that whilst the IOC maximises knowledge transfer and learning opportunities, host cities and stakeholders from the public administration might be missing important opportunities to ensure they maximise knowledge capital through the Games. (Girginov, 2012; Kaplanidou, K. and Karadakis, K., 2010; Halbwirth and Toohey, 2013)

The London 2012 Olympic Games provided a first example on how cities can maximise the learning opportunities from hosting an event such as the Olympic Games through a formal and planned in advanced process. The Olympic Delivery Authority Learning Legacy project aimed to 'to share the knowledge and lessons learned from the London 2012 construction project for the benefit of industry projects and programmes in the future, academia and government' (LOCOG, 2013). An open-access knowledge base available at the Learning Legacy website that combines tools, professional reports and academic research primarily aimed to the construction and event sector.

One of the most relevant learning opportunities in host cities refers to the event industry and in particular to the city event strategy. As hosts of the Olympic Games, cities have an opportunity to develop knowledge to support their event strategy by providing tools, methodologies, lessons learned and best practices that can then be shared with the local event industry. This can also contribute to efficiency in managing public resources allocated by host cities to future events by supporting the decision-taking in bidding process, managing service contracts or setting the city event calendar, as well in maximising the positive impact that sustainable events can have in the city.

Awareness needs to be raised among public administration actors in Olympic host cities about the learning opportunities that exist and the need to secure these as a key part of the intangible legacy. This requires the planning and implementation of a learning strategy based on tailored knowledge management and evaluation processes.

Conclusions

Every event should be viewed as a learning process and as an intensive-knowledge generation and sharing process, not only in relation to the staging and delivery of the event, but to generate added value and competitive advantage as a return for investment for the host city as a key event stakeholder. Knowledge legacy should be considered a key asset for public policy development and event knowledge management should be crucial for accountability, improvement, and learning in the host city. The IOC knowledge management model provides an excellent example of an intensive learning process between the event organiser and the event owner, as well as within the event organising committee when evaluation and improvement tools are implemented. However, it does not provide direct and systematic transfer of knowledge to the host city public sector beyond impact studies.

Until recently, public sector organisations' information and knowledge management practices related to events hosted or organised in the city have mainly focused on impact studies. Far less attention has been paid to develop models with the objective of maximising event capitalisation by using knowledge management to support cities' event strategies. Karadakis, Kaplanidou's and Karlis (2010) research on the Athens 2004 Games began an important discussion on how strategic planning tools can provide knowledge to host cities to properly leverage the event, and this needs to be continued in future research. There is still limited guidance as to how public organisations can develop knowledge management strategies and partnerships with event organisers in order to maximise the capture and exploitation of the knowledge generated from events such as the Olympic Games.

From this perspective, event knowledge management strategies need to be defined from a stakeholder approach. The Olympic Games involve a complex web of organisations that have to work together effectively. It is not just the responsibility of the Games organisers to ensure knowledge transfer and learning occur, but all stakeholders that benefit from the legacy this creates. Each stakeholder would need to define processes for collecting information and generating knowledge, undertaking research and evaluation, as well as determining how to best analyse it, use and disseminate it. But this process needs to be done in a coordinated and integrated way with the current Olympic Games Knowledge Management programme and mechanisms led by the IOC.

Due to the unique context in which an Olympic Games is celebrated, host cities must define their own model in order to reinforce their capabilities to create, capture and disseminate knowledge through the platform (facilities, infrastructure, brand, services or workforce) they provide to the event industry. This knowledge should contribute to an efficient management of public resources and in supporting decision-taking processes regarding the city event policy.

Planning knowledge legacy involves the creation, capture and management of the information and knowledge generated from and about the event This should take place not only as part of the corporate management of an event (pre-event, event and immediate post-event), but as part of a longer-term strategy. Measuring the true legacy of an event such as the Olympic Games might require up to 20 years (Toohey, 2008; Gratton and Preuss, 2008), and includes indicators that go beyond the tangible to include intangible elements such as memory and symbols.

Knowledge is therefore a very valuable resource for event organisers and stakeholders, a competitive advantage. If events are run as learning opportunities, they should follow knowledge management principles and be focused on the long-term knowledge legacy of the event. As stated by Halbwirth 'knowledge is a resource too valuable to leave to chance. It's not what we know but how we share and use [it]'. (Halbwirth 2009: 32).

In our current knowledge economy, the knowledge resulting from events should not only be considered and managed from the event organiser perspective, and should include all event stakeholders and the public authorities in particular. Investing in Olympic Knowledge management capitalises on existing knowledge, contributes to developing event memory and is a key asset for the organisation of future events.

References

Andranovich, G, M. J. Burbank and C. H. Heying (2001). 'Olympic cities: Lessons learned from mega-event politics'. *Journal of Urban Affairs*, 23(2), pp. 113–131.

Barnes, M., J. Newman, A. Knops and H. Sullivan (2003). 'Constituting "the public" in public participation'. *Public Administration.* 81(2), pp. 379–99.

Bianchi, C. (2003). 'The role of archives and documents in the legacy of the Olympic Movement', in Moragas, M. de, C. Kennett and N. Puig (2003). *The Legacy of the Olympic Games 1984–2000: International Symposium.* Lausanne, 14–16 November 2002. [pdf]. Lausanne: International Olympic Committee, pp. 160–167. Available at: http://doc.rero.ch/record/18259 [Accessed 12 May 2014].

Bowdin, G., W. O'Toole, J. Allen, R. Harris and I. McDonnell (2011). *Events Management.* London: Routledge.

Cashman, R. (2003). 'What is "Olympic legacy"?' in Moragas, M. de, C. Kennett and N. Puig (2003). *The Legacy of the Olympic Games 1984–2000: International Symposium.* Lausanne, 14–16 November 2002. [pdf]. Lausanne: International Olympic Committee, pp. XXX. Available at: http://doc.rero.ch/record/18259 [Accessed 12 May 2014].

Cashman, R. (2006). *The Bitter-Sweet Awakening: The Legacy of the Sydney 2000 Olympic Games.* Sydney: Walla Walla Press.

Chalip, L. (2006). 'Towards social leverage of sport events'. *Journal of Sport & Tourism*, 11, pp. 109–127.

Davenport, T. and L. Prusak (1998). *Working Knowledge.* Cambridge: Harvard Business School Press.

Davenport, T.H., D.W. De Long and M.C. Beers (1998). 'Successful knowledge management projects'. *Sloan Management Review,* 39(2), pp. 43–57.

Felli, G. (2003). 'Transfer of knowledge: A Games management tool', in *Architecture and International Sporting Events: Future planning and development. The Second Joint Conference, Lausanne 8th and 9th June 2002*. Lausanne: International Olympic Committee, pp. 121–127.

Flyvbjerg, B. and A. Stewart (2012). *Olympic proportions: Cost and cost overrun at the Olympics 1960–2012*, [pdf]. Oxford: University of Oxford. Available at: <http://ssrn.com/abstract=2238053> [Accessed 12 May 2014].

Getz, D. (2007). *Event Studies: Theory, Research and Policy for Planned Events*. Oxford: Elsevier.

Ghaffar, A. Rahman Abdul, G. Beydoun, J. Shen, W. Tibben and D. Xu (2012). A synthesis of a knowledge management framework for sports event management. In: *ICSOFT 2012. Proceedings of the International Conference on Software Paradigms Trends*, pp. 494–499.

Girginov, V. (2012). 'Governance of the London 2012 Olympic Games legacy'. *International Review for the Sociology of Sport*, 47(5), pp. 543–558.

Gratton, C. and H. Preuss (2008). 'Maximizing Olympic impacts by building up legacies'. *The International Journal of the History of Sport*, 25, pp. 1922–1938.

Halbwirth, S. (2009). 'Knowledge management: Summary of presentation', in World Union of Olympic Cities. Lausanne Summit 2009, 19–21 November, Switzerland: Post-Event Report. [pdf]. Available at: http://doc.rero.ch/record/24663/files/Lausanne_Summit_2009_Event_Report_03–02–10.pdf [Accessed 12 May 2014].

Halbwirth, S. and K. Toohey (2005). 'Sport event management and knowledge management: A useful partnership', in The Impacts of events: proceedings of International Event Research Conference held in Sydney in July 2005. Sydney, Australian Centre for Event Management, pp. 302–314.

Halbwirth, S. and K. Toohey (2013). 'Information, knowledge and the organisation of the Olympic Games', in Frawley, S. and D. Adair (eds.) (2013). *Managing the Olympics*. Basingstoke: Palgrave Macmillan, pp. 33–49.

Hall, C. M. (1997). *Hallmark Tourist Events: Impacts, Management and Planning*. Chichester: John Wiley and Sons.

Hanlon, C. M. and L. Jago (2000) 'Pulsating sporting events: An organisational structure to optimise performance', in H. Allen, L. Jago, and A. Veal (eds.) (2000). *Events Beyond 2000: Setting the Agenda. Proceedings of Conference on event evaluation, research and education*. Sydney: Australian Center for Event Management, pp. 93–104.

IOC (International Olympic Committee) (2009). *2018 Candidature acceptance procedure: XXIII Olympic Winter Games*. [pdf] Lausanne: International Olympic Committee. Available at: http://www.olympic.org/Documents/Reports/EN/en_report_1451.pdf [Accessed 12 May 2014].

IOC (International Olympic Committee) (2012a). *Technical manual on information & knowledge management: 5th updated cycle post Vancouver Winter Games*. Lausanne: International Olympic Committee.

IOC (International Olympic Committee) (2012b). *Guide on Olympic legacy: 5th updated cycle post Vancouver Winter Games*. Lausanne: International Olympic Committee.

IOC (International Olympic Committee) (2013). *Olympic Charter: In force as from 9 September 2013. [pdf.]*. Lausanne: IOC. Available at: <http://www.olympic.org/documents/olympic_charter_en.pdf> [Accessed 12 May 2014].

Jarrar, Y. (2002). 'Knowledge management learning for organizational experience'. *Managerial Auditing Journal*, 16 (2), pp. 322–328.

Kaplanidou, K. and K. Karadakis (2010). 'Understanding the legacies of a host Olympic city: The case of the 2010 Vancouver Olympic Games'. *Sport Marketing Quarterly*, 19, pp. 110–117.

Karadakis, K., K. Kaplanidou and G. Karlis (2010). 'Event leveraging of mega sport events: A SWOT analysis approach'. *International Journal of Event and Festival Management*, 1(3), pp. 170–185.

Kennett, C. and M. de Moragas (2005). 'Barcelona 1992: Evaluating the Olympic legacy', in A. Tomlinson and C. Young (eds.) (2005). *National Identity and Global Sports Events; Culture, Politics, and Spectacle in the Olympics and the Football World Cup*. Albany: State University of New York Press, PAGES.

Lahti, R. and M. Beyerlein (2000). 'Knowledge transfer and management consulting: A look at the firm'. *Business Horizons, 43*(1), pp. 65–74.

Leopkey, B. (2009). *2008 post graduate grant final report: The historical evolution of Olympic legacy*. [pdf] Lausanne: International Olympic Committee. Available at: http://doc.rero.ch/record/12537 [Accessed 12 May 2014].

LOGOC (2013). *Learning Legacy*. [online] Available at: <http://learninglegacy.independent.gov.uk/> [Accessed 12 May 2014].

Mallen, C. and L. Adams (2013). *Event Management in Sport, Recreation and Tourism: Theoretical and Practical Dimensions*. Oxford: Routledge.

Masterman, G. (2009). *Strategic Sports Event Management: Olympic edition*. London: Elsevier.

Moragas, M. and M. Botella (eds.) (2002). *Barcelona: l'herència dels Jocs (1992–2002)*. [pdf]. Barcelona: Centre d'Estudis Olímpics (UAB), Ajuntament de Barcelona, Editorial Planeta. Available at: <http://ceo. uab.cat/2010/docs/C20202.pdf> [Accessed 12 May 2014].

Moragas, M. de, C. Kennett and N. Puig (2003). *The Legacy of the Olympic Games 1984–2000: International Symposium*. Lausanne, 14–16 November 2002. [pdf]. Lausanne: International Olympic Committee. Available at: http://doc.rero.ch/record/18259 [Accessed 12 May 2014].

O'Reilly, N. J. and P. Knight (2007). 'Knowledge management best practices in national sport organisations'. *International Journal of Sport Management and Marketing*, 2(3), pp. 264–280.

Preuss. H. (ed.) (2007). *The Impact and Evaluation of Major Sporting Events*. London: Routledge.

Riege, A. and N. Lindsay (2006). 'Knowledge management in the public sector: Stakeholder partnerships in the public policy development'. *Journal of Knowledge Management*, 10(3), pp. 24–39.

Roche, M. (2000). *Mega-events and Modernity: Olympics and Expos in the Growth of Global Culture*. London: Routledge.

Sherwood, P., L. Jago and M. Deery (2005). 'Unlocking the triple bottom line of special event evaluations: what are the key impacts', in *The Impacts of events: proceedings of International Event Research Conference held in Sydney in July 2005*. Sydney, Australian Centre for Event Management, pp. 16–32.

Shone, A, and B. Parry (2001). *Successful Event Management: A Practical Handbook*. New York: Continuum.

Toffler, A. (1990). *Future Shock*. New York: Bantam Books.

Tookey, K. (2008). 'The Sydney 2000 Olympics: Striving for legacies-overcoming short term disappointments and long term deficiencies'. *The International Journal of the History of Sport*, 25(14), pp. 2098–2116.

Wiig, K. M. (2002). 'Knowledge management in public administration'. *Journal of Knowledge Management*. 6(3), pp. 224–39.

Chapter 10
Perceptions of Legacy: An Educational Perspective

Neil Herrington

Introduction

Legacy is often treated by Games Organising Committees as if it were unproblematic (Cashman, 2006). A precise definition of legacy is not sought. Preuss (2007), however, feels that this leads to a lack of clarity, an unsatisfactory position, especially given the central role that legacy takes in discussion of the Olympics. Thus, whilst only being a relative recent addition to Olympic discourse, (MacRury, 2011), the term 'legacy' emerges from the literature as a multi-dimensional concept. The Olympic Charter (IOC, 2011) makes it clear that a 'positive' legacy from a Games is a key concern for the IOC. There are a number of reasons for this, amongst them as a means to justify expense and as a way of encouraging other cities and nations to bid for future events (Gratton and Preuss, 2008; Poynter, 2009b). The IOC uses the term to encompass the sports facilities and public works turned over to communities and/or sports organisations after the Games (Preuss, 2007). However, the wider literature gives consideration to aspects of legacy that includes: sport infrastructure, regeneration and additional employment, these sitting alongside socially unjust displacements and increases in property prices (Ritchie and Aitkin, 1984; Lenskyj, 2002, 2000; Moragas, Kennett and Puig, 2003; Preuss, 2004; Cashman, 2006; O'Brien, 2006). There is, therefore, a tension in discussions about legacy even when, ostensibly, people are giving consideration to the same phenomenon. These tensions mean that legacy qualifies as a 'wicked problem': a term coined by Rittel and Webber (1973) for those areas of policy which are characterised by scientific uncertainties and high stakes, where there are multiple perspectives on the nature of the problem and the nature of the solution. This study sets out to investigate some of these perspectives.

Investigating 'Wicked Problems'

In order to engage with wicked problems such as legacy, there is a need to examine how individuals and organisations 'arrive at judgments, make choices, deal with information and solve problems' (Bobrow and Dryzek, 1987: 83). While there are a number of interventions that allow such an examination (Baker and Jeffares, 2013), an increasing number of commentators, for example Cuppen (2013), Niemeyer et al (2013) and Gaynor (2013) are using Q methodology in their policy studies. Q methodology was devised and developed by William Stephenson during the 1930s (Stephenson, 1935, 1936a, 1936b). Stephenson was concerned to bring a scientific framework to bear on the elusiveness of subjectivity through the development of an holistic methodological approach. This lead to an approach where:

> Any list of heterogeneous measurements or estimates can be arranged in an order of some kind, or in a scale …
> [in terms of] their … significance for the individual, they may be held to be made homogenous with respect to that individual. (Stephenson, 1936b: 346)

In effect, study participants actively rank order a set of stimulus items, the so called 'Q set'. This is carried out from a first person perspective using as the unit of quantification 'psychological significance' (Watts and Stenner, 2012; Burt and Stephenson, 1939).

The deployment of Q methodology requires a number of different stages: beginning with the generation of the concourse around a specific topic, which might involve a range of methods (documentary analysis, literature review, interviews and focus groups); the construction of the Q set which will form the basis of the Q sort carried out by the participants or P-set; the Q sort itself; the analysis of that Q sort; and the interpretation of the outcomes of the statistical analysis. This is shown in Figure 10.1.

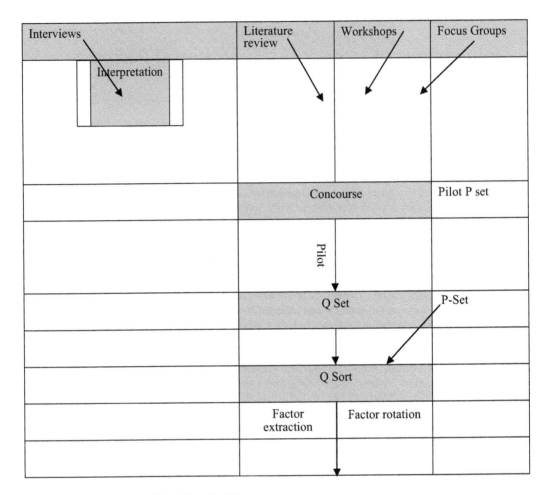

Figure 10.1   Summary of the Q methodology process

Exploring Perceptions of Legacy

The 'Q set' for this study was formed from a concourse generated by literature and document review and a range of focus groups and interviews. A series of pilot studies led to some 57 separate statements (these are attached as Table 10.2, pp. 147–8) which the study participants were asked to rank order in the Q sort. Participants, who are described below, assigned each item of the Q set to a ranking position within a quasi-normal distribution along a simple face valid dimension, defined by a condition of instruction to arrange the statements on the grid along a continuum ranging from 'strongly disagree' (-5) to 'strongly agree' (+5).

Within Q methodology the participant group is termed the P set. As each participant is a variable in the study, some care has to be exercised in the selection of these participants. Q methodology is designed to sensitise the researcher to the existence of certain perspectives, it is not about generalising to the whole population (Brown, 1980), thus a Q study requires sufficient participants:

> to establish the existence of a factor for the purposes of comparing one factor with another. What proportion of the population belongs to one factor rather than another is a wholly different matter and one about which Q technique … is not concerned. (Brown, 1980: 192)

In these terms there are no strict guidelines on the size of the participant group. Stainton-Rogers (1995) indicates that between 30 and 60 is adequate, and Watts and Stenner (2012) offer the advice that the P set size should be less than the size of the Q set. In this study, a Q set of fifty seven statements, the generation of which is described above, were used in a Q sort, described below, that was carried out by a P set of thirty six participants. These participants came from 3 broad groups: secondary school pupils; secondary school teachers; and informal educators, consultants and advisers.

Analysing the Q Sort

Within this study the statistical analysis of the Q sort was carried out using 'PQMethod' (Schmolck, 2002). This software offers a range of options both in terms of factor extraction and factor rotation. This study used centroid factor analysis with the factor's eigenvalue (EV) being used to guide the decision about how many factors to extract. Each Q sort is inter-correlated with every other sort and this generates a correlation matrix (Brown, 1980; Watts and Stenner, 2012) which indicates the extent of the relationship between any two Q sorts. The matrix as a whole describes the relationships between all Q sorts and hence the variability within the study, the so-called study variance. This overall variance can be subdivided into three sub-categories of variance: common variance; specific variance; and error variance (Kline, 1994).

The first describes the amount of variance within the Q sort that is common to the group; the second refers to the variance attributable to individual Q sorts; and the third to those random errors inherent in any methodology. The degree to which each individual Q sort exemplifies a given factor's pattern is termed the factor loading (Watts and Stenner, 2012). The development, through factor extraction and rotation, of a series of viewpoints is the start of the process of interpretation. These viewpoints are prepared by a weighted average of all of the individual Q sorts that load significantly onto the rotated factor (Watts and Stenner, 2012). This leads to a factor array (the arrays for the perspectives discussed here are presented below) for each of the factors. Thus 'a Q factor is not merely a composite of statements but a new generalisation arising from and cutting across individual Q sorts, linking their separate meanings and summarising their communality' (McKeown, 1980: 423). It is from this appreciation of the array that the penportraits presented below are written.

Factor analysis aims to account for the maximum amount of variation possible by looking for 'sizeable portions of common or shared meaning that are present in the data' (Watts and Stenner, 2012: 98). Having identified these patterns, a pattern that constitutes a factor, the portion of common variance that gives rise to that factor is removed. This leaves behind a table of residual variations within which the analysis looks for the next pattern of common meaning and extracts this as a factor. This process continues until all common variance has been removed from the correlation matrix.

Both the variance and the EV indicate the strength of a particular factor in terms of its potential to explain the variation in the correlation matrix. Within this chapter three of the perspectives that emerged are discussed: these have been termed: 'positive', 'ambiguous' and 'marginalised'. The eigenvalues and the percentage of explanatory variance for each of these perspectives are presented in Table 10.1

'Positive'

The first perspective to emerge was generally positive about the Games and its legacy. There was a recognition that the Games were of national significance, providing a lasting legacy of sports facilities and promoting sports education. They also felt that the Games would encourage interest in local volunteering and would raise the self-esteem of local people.

This perspective had a mixture of participants loading onto it. Seven of the eleven were secondary age school students, six of whom were Year 11 students (15–16 years of age) and one was a Year 8 student (12–13 years of age) from five different secondary schools. There were two schools that were attended by two participants each, one of these was more than 15 miles from the Olympic Park, and one was within 10 miles of the Park. The other school-age students each attended a different school, two of which were within 15 miles

Table 10.1   Percentage of explanatory variance and calculated eigenvalue for the rotated factor matrix

Perspective	'Positive'	'Ambiguous'	'Marginalised'
% Explanatory Variance	17	11	10
Eigen Value	5.95	3.85	3.5
Number of Participants Loading onto Factor	11	6	2

of the Park, one was within 5 miles and was within one of the designated Olympic Boroughs. This is also true of one of the adult participants who worked as an Advanced Skills Teacher in a school within 5 miles of the Park. The other three participants came from a wide range of backgrounds within both formal and informal education, but were all based outside of London.

The 'Positive' perspective expressed a strong belief that the Games would have the ability to inspire a new generation of athletes (7: +5), and would provide a lasting legacy of sports facilities (35: +3). The perception that the Games were not just about elite athletes (14:+3) was expressed at the same level as the belief that the Games would lead to an increase in mass participation in sport (12: +3). It was strongly felt that the Games would do much to promote sport education (6: -5) and that the facilities would be used after the Games (25: -5). The view that the facilities will be used is a positive sign for legacy, especially as the majority of participants are students who might be accessing such facilities. However, as the majority of them are based some distance from the Park, the perception might be based on the expected use by others. Indeed, it was noticeable that this factor assigned less significance to the statement about opportunities for 'people like me' to make a contribution to the Games. This could be a reflection of the fact that three of the participants were based outside of London, and only two of the others were within the Olympic Boroughs and an indication that the embodied cultural capital that they feel that they can deploy is limited by location. It was also noticeable that those statements relating to the 'use' of the Games by schools were clustered around the centre of the array, indicating low level significance for this aspect of legacy. These positions demonstrate a low level sense of agency for participants loading onto this factor.

It was felt that the Games would help the regeneration of the area (17: +3), although the extent of the change was coined more cautiously: it may be a catalyst for longer term benefit (44:+1), it may transform the heart of East London (51:+1). It was put a little more strongly in terms of raising the job aspirations of young people (53: +2), and in providing some sustainable jobs (57: -2). The Games were also felt to be likely to have an effect by encouraging people to take part in local volunteering (19: +2), this mirroring the positive perspective of volunteering (24: -3) during the Games held by this factor. These two positions are likely to be linked, as the other perspectives broadly show a reversed pattern, all viewing volunteering as being 'cheap labour' and most not believing that the Games would lead to increases in young people becoming involved in voluntary activity.

This viewpoint might go some way to explain why the 'Positives' also very strongly disagreed with the assertion that the Games were a waste of money (47: -5), having said that, there are others of the distinguishing statements that might also have impacted on this position. Not least amongst these is the view that the Games will lead to health benefits for the community (18: -4). This factor contrasted with the generally tentative view on this matter expressed by the other perspectives. The extraction of the perspective described above from the matrix of correlations removes quite a lot of the common ground that is held by the Q sorts within the data set. Further extractions, and hence the factors discussed below, serve to explore underlying areas of difference.

The Q Sort table (a Q-sort grid with columns ranging from −5 to +5). Each cell contains an item number and a statement.

−5	−4	−3	−2	−1	0	+1	+2	+3	+4	+5
6: The Games will do little to promote sport education	18*: The Games won't lead to any health benefits for the community	5**: Those not directly involved in the Games will feel left out	28: The Games will lead to a huge public debt	23: The Games will inspire people across the country to develop sustainable lifestyles	4: The Games will inspire community development	2: The Games will help to develop an understanding of other cultures	15: Schools should be using the developments around the Games to inform their lessons	10: The Games will provide a lasting legacy of sports facilities	27: The local community must be able to access the Olympic Park facilities after the Games are over	1: The Games will provide opportunities to be involved with people from all over the world
25**: The Games will produce facilities which will not be used after the Games have finished	21: The Cultural Olympiad will exclude local people	8: The voices of local people are being ignored	31: Only certain subjects will be able to use the theme of the Games in their lessons	29: There will be opportunities for people like me to make a direct contribution to the Games	9: The Games will encourage people to gain a greater knowledge about their local area	20: Young people should be involved in deciding what is included within the Cultural Olympiad	19: The Games will encourage young people to take part in local volunteering activity	12*: The Games will lead to an increase in mass participation in sporting activities	35: The Games will give people opportunities to work with people they wouldn't normally meet	3: The Games are an event of national significance
47*: The Games are a waste of money	36: The involvement of young people and schools will only be at a superficial level	22: The Cultural Olympiad will not lead to any long-term benefits to our cultural life	32: The Olympic Park will be disconnected from the surrounding communities	30: Education will be a key strand in the legacy of the Games	11: The Games have diverted money from community projects	34: It is important that school pupils are able to attend events	39: The Games will give the people of East London more self-esteem	14: The Games are not just about elite athletes	40: People are excited about the event	7: The Games will inspire a new generation of athletes
	56: The legacy programme has been thought about in terms of the whole country	24**: The Games will use volunteers as cheap labour	57: The Games won't lead to any sustainable jobs	33: The construction of the Olympic park has caused the destruction of public spaces	13: The Games will lead to a greater understanding of culture in the younger generation	38: The Games provides educational opportunities for cross-curricular work	41: The Games will highlight the good points of East London	17: The Games will help the regeneration of the area	42: The Games will bring people into this part of the city	
		49: There will be affordable homes in the Olympic Park	43: The Games will contribute to the enhancement of the natural environment.	46: The Games will widen the horizons of the local communities	16: The Games will help to connect young people with the UK's artistic communities	44: The Games will act as a catalyst for change eg transport infrastructure for longer term benefit	53: The Games will raise the job aspirations of young people	26*: There will be an increase in personal involvement in activities, sport and volunteering		
			50: The Olympic Park will give people contact with the natural world	52: The legacy programme has been thought about in terms of the whole region	48: The Olympic Park will be a model for future projects in terms of sustainable development	45: The Games will increase community cohesion	37: The Games will be a useful resource for schools.			
				55: Local people will be 'priced out' of their own area after 2012	54: The Games will raise awareness of disability issues	51: The Games will transform the heart of East London				

Figure 10.2   The Q Sort for the 'Positive' Factor

Figure 10.3 The Q Sort for the 'Ambiguous' Factor

-5	-4	-3	-2	-1	0	+1	+2	+3	+4	+5
6 The Games will do little to promote sport education	**13** The Games will lead to a greater understanding of culture in the younger generation	**9** The Games will encourage people to gain a greater knowledge about their local area	**14** The Games are not just about elite athletes	**12** The Games will lead to an increase in mass participation in sporting activities	**4** The Games will inspire community development	**8** The voices of local people are being ignored	**1** The Games will provide opportunities to be involved with people from all over the world	**7** The Games will inspire a new generation of athletes	**15**** Schools should be using the developments around the Games to inform their lessons	**3** The Games are an event of national significance
23 The Games will inspire people across the country to develop sustainable lifestyles	**16** The Games will help to connect young people with the UK's artistic communities	**19** The Games will encourage young people to take part in local volunteering activity	**21** The Cultural Olympiad will exclude local people	**18** The Games won't lead to any health benefits for the community	**5** Those not directly involved in the Games will feel left out	**24** The Games will use volunteers as cheap labour	**2** The Games will help to develop an understanding of other cultures	**11** The Games have diverted money from community projects	**17** The Games will help the regeneration of the area	**27** The local community must be able to access the Olympic Park facilities after the Games are over
29* There will be opportunities for people like me to make a direct contribution to the Games	**56** The legacy programme has been thought about in terms of the whole country	**30** Education will be a key strand in the legacy of the Games	**41** The Games will highlight the good points of East London	**33** The construction of the Olympic park has caused the destruction of public spaces	**31*** Only certain subjects will be able to use the theme of the Games in their lessons	**25** The Games will produce facilities which will not be used after the Games have finished	**10** The Games will provide a lasting legacy of sports facilities	**38** The Games provides educational opportunities for cross-curricular work	**20** Young people should be involved in deciding what is included within the Cultural Olympiad	**28** The Games will lead to a huge public debt
	57 The Games won't lead to any sustainable jobs	**47** The Games are a waste of money	**45** The Games will increase community cohesion	**35** The Games will give people opportunities to work with people they wouldn't normally meet	**39** The Games will give the people of East London more self-esteem	**26** There will be an increase in personal involvement in activities, sport and volunteering	**22** The Cultural Olympiad will not lead to any long-term benefits to our cultural life	**44** The Games will act as a catalyst for change eg transport infrastructure for longer term benefit	**34** It is important that school pupils are able to attend events	
		53 The Games will raise the job aspirations of young people	**50** The Olympic Park will give people contact with the natural world	**40** People are excited about the event	**42** The Games will bring people into this part of the city	**32** The Olympic Park will be disconnected from the surrounding communities	**37** The Games will be a useful resource for schools.	**51** The Games will transform the heart of East London		
			52 The legacy programme has been thought about in terms of the whole region	**46** The Games will widen the horizons of the local communities	**48** The Olympic Park will be a model for future projects in terms of sustainable development	**36** The involvement of young people and schools will only be at a superficial level	**43*** The Games will contribute to the enhancement of the natural environment.			
				54 The Games will raise awareness of disability issues	**55** Local people will be 'priced out' of their own area after 2012	**49** There will be affordable homes in the Olympic Park				

'Ambiguous'

This perspective, whilst also being generally positive was a little ambivalent in some areas. For example, some doubt was expressed about the use of the facilities subsequent to the Games; there was little belief that the Cultural Olympiad would lead to any long term development of the cultural life of the area; nor that the legacy would see health benefits for the community.

This perspective had six participants – all adults – with a range of educational roles loading onto it. Two of the participants were Assistant Head teachers at separate schools, one of which was within an Olympic Borough and within 5 miles of the Park, the other did not work in an Olympic Borough, but was still within 10 miles of the Olympic developments. Two of the participants worked for environmental groups, one national and one London-focused. Both, however, were based in London and were within 15 miles of the Park. Two educational consultants also loaded onto this factor. One was freelance and one worked in the area of widening participation (someone working to increase the number of students from under-represented groups entering higher education) and was based more than 15 miles away from the Park.

The 'ambiguous' perspective strongly agreed that the Games were an event of national significance (3: +5), and disagreed that they are a waste of money (47: -3), but balanced this by fears over the extent to which they will lead to a significant debt (28: +5). This perspective gave clear support for the need for the local community to be able to use the Olympic Park (25: +5), whilst there was a perception that the facilities within the Park might not be used subsequent to the event (25: +1).

There was little confidence that the Cultural Olympiad would lead to any long term developments in the cultural life of the area (22: +2), there being no expectation that it would lead to connections being formed with the wider artistic community (16: -4) or to the development of a greater understanding of culture in the younger generation (13: -4).

Whilst not seeing education as a key strand in the potential legacy of the Games (30: -3), there was a strong feeling that schools should have used the developments around the Games to inform lesson planning (15: 4). There was a belief that the Games would provide opportunities for cross-curricular work (38: +3), but little expectation that people would gain a greater knowledge of the local area through engaging with the Games (9: -3).

Although there was a fairly strong feeling that the Games would operate as a catalyst for change for longer term benefit (44: +3), with the potential to transform the heart of East London (51: +3), there was a strong feeling that the Games had diverted money from existing community projects (11: +3) in order to secure this potential benefit. There was a feeling that there were very few opportunities for people to make a direct contribution to the Games (29: -5). There was also a tentative agreement that the voices of local people were ignored (8: +1). This might go some way to explaining the feeling that the Games would not increase community cohesion (45: -2).

There was a belief that the Games would deliver in providing sustainable jobs (57: -4), although there was little belief that the job aspirations of young people would be raised (53: -3). It did not appear, however, that the role of volunteering in this was recognised, there is a slight agreement with the assertion that volunteers will be used as cheap labour (24: +1), and little belief that this will encourage young people to take part in local volunteering (19: -3).

While expressing a view that the Games was mainly about elite athletes (14: -2), there was an acknowledgment that the Games were likely to inspire a new generation of athletes (7: +3), and would promote sports education (6: -5). This sat alongside a small degree of confidence that the Games would provide a legacy of facilities (10: +2) that these athletes might be able to use. There was less belief that the Games would lead to an increase in mass participation in sport (12: -1), and a similar feeling about the health benefits that might accrue to the wider community (18: -1).

There was a clear belief that schools should be using the developments around the Games to inform lessons. Although there was a recognition of the opportunities for cross-curricular work, the response to the assertion that only certain subjects would be able to use the theme of the Games in their lessons would suggest that this factor acknowledged the difficulties that teachers face in adopting such approaches (Morgan and Williamson, 2008). This constraint imposed by the field of school based accountabilities

-5	-4	-3	-2	-1	0	+1	+2	+3	+4	+5
6 The Games will do little to promote sport education	4 The Games will inspire community development	29 There will be opportunities for people like me to make a direct contribution to the Games	18 The Games won't lead to any health benefits for the community	5 Those not directly involved in the Games will feel left out	2 The Games will help to develop an understanding of other cultures	8 The voices of local people are being ignored	10 The Games will provide a lasting legacy of sports facilities	7 The Games will inspire a new generation of athletes	1 The Games will provide opportunities to be involved with people from all over the world	17 The Games will help the regeneration of the area
43 The Games will contribute to the enhancement of the natural environment.	14 The Games are not just about elite athletes	31 Only certain subjects will be able to use the theme of the Games in their lessons	19 The Games will encourage young people to take part in local volunteering activity	12 The Games will lead to an increase in mass participation in sporting activities	9 The Games will encourage young people to gain a greater knowledge about their local area	11 The Games have diverted money from community projects	24 The Games will use volunteers as cheap labour	20 Young people should be involved in deciding what is included within the Cultural Olympiad	3 The Games are an event of national significance	27 The local community must be able to access the Olympic Park facilities after the Games are over
50 The Olympic Park will give people contact with the natural world	47 The Games are a waste of money	37** The Olympic Park will be a useful resource for schools.	25 The Games will produce facilities which will not be used after the Games have finished	16 The Games will help to connect young people with the UK's artistic communities	13 The Games will lead to a greater understanding of culture in the younger generation	15 Schools should be using the developments around the Games to inform their lessons	36 The involvement of young people and schools will only be at a superficial level	41 The Games will highlight the good points of East London	28 The Games will lead to a huge public debt	33* The construction of the Olympic park has caused the destruction of public spaces
	49 There will be affordable homes in the Olympic Park	46 The Games will widen the horizons of the local communities	30 Education will be a key strand in the legacy of the Games	23 The Games will inspire people across the country to develop sustainable lifestyles	26 There will be an increase in personal involvement in activities, sport and volunteering	21 The Cultural Olympiad will exclude local people	38 The Games provides educational opportunities for cross-curricular work	42 The Games will bring people into this part of the city	34 It is important that school pupils are able to attend events	
		53 The Games will raise the job aspirations of young people	55 Local people will be 'priced out' of their own area after 2012	32 The Olympic Park will be disconnected from the surrounding communities	52 The Games will give people opportunities to work with people they wouldn't normally meet	22 The Cultural Olympiad will not lead to any long-term benefits to our cultural life	40 People are excited about the event	51 The Games will transform the heart of East London		
			56 The legacy programme has been thought about in terms of the whole country	39 The Games will give the people of East London more self-esteem	54 The Games will raise awareness of disability issues	48 The Olympic Park will be a model for future projects in terms of sustainable development	44 The Games will act as a catalyst for change eg transport infrastructure for longer term benefit			
				45 The Games will increase community cohesion		57 The Games won't lead to any sustainable jobs				

Figure 10.4   The Q Sort for the 'Marginalised' Factor

explains both the negative perception of education as a strand of legacy and the belief that the involvement of young people and schools would only be at a superficial level. This indicates that the institutionalised cultural capital (Bourdieu, 1986; Reay, 2004) that is seen in a qualification-driven school sector is a key determinant in the practice of that sector. So while educational engagement might be a feature of the Games, this was seen as ending with the Games rather than an opportunity to develop approaches (Sobel, 2005; Gruenewald and Smith, 2008) around developing positive regard for place, or gaining a greater understanding of the local area.

Table 10.2   The Q Set Statements

No	Statement	No	Statement
1	The Games will provide opportunities to be involved with people from all over the world	16	The Games will help to connect young people with the UK's artistic communities
2	The Games will help to develop an understanding of other cultures	17	The Games will help the regeneration of the area
3	The Games are an event of national significance	18	The Games won't lead to any health benefits for the community
4	The Games will inspire community development	19	The Games will encourage young people to take part in local volunteering activity
5	Those not directly involved in the Games will feel left out	20	Young people should be involved in deciding what is included within the Cultural Olympiad
6	The Games will do little to promote sport education	21	The Cultural Olympiad will exclude local people
7	The Games will inspire a new generation of athletes	22	The Cultural Olympiad will not lead to any long-term benefits to our cultural life
8	The voices of local people are being ignored	23	The Games will inspire people across the country to develop sustainable lifestyles
9	The Games will encourage people to gain a greater knowledge about their local area	24	The Games will use volunteers as cheap labour
10	The Games will provide a lasting legacy of sports facilities	25	The Games will produce facilities which will not be used after the Games have finished
11	The Games have diverted money from community projects	26	There will be an increase in personal involvement in activities, sport and volunteering
12	The Games will lead to an increase in mass participation in sporting activities	27	The local community must be able to access the Olympic Park facilities after the Games are over
13	The Games will lead to a greater understanding of culture in the younger generation	28	The Games will lead to a huge public debt
14	The Games are not just about elite athletes	29	There will be opportunities for people like me to make a direct contribution to the Games
15	Schools should be using the developments around the Games to inform their lessons	30	Education will be a key strand in the legacy of the Games

Table 10.2   The Q Set Statements (*concluded*)

No	Statement	No	Statement
31	Only certain subjects will be able to use the theme of the Games in their lessons	46	The Games will widen the horizons of the local communities
32	The Olympic Park will be disconnected from the surrounding communities	47	The Games are a waste of money
33	The construction of the Olympic park has caused the destruction of public spaces	48	The Olympic Park will be a model for future projects in terms of sustainable development
34	It is important that school pupils are able to attend events	49	There will be affordable homes in the Olympic Park
35	The Games will give people opportunities to work with people they wouldn't normally meet	50	The Olympic Park will give people contact with the natural world
36	The involvement of young people and schools will only be at a superficial level	51	The Games will transform the heart of East London
37	The Games will be a useful resource for schools.	52	The legacy programme has been thought about in terms of the whole region
38	The Games provides educational opportunities for cross-curricular work	53	The Games will raise the job aspirations of young people
39	The Games will give the people of East London more self-esteem	54	The Games will raise awareness of disability issues
40	People are excited about the event	55	Local people will be 'priced out' of their own area after 2012
41	The Games will highlight the good points of East London	56	The legacy programme has been thought about in terms of the whole country
42	The Games will bring people into this part of the city	57	The Games won't lead to any sustainable jobs
43	The Games will contribute to the enhancement of the natural environment.		
44	The Games will act as a catalyst for change e.g. transport infrastructure for longer term benefit		
45	The Games will increase community cohesion		

The perception that there would be limited networking opportunities is somewhat at odds with the variety of people who have loaded onto this perspective. The potential for the development of bridging social capital does not spontaneously lead to its generation. This indicates a low level of embodied social capital, certainly less than the significance given to the institutionalised cultural capital discussed above. This is in tune with the low significance that was given to the statement about the likelihood of increasing 'personal involvement in activities, sport and volunteering'. It also resonates with the very strong disagreement with the assertion that there would be opportunities for people like them to make a direct contribution to the Games. This is similar to the 'positive' perspective, but differs in as much as none of participants loading onto the 'ambiguous' perspective were geographically distant from the Olympic site. It would appear that the potential for 'disconnect' is not merely a function of distance.

'Marginalised'

A further perspective, with a number of school students loading onto it, was that spending on the Games diverted money from community projects, that public space was destroyed in the construction of the Park and in essence they felt marginalised by the developments around the Games.

The two participants loading onto this perspective were both students at the same school. One was a Year 9 student (13–14 years of age), the other a Year 11 student (15–16 years old). The school was within one of the Olympic Boroughs and within 5 miles of the Olympic Park.

The Games were seen as being a strong facilitator in the regeneration of the area (17: +5), transforming the heart of East London (51: +3) and highlighting its good parts (41: +3). This was set alongside the view that in developing the site, existing public space had been destroyed (33: +5). Whilst the Games, it was believed, would incur large public debt (28: +4), they were not viewed by participants loading onto this perspective as being a waste of that money (47: -4). Thus, the benefits of the infrastructure legacy were recognised with the Games being seen as helping in the regeneration of the area and were thought likely to transform the heart of East London, acting as a catalyst for longer term benefit. However, this factor very strongly believes that the construction of the Olympic Park has caused the destruction of public space. This perception is different to the majority of other perspectives that emerged from this study. This indicates a difference in the way in which the changes in the cultural capital embodied in the places around the Park are perceived, probably based upon its use value. Some of this value which was perceived to have been lost might be reclaimed if, as this factor believes, the sports facilities within the Park are used after the Games.

The Games were seen as being an event of national importance (3: +4). Whilst it was felt to be very important that pupils were able to attend events (34: +4), their actual involvement and that of their schools was thought likely to be only at a superficial level (36: +2), with the use of the Olympics as a resource being called into question (15: +1). As such, education was not seen as being a key legacy flowing from the Games (30: -2). It was hoped, however, that opportunities for young people to become involved in the Cultural Olympiad would be forthcoming (20: +3). There was disagreement that the Games would be a useful resource for schools, and a perception that any engagement of schools and young people with the Games would be shallow. This is a position indicative of a feeling of marginalisation. This is especially likely when taken alongside the perception that the voices of local people were ignored and the significance given to the statement that local people will be excluded from the Cultural Olympiad.

While recognising that the Games presented opportunities to be involved with people from all over the world, the beneficial outcomes that one might expect to flow from the opportunity are called into question by the perception that there will be limited opportunities to make a direct contribution to the Games and that the Games are unlikely to widen the horizons of local communities. Similarly low expectations are evident for the encouragement of young people into local volunteering and other activities. Taken together these indicate low levels of embodied cultural capital and resonate with the links between socio-demographics and the likelihood of taking part in volunteering activity (Anheier and Salamon, 1999; Lammers, 1991; Pearce, 1993; Wardell, Lishman and Whalley, 2000).

Conclusion

This chapter has revealed the range of perspectives on the legacy of the London Olympic Games held by educational stakeholders, and thus addresses the second of this study's research questions. It has done so using a statistically robust methodology that sets individuals as variables and allows a perspective to be developed across a number of aspects of potential legacy.

The perspectives revealed through the Q sort demonstrate a general feeling of warmth towards the Games, but also showed that there was an underlying pattern that indicated levels of marginalisation and the potential for further marginalisation. This was often located in low levels of embodied cultural capital. The perspectives should sensitise us to the primacy given to the institutionalised cultural capital of the formal curriculum and that this is likely to work against engagement with the event. In this is exposed the internalised arbitrary of

Bourdieu and Passeron (1977), a key part of their argument about how the values of the dominant class are transmitted through the curriculum. This helps to develop an understanding of how practices are perpetuated across time through the habitus of individuals and the operation of this 'internalised arbitrary' (Bourdieu and Passeron, 1977; Marsh, 2006). In terms of this analysis this understanding is important, as is an appreciation of how these perspectives are received by, and responded to, in a dialectic development of practice and policy,

This study did not use the perspectives that emerged in this dialectic, but it would have been possible to combine the outcomes of the Q sort with a deliberative process in a manner described by, for example, Niemeyer et al (2013). They feel that such an approach provides an opportunity to engage with the issue and to consider the implications of different alternatives, providing a potentially powerful tool allowing one to move beyond the pre-deliberative situation which is often characterised by the strongly competing claims that the Q reveals. Q offers an opportunity to systematise interpretative enquiry and analysis without the need for resource intensive qualitative interviewing (Baker and Jeffares, 2013), with the quantification and factor analysis offering an, admittedly not universally accepted (Yannow, 2007), enhanced sense of legitimacy. The further exploration of perspectives in the post-event phase and the development of a mechanism for deliberative engagement with those perspectives is the next stage of this research.

References

Anheier, H. and L. Salamon (1999). 'Volunteering in cross-national perspective: Initial comparisons'. *Law and Contemporary Problems*, 62(4), pp. 43–65.

Baker, R and S. Jeffares (2013). 'Introduction to the special issue: Public policy', *Operant Subjectivity*, 36(2), pp. 69–72.

Bobrow, D. and J. Dryzek (1987). *Policy Analysis by Design*. Pittsburgh, PA: University of Pittsburgh Press.

Bourdieu, P. (1986). 'The forms of capital', in Lauder, H., P. Brown, J. Dillabough and A. H. Halsey (eds.) (1986). *Education, Globalisation and Social Change*. Oxford: OUP.

Bourdieu, P. and J. C. Passeron (1977). *Reproduction in Education, Society and Culture*. London: Sage.

Brown, S. (1980). *Political Subjectivity: Applications of Q Methodology in Political Science*. Newhaven, CT: Yale University Press.

Burt, C. and W. Stephenson (1939). 'Alternative views of correlations between persons'. *Pyschometrika*, 4(4), pp. 269–281.

Cashman, R. (2006). *The Bitter-Sweet Awakening*. Walla Walla Press.

Cuppen, E. (2013). 'Q methodology to support the design and evaluation of stakeholder dialogue'. *Operant Subjectivity*, 36(2), pp. 135–163.

Devine-Wright, P. (2005). 'Local aspects of UK renewable energy development: exploring public beliefs and policy implications'. *Local Environment*, 10(1), pp. 57–69.

Dunn, W.N. (2001). 'Using the method of context validation to mitigate type III error in environmental policy analysis', in Hisschemoller, M. R. Hoppe, W.N. Dunn and J. R. Ravetz (eds.) (2001). *Knowledge, power and participation in environmental policy analysis*. New Brunswick and London. Transaction Publishers, pp. 417–436.

Frederickson, H. G. (1991). 'Toward a theory of the public for public administrators'. *Administration and Society*, 22(4), pp. 395–417.

Gaynor, T. (2013). 'Building democracy: Community development corporations' influence on democratic participation in Newark, New Jersey'. *Operant Subjectivity*, 36(2), pp. 93–113.

Gratton, C. and H. Preuss (2008). 'Maximising Olympic impacts by building up legacies'. *International Journal of the History of Sport*, 25(14), pp. 1922–1938.

Gruenewald, D. and G. Smith (2008). *Place-based Education in the Global Age: Local Diversity*. New York: Routledge.

International Olympic Committee. (2011) Olympic Charter. Lausanne: IOC.

Kline, P. (1994). *An Easy Guide to Factor Analysis*. London: Routledge.

Lammers, J. (1991). 'Attitudes, motives and demographic predictors of volunteer commitment and service duration'. *Journal of Social Science Research*, 14(3/4), pp. 125–140.

Lenskyj, H. J. (2002). *The Best Olympics Ever? Social Impacts of Sydney 2000*. Albany: State University of New York Press.

MacRury, I. (2011). Framing legacies and governing development: Urban experience, transformation and narratives of change after 2012. Paper presented at the Legacy Symposium.

Marsh, J. (2006). 'Popular culture in the literacy curriculum: A Bourdieuan analysis'. *Reading Research Quarterly*, 47(2), pp. 160–174.

Moragas, M., C. Kennett and N. Puig (eds.) (2003). *The Legacy of the Olympic Games 1984–2000*. Lausanne: IOC.

Morgan, J. and B. Williamson (2008). *Enquiring Minds: Schools, Knowledge and Educational Change*. Bristol: Futurelab.

Niemeyer, S, S. Ayirtman and J. HartzKarp (2013). 'Understanding deliberative citizens: The application of Q methodology to deliberation on policy issues'. *Operant Subjectivity*, 36(2), 114–134.

O'Brien, D. (2006). 'Event business leveraging the Sydney 2000 Olympic Games'. *Annals of Tourism Research*, 33(1), pp. 240–261.

Pearce, J. (1993). *Volunteers: Organisational Behaviour of Unpaid Workers*. London: Routledge.

Poynter, G. (2009). 'The 2012 Olympic Games', in Imrie, R., L. Lees and M. Raco (eds.) (2009). *Regenerating London: Governance, Sustainability and Community in a Global City*. London: Routledge, pp. 132–148.

Preuss, H. (2004). *The Economics of Staging the Olympics: A comparison of the Games 1972–2008*. Cheltenham: Edward Elgar.

Preuss, H. (2006). 'Impact and evaluation of major sporting events'. *European Sport Management Quarterly*, 6(4), pp. 313–316.

Preuss, H. (2007). 'The conceptualisation and measurement of mega sport event legacies'. *Journal of Sport & Tourism*, 12(3/4), pp. 207–228.

Reay, D. (2004). 'Education and cultural capital: The implications of changing trends in education policies'. *Cultural Trends*, 13(50), pp. 73–86.

Ritchie, J. and K. Aitkin (1984). 'Assessing the impact of the 1988 Olympic Winter Games: The research programme and initial results'. *Journal of Travel Research*, 22(3), pp. 17–25.

Rittel, H. and M. Webber (1973). 'Dilemmas in general theory of planning'. *Policy Sciences*, 4, pp. 155–169.

Schmolck, P. (2002). PQmethod (version 2.11). Retrieved 6th October, 2008, from www.lrz.de/-schmolk/qmethod

Sobel, D. (2005). *Place-based Education*. Barrington MA: The Orion Society.

Stainton-Rogers, R. (1995) 'Q methodology', in Smith, J., Harre, R. and Van Langanhove, L. (eds.) (1995). *Rethinking Methods in Psychology*. London: Sage, pp. 178–192.

Stephenson, W. (1935). 'Technique of factor analysis'. *Nature*, 136, p. 297.

Stephenson, W. (1936a). 'The Foundations of Psychometry: Four factor systems'. *Psychometrika*, 1(3), pp. 195–209.

Stephenson, W. (1936b). 'The inverted factor technique'. *British Journal of Psychology*, 26(4), pp. 344–361.

Van Eeten, M. (2001). 'Recasting intractable policy issues: The wider implications of the Netherlands civil aviation controversy'. *Journal of Policy Analysis and Management*, 20, pp. 391–414.

Wardell, F., J. Lishman and L.Whalley (2000). 'Who volunteers?' *British Journal of Social Work*, 30, pp. 227–248.

Watts, S. and P. Stenner (2005). 'Doing Q methodology'. *Qualitative Research in Psychology*, 2, pp. 67–91.

Yannow, D. (2007). 'Qualitative-interpretive methods in policy research', in Fischer, G., G. Miller and M. Sidney (eds.) (2007). *Handbook of Public Policy Analysis, Theory, Politics and Methods*. London: CRC, pp. 405–416.

Chapter 11
Legacy as Knowledge

Ailton Fernando S. De Oliveira, Celi Nelza Z. Taffarel, Cristiano M. Belem and
Lamartine P. DaCosta

Introduction

The objective of this chapter is to review the concept of legacy, as it relates to sport mega-events and to their host cities, focusing on knowledge, a concept that, until recently, has not been analysed in this context. The legacy of mega-events is normally discussed in terms that are geographically bound by the borders of the host countries. Knowledge, however, has the capacity to overcome these geographical boundaries and provide a broader framework for the understanding of sport in general and of mega-events in particular. Moreover, legacy 'as knowledge' might also be applied to comparative research within the tradition of Olympic Studies with such an approach referring to the knowledge-sharing legacies that originate from national, regional and local analyses of specific sporting mega-events or sports in general.

Approaching legacy as knowledge requires the conceptualisation of its different dimensions. This involves distinguishing between explicit knowledge, including technical and scientific knowledge, and tacit knowledge, which is usually acquired through experience (Polanyi 1958). The intermediation between knowledge production and its application is provided by *knowledge management*, usually defined as a process of capturing, distributing and effectively applying knowledge (Davenport 1994). This allows us to configure the legacy knowledge produced in the past and present in order to apply it to the construction of future legacy.

To illustrate our methodology and explore the comparison between sport in general and the Olympic Games in particular we examine two case studies. The first is DIESPORTE[1], a project of the Ministry of Sports of Brazil, which aims to identify and classify the problems faced by the institutions which constitute the Brazilian system of sport, and suggest strategies for their solution that will, in turn, inform the development of public policies. It incorporates data, information and knowledge from sports in general and their effects – mega-events included. DIESPORTE illustrates the broad use, circulation and maintenance of knowledge as applied to sport. The second case study illustrates aspects which are specific to mega-events such as the Olympic and Paralympic Games. Here we analyse the knowledge-legacy relationships proposed and implemented by the International Olympic Committee (IOC).

Over recent years the IOC-sponsored Olympic Games Impact Studies (OGI) have carried out research on local impacts and legacies of the Olympic Games, covering local, regional and national impacts over a period of 11 years and producing relevant data for the periods before, during and after the Games (Brimicombe 2010). This IOC-inspired attempt to produce knowledge from sports mega-events demands research of broad scope and duration. DIESPORTE adopts a similar approach, but seeks to encapsulate several kinds of broader developments within sports and leisure in Brazil. In the first stage of DIESPORTE (2011–2013), the approaches adopted were experimental. They included geo-referential studies of selected urban areas, among which was a survey of the area of Barra da Tijuca in the city of Rio de Janeiro, where the Olympic Park of the 2016 Olympic Games is located (Belem 2013a). DIESPORTE and the Olympic Games impact studies display

---

1 DIESPORTE is short for Diagnostico Nacional do Esporte (National Sports Diagnostics). It comprises a range of research projects promoted by the Ministry of Sports and supported by research funding agencies of the Ministry of Science and Technology – namely the Brazilian Innovation Agency (FINEP) and the National Council for Scientific and Technological Development (CNPq). The research is taking place across all Brazilian territories and has been conducted since 2011 by a network involving six Brazilian federal universities: UFBA – state of Bahia; UFS – state of Sergipe; UFRJ – state of Rio de Janeiro; UFRGS – state of Rio Grande do Sul; UFAM – state of Amazonas; UFG – state of Goiás (Taffarel and Oliveira 2013).

similarities and differences in relation to knowledge production. Both generate hypotheses about the concept of legacy that comes from different modes of production, one specific and the other much broader in focus.

In the remaining sections of this chapter we examine the way in which knowledge production contributes to the concept of legacy, illustrated by reference to various examples of academic knowledge produced and to the two case studies in particular. We then explore Knowledge Transfer and Knowledge Management Practices and, in the conclusion, we review our conception of Knowledge Legacy.

Knowledge Production

Cashman and Toohey (Cashman and Toohey 2002) pioneered the academic discussion of legacy as knowledge when they surveyed the first planned and managed initiative of knowledge production that took place in relation to the Sydney Olympic Games (2000). They identified the unique role played by universities with regards to the production and dissemination of knowledge, e.g. through publications, including books, scholarly articles, seminars and workshops. From these several national and international research partnerships emerged. The authors argued that the knowledge produced as a result of the 2000 Olympic Games could itself be viewed as a legacy, once it was submitted for publication, made available to the general public and shared with sports and government entities (Cashman and Toohey 2002: 70–72). Subsequent Olympic Games followed the tradition of knowledge production by universities started by Sydney 2000. The culmination of this process took place at the Sochi 2014 Winter Games when the Games led to the creation of a sports university which will use the local installations and competition venues created for the Games (Sochi OCOG 2013).

However, the tradition has not been maintained in relation to the 2016 Rio Olympic Games, where the Local Organising Committee 'Rio 2016', has not to date involved itself in knowledge production in collaboration with universities. This omission – or possible delay – regarding the involvement of universities with Olympic events in Brazil, constitutes a break with tradition. The Brazilian National Olympic Committee (COB), whose leaders formed the main nucleus of Rio's OCOG in 2009 (the year in which the city of Rio de Janeiro was chosen as the host site for the 2016 Games), has a long history of organising partnerships with universities that commenced in 1968 during the Mexico Olympic Games. On that occasion, COB sponsored a study of the urban and technological changes in the city that hosted the 1968 Games (DaCosta et al. 1969); also COB gave support to a book that developed scientific and management guidelines for the participation of the Brazilian Olympic team at the high altitude of 2250 meters in which the Games took place (DaCosta 1968a). In both cases, important university sports centers were called upon to participate in these studies – Gama Filho University (Rio de Janeiro) and the University of São Paulo – creating an initiative similar to the one described by Cashman in Sydney three decades later.

The research on Olympic themes that was undertaken in the 1960s in Brazil was innovative and gave rise to several original proposals. Among these was the so-called 'Altitude Training', a new training method that achieved international recognition (DaCosta 1968b). As Cashman and Toohey (2002: 48–68) highlighted, a similar development occurred in Australia in the years that followed the 2000 Games. In short, Sydney 2000 was more than a simple incubator of innovations. As far as *post hoc* studies are concerned, Halbwirt and Toohey (2001) have shown that the knowledge produced on behalf of the Games circulated with greater efficiency within the Sydney OCOG (SOCOG) than it did within its associated universities.

In contrast to Rio 2016, the current DIESPORTE Project involves its associated universities in relationships of exchange of data and information and this exchange also includes several other institutions, especially government ones. The lack of continuity concerning knowledge management in the current Olympic environment in Brazil suggests that a form of inappropriate governance is taking place which could jeopardise the creation of positive legacies for the nation.

Knowledge Transfer and Knowledge Management

Knowledge transfer is integral to the IOC's relationships with NOCs and OCOGs and the knowledge produced and preserved by each host site of summer and winter Olympics is an essential component of this process of transfer. The movement of the Games every two years from one host city to the next imposes a tight timetable for the learning process of each new OCOG. In research designed to trace knowledge transfer and knowledge management in the case of London 2012, Girginov and Gold (Girginov and Gold 2013) characterised the process of knowledge transfer as occurring at three levels: cognitive, organisational and societal. The cognitive level involves knowledge residing with the individuals involved in the planning and operation of the Games; the organisational level concerns knowledge embedded in the culture of the institutions involved and the societal level refers to knowledge possessed by institutions and agencies. In the case of the London Organising Committee of the Olympic Games (LOCOG), at the cognitive level the most important elements were the skills and experience of the key personnel who worked for LOCOG and the means by which their acquisition of new knowledge would be facilitated. At the organisational level knowledge was mainly transferred by LOCOG's own management system, through its decision-making and communications structure and the societal level involved LOCOG's interaction with some 150 organisations external to the Olympic system. Knowledge management at LOCOG fostered the relationships between the various levels and promoted the development of management innovation at all levels.

Bovy (2013), a source institutionally related to the IOC, argues that the knowledge transfer that takes place within the operational relationships of the IOC constitutes a management function just like finance, sustainability, marketing and so forth. In other words, knowledge management in the framework of IOC-NOC-OCOG is a specialist function for management, broadly similar to that which happens within companies when dealing with the transfer of knowhow. However, the full public disclosure of this information in relation to the Olympic Games has not yet occurred, making the knowledge transfer system a semi-closed one.

The IOC-NOC-OCOG semi-closed system, of which Rio 2016 is an extreme case, contrasts with the more open systems of universities. Top sports administration organisations in Brazil, which include the current Brazilian NOC and the Rio 2016 OCOG, possess a tradition of autonomy in relation to local, state and federal governments (DaCosta 1996). While in Brazil all government organisations are obliged by law to make information they generate available to all citizens, the special autonomous status of the Brazilian NOC and of the Rio 2016 OCOG means that they are absolved from this responsibility, which has resulted in a lack of public transparency. Private companies and concerns that receive government funding are only required by law to account publicly for the spending of government-provided funds. This lack of external openness in the Brazilian case reduces and even eliminates knowledge production as legacy.

A programme designed to manage the transfer of knowledge between the IOC, the NOCs and the OCOGs was established in the run up to the Sydney 2000 Olympics. This constituted a platform of services that emphasised good practices. Knowledge Management was used to identify the present and future legacies of the Olympic Games within the host city and nation. It was originally named the *Olympic Games Knowledge Management* (OGKM), an entity made up of consultants and formally included in the structure of the IOC in Lausanne, Switzerland. The OGKM programme consists of three main sources: information, services and personal experience. Information includes the Official Games Reports, technical manuals, knowledge reports and a range of other documents and publications made available on a dedicated extra-net. Services include workshops, seminars and a network of experts with Games experience on a range of Olympic topics. Personal experience of Games preparations and operations is made available through the Games-time Observer Programme, the official Games debriefing and the secondment programme. This enables personnel of future OCOGs to work on the current Games in order to gain first-hand experience of Games operations. The operation of the OGKM is discussed in Clark (2011). Clark characterises the IOC's knowledge transfer programme as a learning system. According to the head of OGKM, Phillipe Furrer, it is concerned with the study of legacies through managed knowledge. Explicit knowledge is transferred by making available the material collected through the information component of the OGKM. The briefing and debriefing of technicians and managers who receive support from OGKM, organised through the personal experience component of the programme, is aimed at transferring the tacit knowledge acquired by the actors. OGKM consultants carry out knowledge transfer operations that include both tacit and explicit aspects. The knowledge generated by universities involved in the study of Olympic Games can essentially be characterised as explicit knowledge.

There are similarities between OGKM and DIESPORTE in relation to the explicit and tacit knowledge model[2]. DIESPORTE uses a set of definitions that have emerged through knowledge transferred between the various triennial National Sports Conferences held in Brazil. These events bring together community representatives who possess tacit knowledge, university researchers who have generated explicit knowledge and government officials who possess tacit knowledge and have also produced explicit knowledge. However, in contrast with the OGKM, the sports legacy developed by DIESPORTE often takes itself the form of knowledge rather than being a subsidiary end-product. This evidence-based knowledge has been primarily an explicit knowledge legacy which has informed public policies on sport developed by government at different levels.

An analysis of the activities carried out as part of DIESPORTE from 2011 to 2013 shows that its knowledge legacy arises in the form of the results of research on the situation of sport and leisure in the country. Collectively they have been referred to as the National Sports Diagnostics. These include the results of sectoral and local research undertaken in response to specific demands and in response to policies emanating from the Federal Government. The studies of impacts of sport mega-events in the city of Rio de Janeiro that have been recently made available constitute examples of this (Belem 2013b). The 2011–2013 Diagnostics is under preparation. A preliminary methodological discussion has been produced by Oliveira (2013).

DIESPORTE, in its initial phase (2011 to 2013), has already provided a contribution to the legacy of knowledge by collecting, as a background to the project, previous versions of the data and information it is seeking to update produced in Brazil between 1971 and 2010. This aspect of DIESPORTE suggests a similarity with the procedures of OGKM in terms of continuous improvement carried out in the context of a long-term perspective. In both cases, the knowledge legacy is explicit and stable, and thus consistent with the approach adopted by both the IOC-NOC-OCOG system and the national diagnostics of DIESPORTE.

DIESPORTE has produced partnership research with universities that has developed explicit knowledge. In this respect, an agreement made with the University of Rome in Italy which has resulted in a first publication of international collaborative research on a Brazilian theme, deserves special note (Mussino, Taffarel and Oliveira 2013). This international collaboration highlights one of the differences previously mentioned between DIESPORTE and OGKM, in that the knowledge generated by the Brazilian project is much more open to external access and is, therefore, more likely to be understood as a legacy because of its greater social and cultural impact.

Summarising, both OGKM and DIESPORTE promote knowledge transfer but the management of the former is institutional while the latter operates through partnerships. OGKM is guided mainly by the provision of services, while DIESPORTE is based more on personal and institutional experience. This suggests that OGKM uses knowledge to produce legacies and DIESPORTE transforms knowledge into legacy.

The international exchange of knowledge between organisations of the Olympic Movement, particularly NOCs, has been mainly based on the International Olympic Academy (IOA). This started life in the 1960s. The IOA is headquartered in Olympia, Greece, and until recently led a proactive network of National Olympic Academies (NOAs). The main financial and management support comes from the Government of Greece. Additional finance is provided by the IOC. This collaborative knowledge production between the IOA and NOAs has traditionally involved universities. At the time of writing this chapter, there was a reduction in their activities because of the European economic crisis and local socio-economic problems.

The IOC has recently also attached less importance to the IOA and its national counterparts and to top universities in the field of Olympic Studies due to the priority given to its own system of knowledge transfer from the year 2000 on. This autonomous development of the so-called 'Olympic family' has often been presented as self-learning and experience-based. These features have been highlighted by Pollard (Pollard 2012). Pollard also argues that the OGKM system is organised by briefing and debriefing, focusing on best practices and local contexts, inspired by business experiences.

---

2 Griginov and Gold (2013), following Nonaka and Toyama (2003), characterise the nature of the relationship between tacit and explicit knowledge in the Olympic knowledge production and transfer system as dialectic in that interactions between the two types of knowledge, including contradictions between them, generate new knowledge. They also argue that the same dialectical relationship applies to the dichotomy between individual (cognitive) and collective (institutional) knowledge.

This business-focused option in conducting knowledge transfer in the Olympic environment may also explain the recent neglect of universities by the Brazilian NOC. Figure 11.1 outlines the relationships of the Brazil NOA in the 2000s when it was seen as a relevant exchange between universities and NOAs both nationally and internationally. However, this Brazilian-based Olympic alternative of knowledge exchange is today reduced to a minimum, as opposed to what is happening in other Olympics-related initiatives in the country.

Knowledge Management has become an operational function of the IOC management system. In analysing these developments, Werner (2012) concluded that IOC management is atypical in that it has transformed its own management system into a tool for innovation when it shares its knowledge with its network of related organisations. Werner also demonstrated that this approach to knowledge exchange adopted by IOC-OGKM enabled the integration of other Olympic stakeholders into its own management and confirmed its semi-closed status.

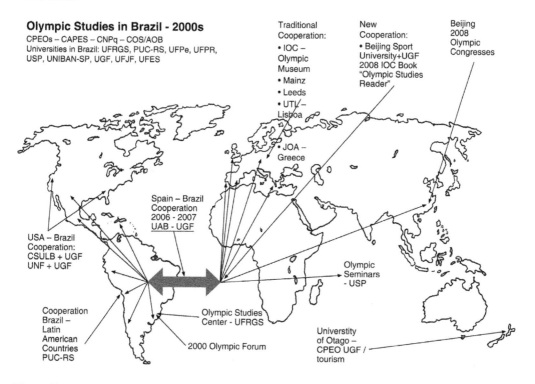

Figure 11.1   Olympic Studies Exchange of Knowledge between Brazil and other countries with the participation of National Olympic Academies and Universities during the 2000s

Parent, Macdonald and Goulet (2013) found in relation to Vancouver 2010 that the degree of openness – characterised as 'socialisation' – to the general public and to universities depended more on the performance and interest of the Olympic stakeholders than on the activities of the international consultants from OGKM. The communication between parties, in this case, flowed from explicit to tacit knowledge, since the reverse was not understandable to local recipients. In the end, local culture was found to be more receptive to management practices that leveraged the explicit knowledge made available.

The cases of Vancouver 2010 and London 2012 suggest that the IOC/OGKM system of knowledge transfer has reached maturity as far as universities are concerned in that knowledge produced by them has been incorporated into the IOC network. The explicit knowledge obtained from the universities and the information and data produced by internal discussion are consolidated into a legacy which benefits in the first instance the IOC's own Olympic network and is then offered to external audiences often in summary formats.

The IOC has used the expertise of university researchers particularly when faced with external threats, as happened in the confrontation with the European Parliament in 1992 over issues of environmental protection in the Winter Olympics (DaCosta 2002). Flores (2014) has termed this ability of the IOC to defend itself from external threats and to use them as a springboard for extending its reach, the IOC's 'multiplier face'. This ability has been institutionalised as knowledge in the governance of the IOC from its inception by Pierre de Coubertin, founder of the modern version of the Olympic Games, and its first test was the challenge to it created by the formation of the League of Nations after World War I which was turned into an opportunity to extend its international reach.

Comparing the Knowledge Management systems of the IOC network with that of DIESPORTE from the point of view of the international reach of the knowledge generated and managed, it should be noted that the IOC system is by its nature international insofar as the network of knowledge generation and transfer includes as a minimum all the 204 countries which possess a NOC, whilst DIESPORTE has tried to internationalise knowledge generation and management by investing in collaborative agreements with foreign universities such as the one with the University of Rome already previously mentioned. In the case of the IOC network, internationalisation occurs naturally and operates by internal knowledge transfer between the constituent entities and individuals. In the case of DIESPORTE internationalisation is actively sought and operates by collaborative agreements. This contrast is of some analytical importance for the study of other knowledge management systems which have an international dimension. It also has significant implications for the international dimension of knowledge as legacy.

Conclusions

The preceding review of the concept of legacy as knowledge in relation to both sports mega-events and sports in general is consistent with the idea of knowledge as a means of governance which has historically been developed in the preparation of Olympic Games, and in the creation of their legacies (Flores 2014). Recently this understanding has been achieved mainly through the operations of the Olympic Games Knowledge Management (OGKM) system of the IOC, a system that we have argued is semi-closed in the sense that knowledge is first generated and managed internally in the Olympic organisations most closely associated with the IOC before being made available to universities and to the general public. There has been a tendency for the system to become more closed as the IOC has adopted a business approach to knowledge management. Brazil has a tradition of openness in knowledge creation and distribution in relation to sport through the involvement of universities in association with government and eventually with other sports-related entities. However, this tradition so far appears to have been broken in relation to the preparations for the Rio 2016 Olympic Games as the universities have been kept at a distance from the process. The adoption of business models may once again have been a factor in this.

In the case of DIESPORTE, the Brazilian research programme that seeks to develop knowledge about sport in general, including the Olympic Games, an examination of the activities of the universities involved, and of their relations with the government and other sport-related entities which have supported them, suggests that, in this instance, knowledge is an end in itself. In this context, legacy is the explicit knowledge in the form of data and information that is captured, managed and distributed for future use. DIESPORTE, in the way that it was conceived and has operated, has generated a system of knowledge creation, management and distribution which is open and available to the general public via the Internet (Oliveira 2013).

Knowledge that circulates within the system of Olympic entities also often becomes an end in itself. According to Girginov and Gold (Girginov and Gold 2013) and Stewart (Stewart 2012), the legacy of Olympic events is becoming dependent on the level of transparency of the knowledge generated and subjected to successive revisions and syntheses in the context of the interrelations between the entities involved. These requirements also confirm the view that the IOC's insistence on the need for legacy to arise from future Olympic Games is part of the IOC's own search for legitimacy (DaCosta 2013). From this perspective, the main conclusion of this study is the proposition that the knowledge transferred within the system of Olympic institutions should not be semi-closed. It should be accessible to the public, and so should, as DIESPORTE illustrates, the knowledge produced in the context of examining government policies in relation to sport in general.

References

Belem, C. M. (2013a). A Ocupação Urbana no Entorno do Parque Olímpico 2016 no Rio de Janeiro: Impactos dos Jogos Pan-Americanos 2007 e dos Preparativos dos Jogos Olímpicos 2016. Tese de Doutorado – Universidade Gama Filho, Rio de Janeiro (unpublished).

Belem, C. M. (2013b). The Urban Environment and Occupation of the Future Olympic Park Region: A case study of the 2007 Pan American Games towards Rio 2016 Olympic Games. Paper presented at the International Conference Olympic Legacies, 4–6 September 2013, University of East London.

Bovy, P. (2013). *Mega-event Transport Planning Legacy and Sustainability*. Lausanne: Swiss Federal Institute of Technology.

Brimicombe, A. (2010). Olympic Games Impact Study – London 2012, Pre-Games Report October 2010. London: University of East London.

Cashman, R. and K. Toohey (2002). The Contribution of the Higher Education Sector to the Sydney 2000 Olympic Games. Sydney: University of South Wales – Centre for Olympic Studies.

Clark, S. (2011). Learning from the past, getting ready for the future. Lausanne: International Olympic Committee Website. http://www.olympic.org/news/learning-from-the-past-getting-ready-for-the-future/143493 (accessed 25 October 2011).

DaCosta, L. P. (1968a). Planejamento Mexico. Brasilia: Divisao de Fducacao Fisica – Ministerio da Educacao e Cultura.

DaCosta, L. P. (1968b). 'Altitude training', in DaCosta, L. P. (ed.) (1968). *Introducao a Moderna Ciencia do Treinamento Desportivo*. Brasilia: Divisao de Fducacao Fisica – Ministerio da Educacao e Cultura, pp. 239–284.

DaCosta, L. P. (1969). Mexico 68 Aspectos Tecnicos Evolutivos. Rio de Janeiro: Comite Olimpico Brasileiro/ SEED-Ministerio da Educacao e Cultura.

DaCosta, L. P. (1996). 'The state versus free enterprise in sport policy: The case of Brazil', in Chalip, L., A. Johnson and L. Stachura (eds.) *National Sports Policies*. Westport: Greenwood Press, pp. 23–38.

DaCosta, L. P. (2002). Olympic Studies – Current Intellectual Crossroads. Los Angeles: LA84 Foundation, pp. 91–106. Available online at: library.la84.org/SportsLibrary/Books/OlympicStudies.pdf (accessed 11 November 2013).

DaCosta, L. P. and A. Miragaya (2008). 'The state of the art in legacies of sport mega-events', in DaCosta, L. P. (ed.) (2008). *Legacies of Sport Mega-events*. Brasilia: Ministerio do Esporte/ CONFEF, pp. 33–39.

DaCosta, L. P., A. Miragaya and V. Bitencourt (2012). Epistemological Experiments in the Perspective of Sport in the Global Era. Paper presented at the 40th Annual Conference of the International Association for the Philosophy of Sport. Porto: University of Porto.

DaCosta, L. P. (2013). Future Mega-event Cities. Paper presented at the International Conference Olympic Legacies, 4–6 September 2013, University of East London.

Davenport, T. H. (1994). 'Saving IT's soul: Human centered information management'. *Harvard Business Review*, March–April, 72 (2), pp. 19–131.

Flores, M. (2014). 'Conhecendo o Comite Olimpico Internacional', in Flores, M. (ed.) (2014). *Sustentabilidade, Governanca, e Megaeventos: Estudo de Caso dos Jogos Olímpicos*. Rio de Janeiro: Elsevier, pp. 32 – 36.

Girginov, V. and J. Gold (2013). *London 2012 Olympic and Paralympic Games Knowledge Transfer*. London: Podium Report.

Halbwirth, S. and K.Toohey (2001). 'The Olympic Games and knowledge management: A case study of the Sydney organising committee of the Olympic Games'. *European Sport Management Quarterly*, Volume 1, Issue 2, pp. 91 – 111.

Mussino, A., A. F. S. Oliveira and C. Z. Taffarel (2013). Il Diagnostico Nacional do Esporte e Lazer: Conoscere per Governare un Sistema Sportivo. Rivista Trimestrali di Scienza dell Amministrazione. V.1, Roma – Itália, 2013, pp. 45–64.

Nonaka, I. and R. Toyama (2003). 'The Knowledge-creating system revisited: Knowledge creation as a synthesising process'. *Knowledge Management Research and Practice*, 1, pp. 2–10.

Oliveira, A. F. S., C. N. Z. Taffarel, C. M. Belem and L. P. DaCosta (2013). Diagnosis of Sport in Brazil: a Legacy for Assessment and Planning Policies of Sport. Paper presented at the International Conference Olympic Legacies, 4–6 September 2013, University of East London.

Oliveira, A. F. S. (2013). Diagnostico esportivo no Brasil: Desenvolvendo Metodos e Tecnicas. Tese de Doutorado. Universidade Federal da Bahia, Salvador (unpublished).

Parent, M. M., D. MacDonald and G. Goulet (2013). The theory and practice of knowledge management and transfer: The case of the Olympic Games. *Sport Management Review,* in press. Available online 12 July 2013 at: http://www.sciencedirect.com/science/article/pii/S1441352313000351

Pollard, C. (2012). OGKM: Learning from experience. Lausanne: International Olympic Committee Website. http://www.olympic.org/news/ogkm-learning-from-experience/170562 (accessed 3 August 2012).

Polanyi, M. (1958). *Personal Knowledge: Towards a Post-Critical Philosophy.* Chicago: University of Chicago Press.

Sochi OCOG (2013). *The Olympic Games Impact – Summary Report.* Moscow: Moscow State University, pp. 44–45.

Taffarel, C. N. Z. and A. F. S.Oliveira (2013). 'Sport diagnosis in Brazil: Indepth analyses'. *FIEP BULLETIN, V.83,* Article 1, pp. 358–36.

Werner, K. (2012). The Impact of a Nationwide Mega-event on Tie Strength, Collaborative Capacity and Knowledge Transfer Dynamics within Regional Destination Marketing Networks. Doctoral thesis. Auckland University of Technology. Available 7 December 2012 at http://hdl.handle.net/10292/5338

PART IV
Sustainability and Mega-event Legacy

Part IV – Introduction
Sustainability and Mega-event Legacy

Ozlem Edizel and Ralph Ward

Mega-events and their legacies have become increasingly attractive to cities as an urban development tool, and the competition to host mega-events continues to widen and deepen. Mega-events are seen as a unique mechanism to improve a city's image and to trigger physical transformation which together can enhance their international competitiveness, and promote growth and regeneration

In parallel, 'sustainability' has become established as a necessary component of urban development strategies. Sustainability has thus become an essential part of the language of mega-event planning and delivery, and of mega-event legacy, particularly at the local level and especially following the International Olympic Committee's (IOC) adoption of the theme. This dates from the Centennial Olympic Congress in 1994, where concern for the environment was reflected in the creation of the IOC's Sport and Environment Commission whose work ensured the incorporation of the twin themes of environment and sustainability in the Olympic Charter in 1996, as the third Pillar of Olympism. The IOC defined its role in the Olympic Charter as 'to encourage and support a responsible concern for environmental issues, to promote sustainable development in sport and to require that the Olympic Games are held accordingly' (IOC, 1996:11). It also states that it is an IOC responsibility 'to promote a positive legacy from the Olympic Games to the host cities and host countries' (IOC, ibid.).

FIFA as organisers of the World Cup have followed. After the 2010 FIFA World Cup, which prompted concerns about the sustainability of the event, FIFA began to develop the 2014 FIFA World Cup Sustainability Strategy with discussions and meetings with the Local Organising Committee (LOC) and government representatives in Brazil in 2011.

The Challenge of Sustainability

Sustainability presents challenges for the mega-event. The mega-event and its legacy tend to centre on a property-led development project, and historically their appeal to cities has derived from the positive economic impact they are perceived to generate. By the early 2000's, the need to resolve potential tensions between sustainability and economic development was beginning to be acknowledged in national and international policy documents[1].

To deliver major property-led regeneration projects in designated locations with minimum delay, cities have popularly used powerful quangos[2], undertaking 'fast track' initiatives with uncomplicated objectives and limited local involvement or accountability, (Cochrane, 2007; Raco and Tunney, 2010). Sustainable development, in contrast, implies a carefully considered and calibrated process which engages with social and environmental as well as economic agendas, and involves local participation. In the UK, the term 'sustainable' fits in the 'agendas of inclusiveness, multiagency partnerships, and the shift from government to governance' (Jones and Watkins, 2008: 1416)

Nevertheless the successful delivery of the physical mega-event to brief, time and budget, remains its non-negotiable top priority, which in extremis has to trump all other considerations. Perhaps it is not surprising to find that London chose to create a delivery mechanism for London 2012, the ODA, or Olympic Development Agency, which was directly modeled on earlier conventional development corporations, such as the London

---

1   Tony Blair, British Prime Minister at the time is quoted in the UK Government's 'Securing for the Future' Report:'Development, growth, and prosperity need not and should not be in conflict with sustainability' (DEFRA, 2005).

2   Quasi-autonomous non-governmental organisations', separate to Government but funded and accountable to it.

Docklands Development Corporation. The organization set up by the London mayor to manage the legacy of 2012, the LLDC or London Legacy Development Corporation, has even greater local autonomy and powers than the ODA.[3]

The appraisal of the impact of mega-events is assisted by the new 'paradigm' of legacy which focuses on positive long-term impacts of events (Smith, 2012). Poynter and MacRury (2009b: 5) suggest that Olympic legacy 'offers bridges between two potentially divergent narratives setting the practical accountancy (and financial and political accountability) of city planning, against the 'creative' accounting that underpins Olympic dreams and promises'. The concepts of legacy and sustainability thus overlap in mega-event literature, and legacy creates the essential event hinterland in which sustainability becomes possible. The Commission for a Sustainable London (CSL), set up by the Mayor of London to provide assurance to the Olympic Board and the public on how the London 2012 event and legacy were meeting their sustainability commitments, began their 2010 Annual Review by saying 'We have always maintained that, taken in isolation, delivering and Olympic and Paralympic Games is an inherently un-sustainable thing to do. We therefore cannot call the programme truly sustainable unless the inspirational power of the Games can be used to make a tangible, far reaching difference' (CSL, 2010: 3). The forward to their final 2013 report 'Making a Difference' says 'On balance we believe there is sufficient evidence to conclude that sustainable practices inspired by the Games should outweigh the inevitable negative impacts of the Games over time' (CSL, 2013: 2).

Delivering Sustainability

Sustainability is now acknowledged as having not only an environmental dimensions but also social and economic dimension as well. It is about the integration and reconciliation of competing goals; the search for common ground which brings these issues into balance. As yet, however, the predominant practical focus of the policy documents that relate directly to mega-events has been environmental. The IOC's 2013 publication 'Sustainability through Sport' talks of sport presenting 'broad opportunities to promote environmental awareness, capacity building and far reaching action for environmental social economic development across society' (IOC, 2013: 5). How these concepts can be translated into operational and practical policy is elusive. The new international standard ISO 20121 now provides guidance to help event organisers map the events' economic, environmental and social impacts.

Perhaps it is not surprising then that, prior to 2012, sustainability has not been an explicit highlight of past host city strategies, and past Olympic performance and achievement in the field of sustainability is patchy (CSL, 2013). Relevant targets, in so far as they were set at all by organizers, have tended to centre on fairly obvious environmental outcomes. One of the most frequent sustainable goals, shared by most if not all events of all kinds, is the avoidance of 'white elephants' – infrastructure that has no further use after the event has finished. This relatively uncomplicated goal has proved disappointingly difficult to achieve. Horne and Manzenreiter's (2004) study on the World Cup 2002 showed that both South Korea and Japan ended up with under-utilised football stadia since interest in football is very low in these countries and they had not assigned any other use to the stadia. Burton (2003) observes that Olympic venues are attractive during the event time, but they can turn into ghost towns after the Olympics has left the city. A visit to the site of the Seville Expo is a dismal experience. As we write, conversation in Brazil, now the party is well and truly over, is returning angrily to the cost of stadia widely seen as superfluous to the countries far more visceral social needs.

London 2012 is the first summer host city to embed sustainability within its Games and its legacy programme from the beginning. This added complexity and potentially cost to the planning and development process, but nevertheless was not seriously questioned inside or outside the project. The UK Government even adopted 'The development of the Olympic Park as an exemplar of sustainable living' as one of its five 'Legacy Promises'.

CSL caution that 'Sustainability is driven by context and the Olympic and Paralympic Games taking place in different parts of the world will have different contexts, and therefore a different approach to sustainability'

---

3   The LLDC has additional powers to make local policy and plans, as well as simply take planning decisions, which in the UK's 'plan-led' system is very significant.

(CSL, 2013). What London however perhaps does demonstrate is how sustainability can be incorporated into the heart of the mega- event project by building around it a sustainable long term goal – in London's case, the social environmental and economic renewal of East London – that had been a policy priority, agreed and widely supported locally, for decades.

As the following chapters in this section demonstrate, green development goals that we would now describe as sustainable have, in practice, been features of the strategies for past mega-events. They demonstrate how hosting mega-events has enabled cities to plan for and achieve particular sustainable goals at the strategic level. Munich, in particular, is clearly overdue a reappraisal of its urban legacy.

References

Burton, R. (2003). 'Olympic Games host city marketing: An exploration of expectations and outcomes'. *Sport Marketing Quarterly,* 12, pp. 37–47.

Cochrane, A. (2007). *Understanding Urban Policy: A Critical Approach.* Oxford: Blackwell.

Commission for a Sustainable London (2010). Annual Review, http://www.cslondon.org/wp-content/uploads/downloads/2011/04/CSL-Annual-Review-20102.pdf; accessed May 12, 2014.

Commission for a Sustainable London (2013). Making a Difference, http://www.cslondon.org/wp-content/uploads/downloads/2013/03/CSL-Making-a-Difference-2013.pdf; accessed May 14, 2014.

DCLG (2008). Previously-developed land that may be available for Development: England 2007.

DEFRA (2005). *Securing the Future – UK Government Sustainable Development Strategy.* London: DEFRA.

Hall, C. M. (1992). *Hallmark Tourist Events: Impacts, Management and Planning.* London: Bellhaven Press.

Horne, J. D. and W. Manzenreiter (2004). Accounting for Mega-Events: Forecast and Actual Impacts of the 2002 Football World Cup Finals on the Host Countries Japan/Korea. *International Review for the Sociology of Sport,* 39, pp. 187–203.

International Olympic Committee (1996). Olympic Charter, Lausanne: IOC, http://www.olympic.org/Documents/Olympic%20Charter/Olympic_Charter_through_time/1996-Olympic_Charter.pdf; accessed May 12, 2014.

Jones, C. and C. Watkins (1996). 'Urban regeneration and sustainable markets'. *Urban Studies,* 33, pp. 1129–1140.

Newman, P. and J.R. Kenworthy (1999). *Sustainability and Cities: Overcoming Automobile Dependence.* London: Island Press.

Pacione, M. (2005). *Urban geography: A Global Perspective, Second. ed.* London: Routledge.

Poynter, G. and I. MacRury (2009). London's Olympic Legacy A 'Thinkpiece' report prepared for the OECD and Department for Communities and Local Government. London East Research Institute.

Raco, M. and E. Tunney (2010). 'Visibilities and invisibilities in urban development: Small business communities and the London Olympics 2012'. *Urban Studies,* 47, pp. 2069–2091.

Richards, G. and R. Palmer (2010). *Eventful Cities: Cultural management and Urban Revitalisation.* London: Butterworth-Heinemann.

Robinson, J. A. and R. Torvik (2005). 'White elephants'. *Journal of Public Economics,* 89, pp. 197–210.

Smith, A. (2012). *Events and Urban Regeneration: The Strategic Use of Events to Revitalise Cities.* London, New York: Routledge.

Tallon, A. (2013). *Urban Regeneration in the UK.* New York: Routledge.

Chapter 12
Creating Sustainable Urban Legacies? Olympic Games Legacies in Munich and London

Valerie Viehoff

Introduction

Creating sustainable legacies for the host city has become an increasingly important aspect for any city intending to host the Olympic Games, especially since positive legacies have been included as one of the core aims of the Olympic Movement, enshrined in the Olympic Charter. To justify significant public investment many host cities have started to develop holistic planning strategies to capture some of the value created through the hosting of the Olympic Games. Since creating lasting legacies has become essential for host cities, research on legacies of mega-events has burgeoned. Studies of Olympic legacies suffer, however, a bias in favour of a handful of case studies, e.g. Barcelona 1992, as the blueprint for successful urban regeneration. Other Olympic cities, such as Munich 1972, might also provide useful insights even though their urban development plans might not have employed the now so ubiquitously used labels "legacy" or "sustainability".

Following a brief introduction to the development of the ideas of sustainability and legacy in the context of the Olympic Games, we take a look back at the Munich Olympic Games of 1972 and the more recent London 2012 Olympic and Paralympic Games, analysing their promises, preparation and legacies. A particular focus lays on the Olympic Park and the Olympic village as the key sites of Olympionism, which have to be integrated into the existing urban fabric post-Games. The final part will outline some surprising similarities and fundamental differences between Munich 1972 and London 2012.

Sustainability and Legacy

While the IOC's definition comprises of five legacy categories: sporting, social, environmental, urban and economic (e.g. IOC, 2012, 2013a, 2013b) legacy discourses have been criticised for often focussing on purely economic issues, e.g. on the "returns on investments" of an event. This bias might stem from (event) management studies' rather narrow focus on economic legacies as criticised by Leopkey (Leopkey, 2008; Leopkey and Parent, 2012) or from an intentional discursive replacement of the term "Olympic brand" with the more palatable term "Olympic legacy", because of the latter's less obvious roots in marketing parlance (MacAloon, 2008). Community organisations (NEF, 2008) and urban researchers (e.g. Poynter, 2009) have warned that the predominant public (and academic) focus on the financial component of legacy poses the risk that other, less easily quantifiable or less tangible legacies (environmental, social, urban, sporting) will be marginalised, especially in the current climate of economic crisis and fiscal austerity (see also Scherer, 2011; Armstrong et al., 2011).

Although the urban legacy plans of London 2012 were amongst the most ambitious regeneration plans of any Olympic city yet, the idea of using mega-events as a catalyst for brown field regeneration is not new and was already applied in Munich 1972, Barcelona 1992, and Sydney 2000, to name just a few (e.g. Cashman, 2003, 2005; Chalkley and Essex, 1999; Essex and Chalkley, 1998, 2003; Toohey, 2008, 2010, 2012). A large amount of literature concerns Barcelona, which has become a role model for successful brownfield and waterfront regeneration (e.g. Abad, 1996; Brunet, 2009; Nello, 1997) although not uncontested (e.g. Blanco, 2009; Garcia-Ramon and Albet, 2000).

The legacies of Munich 1972 are seldom discussed, unless in the context of security issues (Cottrell, 2003) or from an architecture or urban design perspective (Heger, 2014; Modrey, 2008; Wimmer, 1976; Traganou, 2012) even though Munich's Olympiapark and former athletes' village are enjoying continuous popularity.

That Munich might represent a neglected example for the creation of sustainable urban legacies was already suggested by Geipel, Helbrecht and Pohl in 1993. In their opinion, Munich 1972 demonstrates "how an international mega-event can be successfully channelled towards local needs and interests and exploited to realise urban planning policies" (Geipel, Helbrecht and Pohl, 1993: 279). As early as 1979, Willi Daume, President of the Munich Olympic Organising Committee, had felt able to claim: "In Munich there are no 'Olympic ruins'" (Schiller and Young, 2010b: 227).

For the future of the Olympic movement it is of crucial importance that each successive Olympic Games is perceived a "success", in financial terms (see also: Preuss, 2003) as well as in terms of positive legacies left behind for the host city, region and country, in order to encourage sufficient numbers of candidate cities willing and capable of hosting the event for the IOC in the future. Therefore, the IOC had to continuously renew its appeal and adapt to changing social, political and economic circumstances. Following the Earth Summit in Rio de Janeiro in 1992, the environment was included as third pillar of Olympism in addition to sport and culture in 1994 (Chappelet, 2008) and in 1999 the IOC adopted its own Agenda21 with the aim to "encourage the members of the Olympic Movement to integrate sustainability principles into their operations" (IOC, 2014: 1).

At the beginning of the 21st century, "legacy" emerged as the new dominant agenda (see Leopkey, 2008; Leopkey and Parent, 2012), "which would rival and surpass sustainability as the guiding framework for considering urban outcomes" (Gold and Gold, 2013: 3530). Even though neither "legacy" nor "sustainability" was used by the organiser of Munich 1972, we contend that in essence, the motivations of Munich's mayor Dr Hans-Jochen Vogel and former mayor of London Ken Livingstone were similar: to mobilise (private) financial capital and tap into (public) funds for the long-term development of their cities that would not have been available without the mega-event (Vogel, 1972).

Munich 1972: Green, Compact and Joyful Games

The concept of the Olympic Games 1972 in Munich and the aims and goals it sought to achieve were centred on three themes:

1. *Spiele der kurzen Wege* – "Compact Games"
   In contrast to the Games of Tokyo 1964 and Mexico 1968 and in line with the ancient Greek model, Munich proposed setting the Olympic village and most sport facilities in a newly created Olympic Park with short distances to most venues.
2. *Spiele im Grünen* – "Green Games"
   For the Games a former airfield was transformed into a new park with a variety of sporting facilities. By connecting it with existing green spaces an uninterrupted chain of green parkland was created, providing centrally located, accessible green spaces for sporting and recreational purposes.
3. *Fröhliche Spiele* – "Joyful Games"
   The ambitious aim of the 1972 Olympic Summer Games was to televise the image of a "new Germany" around the world. The image of a nation that had changed since 1945 from a threatening, warmongering, fascist nation into a stable democracy, an economic powerhouse, a reliable partner in international politics and a friendly, joyful and open country to visit and live.

The Games helped to accelerate the implementation of a new urban development strategy (Heger, 2014) focussed on an extension of the public transport system, the creation of a car-free pedestrian zone in the city centre and the construction of urgently needed new housing. These improvements were aimed at increasing the local residents' quality of life rather than attracting foreign investment or visitors (Schiller and Young, 2010; Stadt München, 1970). The Games also functioned as catalyst for the creation of a completely new department for urban development planning in the city's administration.

"Legacy" Goals

Even though the word legacy never appears in the tendering documents, the organising committee still had a clear vision to create something that would last. From the very beginning of the bidding and application process plans were developed for the post-Games use of most venues:

> "In keeping with the Olympic Idea and the intentions of the IOC, nearly all the investments I have mentioned will be of lasting value even after the Games. For instance, the Olympic Village and the Press Village will be used to house families and students, the Press Centre will be turned into a department store, and the Radio and Television Centre will serve as a university sports centre". (Vogel, 1969: 3)

Oberwiesenfeld, a former airfield and brownfield site, provided the perfect location for the new Olympic Park. Measuring 280 hectares, the site featured some playing fields, a pile of debris from buildings destroyed in WWII, a television tower and an ice rink. Already in Munich's first urban development plan in 1963 the site had been designated as an area for recreational use and sport activities (Internationaler Arbeitskreis Sportstättenbau, 1967: 608; see also: Schiller and Young, 2010; Stadt München, 1970).

Apart from being "empty" and relatively close to the city centre, Oberwiesenfeld also had the advantage of being in public ownership. Munich's mayor recognised the opportunity to use the Olympic Games as a catalyst for the realisation of other projects and to advance his ideas of a more sustainable urban development, focused not only on economic growth, but also on quality of life. The main goals were (1) to extend and upgrade the public transport network, (2) to build new sport facilities, (3) to develop Munich's neglected (industrial) north, (4) to provide a new public park for sport, recreation and events and (5) to create a continuous ribbon of connected green spaces across the city

By aligning local (city of Munich), regional (Bavaria) and national (FRG) interests, the Organising Committee managed to reduce the financial burden for Munich. The Games hence became a joint project of the local, regional and national governments with equal financial contributions agreed by all three parties and representatives of all three tiers of government in the Olympic Organising Committee (Stadt München, 1972).

Urban Planning Ideals

The Games reflect urban planning ideals of 1960s Germany, in representing a final high point of modernist utopian beliefs in planning, technology, and architectural solutions to social problems, yet, the first influences of a new paradigm can already be detected that would start to cast doubt on the sustainability of the post-war economic system. In 1972 the Club of Rome published *The Limits to Growth*, initiating the first sustainability debates (Meadows et. al., 1972), and with the oil crisis of 1973 the previous growth-focused paradigm would come to a crushing halt.

After the end of WWII urban planning in Germany had been characterised by a period of historical reconstruction (*Wiederaufbau* 1945–1955), mainly focused on the alleviation of housing needs and strongly influenced by the concept of the "structured and low-density city" (*Gegliederte und aufgelockerte Stadt* 1950–1960). Inspired by the *Charta of Athens* and the concept of the *functional city* it was based on a separation of functions combined with extensive green spaces especially in residential areas.

New, functionally structured, low-density neighbourhoods with ample green-space were built at the urban fringe. This suburbanisation resulted in rapid urban sprawl, the loss of *urbanity* and an increase of individual road traffic (Heineberg, 2001: 122). The number of commuters from Munich's suburbs and the wider region into the centre, doubled between 1961 and 1970, while the use of public transport dropped from 56 per cent to 33 per cent (Dheus, 1972: 268). With the new concept called "car-adapted city" (*autogerechte Stadt)* Planners in Germany sought to adapt the traffic infrastructure to the increasing demand and to optimise traffic flows (1) by building new concentric ring roads with radial outlets and (2) by separating different road users (e.g. pedestrians and cars). Between 1960 and 1973 this new urban ideal of the car-friendly city was complemented by a new concept called *Urbanität durch Dichte* (urbanity through density) with new high-density developments surrounded by vast public green spaces being constructed on existing (peri)urban green spaces (see Heineberg, 2001).

Munich's Olympiapark and Olympic Village strongly refer to these planning ideals. The athletes' village in the northern part of the Olympic Park is a show-case example of a neighbourhood built according to the urban planning ideal of *urbanity through density* with clusters of high-rise blocks of 8 to 25 floors. The buildings are turned away from the busy ring road in to the north, opening up to the vast green spaces of the *Olympiapark*. A small shopping centre, a nursery, doctors' surgeries and a pharmacy complete the infrastructure of this high-density neighbourhood.

While housing was the key priority in the 1960s and 1970s, improving quality of life was also an important public aim and this included a detailed strategy to improve provision of sport facilities. The so called *Golden Plan*, invented in the early 1960s (Deutsche Olympische Gesellschaft, 1962) empowered municipalities and local authorities to calculate their need for sport facilities, e.g. public swimming pools or sports grounds.

The Olympic Park and Village

Based on a series of guiding principles (e.g. designated legacy uses for all venues and the park itself, short distances between the athletes' accommodations and the venues, etc.), a competition for the Olympic Park and Village was opened in 1967. The winning bid by Günter Behnisch and Partner from Stuttgart was then, in an unusual and legally contested way, combined with the plans for the Olympic village by third-ranked architects Heinle, Wischer and Partner.

The three outstanding elements of the final proposal were (1) a new landscaped park (with an artificial lake and a man-made Olympic hill) into which the various sporting venues would be embedded, (2) a tent-like roof construction over parts of the Olympic stadium and (3) a high-density Olympic village that combined a variety of different building shapes, including small cubic cottages, large court-yard bungalows, a central high-rise tower and several rows of staggered mid-rise buildings – providing mixed-use housing after the Games (see also: Krämer, 1970, 1972).

The concentration of Olympic venues within a compact Olympic Park, with walkable distances between the majorities of venues, was one of the innovative contributions of the 1972 Olympics (see Table 12.1).

The Olympic village was planned for the north-eastern corner of the park within easy walking distance of most sporting venues. The Olympic stadium and various new venues were embedded into a new artificial landscape in the southern part of the park with an open space near the lake in the southern part intended as an open plaza and event space for "manifold activities". The most famous and most contested detail of the winning proposal by was the technically challenging, avant-garde, tent-like roof construction (see Figure 12.1).

What is colloquially called the "Olympiapark" actually consists of various distinct parts: the central Olympic compound containing the Olympic stadium, the Olympic village in the north-east, the former press centre in the far north-west (separated from the rest of the park by a road and railway tracks), the university sport fields and halls (including the former radio and television centre) in the north, and the actual "parkland" in the south. The entire site of the former Oberwiesenfeld measures 280ha, yet the Olympic park as such (excluding the Olympic village and the university sport fields) covers only 140 ha (Landeshauptstadt München, 2008).

Legacies of Munich 1972

The Olympic Park and its Sport Venues

Even before the first construction work started on the Olympic site, plans had been developed for the post-Olympic use of the park, the Olympic village, the press and media centres and most sporting venues. The key concept for the Olympic Park in its post-Olympic phase or – in today's terms – "legacy mode", was the creation of a public, green, and accessible space for sport, leisure, recreation, relaxation and entertainment.

Table 12.1 **"Games of short walks" – Distances of Olympic venues from the Olympic Stadium**

Facility/Venue	Distance from Stadium
Sports hall (Sporthalle)	250 m
Aquatics Centre (Schwimmhalle)	300 m
Velodrom	440 m
Radio and TV Centre	570 m
Volleyball hall (Volleyballhalle)	650 m
Olympic Village	700 m
Hockey fields	900 m
Subway Station	1000 m
S-Bahn Station	1000 m
Water Polo (Dante-Freibad)	1000 m
Media Centre (Pressezentrum)	1050 m
Tram Station	1700 m
Archery (Bogenschießanlage)	2500 m
Dressage (Schloß Nymphenburg)	4000 m
Olympic Centre at Munich Exhibition Centre	5000 m
Rowing (Regattastrecke Oberschleißheim)	7000 m
Show Jumping (Reitanlage Riem)	12000 m
Sailing (Kiel)	900 km

Source: Own table, based on information retrieved from Harrenberg 1971.

Designed by the landscape designer Günther Grzimek, the park was intended as a "democratic green", as a distinctly urban park that could be freely appropriated as Grzimek explained: "The Olympic landscape is different from traditional parks (…). It intends and tolerates the visual inclusion of the city. It provokes an urban feeling and attitude" (Haus der Bayrischen Geschichte, 2010: 45).

According to the *Konsortialvertrag* (Olympic agreement) signed in July 1967 the city of Munich would become the official owner of the Olympic park and the key sporting venues and would hence also be responsible for any legacy funding. In 1970, two years before the start of the Games, the *Münchner Stadion GmbH* (later renamed Olympiapark GmbH) was created with responsibility for the post-Games management of the park and its venues. Officially a "limited company" the Olympiapark GmbH benefits nevertheless from public backing and financial support. It continues to manage the Olympiapark – especially the acquisition of sponsoring contracts and income generating events. According to the initial Olympic agreement the legacy costs should be borne by all three signatory parties, yet, to avoid any long-term involvement the national government contributed a one-off payment of 130 million DM, based on their calculation of the "additional

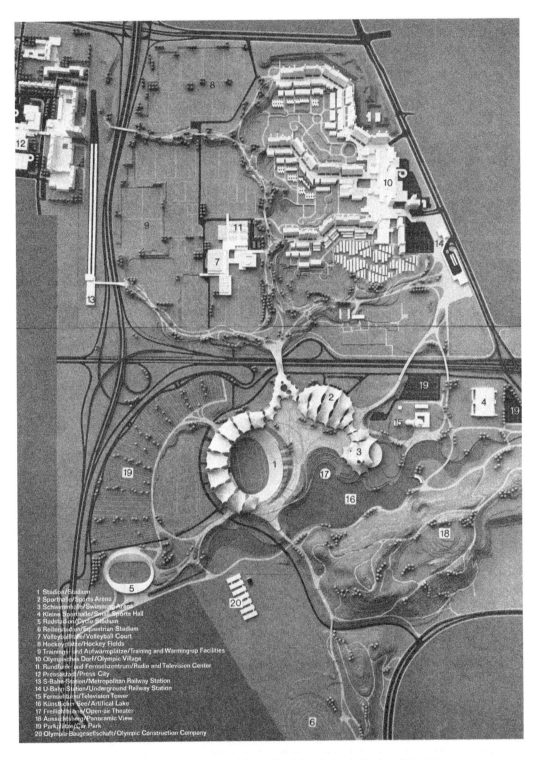

1 Stadion/Stadium
2 Sporthalle/Sports Arena
3 Schwimmhalle/Swimming Arena
4 Kleine Sporthalle/Small Sports Hall
5 Radstadion/Cycle Stadium
6 Reiterstadion/Equestrian Stadium
7 Volleyballfläche/Volleyball Court
8 Hockeyplätze/Hockey Fields
9 Trainings- und Aufwärmplätze/Training and Warming-up Facilities
10 Olympisches Dorf/Olympic Village
11 Rundfunk- und Fernsehzentrum/Radio and Television Center
12 Pressestadt/Press City
13 S-Bahn-Station/Metropolitan Railway Station
14 U-Bahn-Station/Underground Railway Station
15 Fernsehturm/Television Tower
16 Künstlicher See/Artificial Lake
17 Freilichtbühne/Open-air Theater
18 Aussichtsberg/Panoramic View
19 Parkplätze/Car Park
20 Olympia-Baugesellschaft/Olympic Construction Company

Figure 12.1   Photograph of the architectural model of the Olympic Park in Munich
Source: Archiv Heinle, Wischer und Partner, Freie Architekten.

Olympic costs" over a period of 25 years. Bavaria's contribution was made in the form of land transfers worth around 31 million DM (Deutscher Bundestag, 1975).

By leveraging and investing these initial funds prudently and thanks to the steady income from the football clubs (10 per cent of their revenues from ticket sales, plus income from sideline advertising), the Olympiapark GmbH was able to manage the park without any subsidies until the mid-1990s, aided, however, by restrictive laws minimising competition from other venues. The new football stadium *Allianz Arena* is, for instance, only allowed to host football, hence guaranteeing the Olympic stadium a monopoly position for hosting concerts or other non-football events.

The move of Munich's Bundesliga football clubs to a new venue has had a significant financial impact leaving a large deficit in the Olympiapark GmbH's budget. In spring 2013 the city of Munich increased its annual contribution to 40 million € per year (*Süddeutsche Zeitung*, 12 October 2013). The city's hopes of generating funds by hosting the 2018 or 2022 Winter Games in Munich (Landeshauptstadt München, 2008) were dashed when the Olympic candidacy plans were rejected in a public referendum. It remains hence unsolved how the estimated 460 million € required for refurbishments over the next 20 years will be financed.[1]

The key factors for the successful legacy management of the Olympiapark until the mid-2000s were, firstly, the early creation of the legacy body, secondly, the comfortable start-up budget of the legacy institution, and thirdly, the steady income from long-term lease contracts with premier league football clubs.

Munich's Olympic Village

The Olympic village, designed by Heinle, Wischer & Partner had been planned with clear post-Games use in mind. The "village" consists of three rows of medium high-rise buildings with staggered balconies, spreading out "like fingers on a hand", a cluster of high-rise buildings (19 floors high) at the centre and a mix of different court-yard bungalows, townhouses, terraced houses and mini-apartment cubes, scattered between the three main blocks. Spaces between the buildings were car-free. A network of underground access roads and car parking provided direct car access to each home. The nearby tube station and the provision of safe spaces for cyclists and pedestrians meant to incentivise a car-free life – a rather novel idea in car-loving Germany in the 1970s.

Initially the Olympic village seemed to be a financial disaster for the six consortia (a mix of housing associations and commercial developers), who had financed and owned the Olympic village. Only about 800 of the 3,000 homes had been sold at the end of 1972. Yet, the mood soon changed, people discovered the child-friendliness of the car-free Olympic village and by the end of 1975 all homes had been sold (Heinle, Wischer and Partner, 1980: 27). A different case altogether were the student accommodations located (a) in one of the high-rise towers and (b) in the small cubic 1-up-1-down bungalows of the former female Olympic village, which had been built by the *Studentenwerk* of the University of Munich before the Games. The "Ollydorf" remains Munich's most favourite student residence, with long waiting lists for the 1853 flats (801 in the high rise and 1052 in bungalows).

Although family-friendly, the new homes in the Olympic village (in their post-Games use) provide, however, no social housing. In 1972 the Olympic village was even criticised for becoming a "ghetto for the rich" (SPIEGEL, 17 February 1972: 65). Already during the planning process the architects were reluctant to meet the then existing norms regarding minimum spatial standards for social or subsidised housing.[2] They worried about increasing the size of the Olympic village and thereby dwarfing their prestige projects, the Olympic park and the sporting venues (Heger, 2014: 94–95).

A legacy of social "housing for low- and middle class families" (Bernstock, 2014: 5) was created, in the former press and media village, which was converted into 760 subsidised flats after the Games. However, in 1998 the tower blocks were sold by the housing association to a private company and subsequently converted into private ownership flats, with only very few preserving a rent-cap guarantee or a secure-tenancy agreement.

---

1   It still remains difficult to obtain details of the Olympiapark financial details and the company's financial operations, despite the fact that the Olympiapark GmbH was ordered to reveal certain details of their budget in reply to a freedom of information request by a journalist in 2006.

2   e.g. DIN 283, which prescribed min. 16m² per double bedroom.

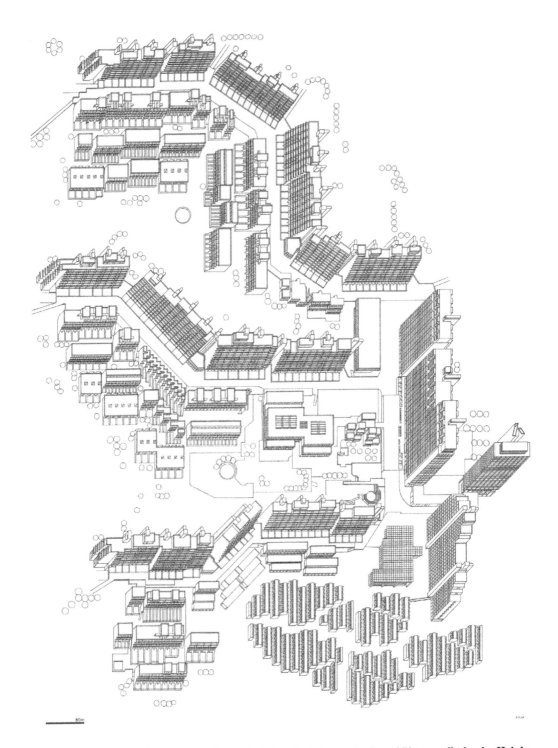

80m

Figure 12.2   Isometric representation of the Olympic Village, Munich 1972; overall plan by Heinle, Wischer und Partner

Source: Archiv Heinle, Wischer und Partner, Freie Architekten.

Table 12.2 **Number of bedrooms and distribution of housing units in the Olympic village post-Games**

Number of bedrooms per housing unit	Percentage of housing units built
1 bedrooms	5%
2.5 bedrooms	25%
3 bedrooms	25%
4 bedrooms	40%
4.5 bedrooms	4%
5 bedrooms	1%

Source: Own table, based on information from Heinle, Wischer & Partner 1980.

Olympic and Paralympic Games, London 2012

Legacy Promises

The initial legacy promises of London's bid, which was prepared under the aegis of the then national government of the UK (Labour party), the mayor of London, the national Olympic Committee and various sporting bodies, contained a strong focus on sustainability and on creating a lasting legacy from the Olympic and Paralympic Games. When a new national government (coalition of Conservatives and Liberal Democrats) came to power in 2010, the legacy promises were amended (DCMS, 2010: 1) and sustainability was dropped – with reference to the Olympic Park as a "blueprint for sustainable living" as well as in the broader sense of delivering sustainable Games (see Table 12.3).

Planning for the Games and for Legacy

London created a body with responsibility for delivering the promised legacies about two years before the Games. The Olympic Park Legacy Company (OPLC) was jointly owned and reported to the Mayor of London and the UK national government (GLA, 2010: 15). Even though London's legacy institution was created early, its work was initially complicated by the fact that it existed "within a complex network of organisations, one that includes overlapping responsibilities and lines of accountability" (GLA, 2010: 19). Soon after the *Localism Act* (November 2011) was introduced by the new coalition government, granting the Mayor of London additional powers, the mayor Boris Johnson identified east London as "London's single most important regeneration project for the next 25 years" (GLA, 2012) and converted the OPLG into the London Legacy Development Corporation (LDDC) under his direct control.

Since October 2012 the LLDC has assumed planning powers from local authorities, from the LTGDC and from the Olympic Delivery Authority (ODA) for an area covering and extending beyond the borders of the Olympic Park.

The Olympic Park

The Queen Elizabeth Olympic Park in east London stands in the tradition of both, Olympic Parks worldwide (including Sydney 2000, Barcelona 1992 and Munich 1972) and public parks in Great Britain, which, since their earliest days in the 1840s, have embraced "social, health and amenity objectives" (Hopkins and Neal, 2013: 26). Their purpose was to improve the health and quality of life especially of the urban poor, living in overcrowded accommodation, considered "unhygienic", while at the same time providing a new "asset" to a

Table 12.3   Legacy promises related to London 2012

Questionnaire for cities applying to become Candidate cities ("pitch book", January 2004)	London 2012 Bid (presentation in Singapore, 6th July 2005)	Tessa Jowell, Minister for Culture, Media and Sport (DCMS, 2007)	Ken Livingston, Mayor of London (January 2008) endorsed by new Mayor	New Coalition Government (DCMS, 2010)
• The Olympic Games (…) would enhance sport in London and the UK forever. (…) The nation will be healthier, happier and more active. • (…) will be the most powerful catalyst imaginable for the regeneration of one of our most underdeveloped areas. It will accelerate the most extensive transformation seen in London for more than a century. Tens of thousands of lives will be improved by new jobs and sustainable new housing, new sports venues and other facilities (…) There will be a real and long-lasting legacy (p. 1). The village (…) [will be built] as a sustainable long-term development in line with London-wide housing strategies, it will have guaranteed legacy use as affordable housing for key occupations such as teachers and medical personnel (p. 12). • Environmental quality and sustainability will be the cornerstones of London 2012 (p. 23).	• (…) our legacy will be immediate. The Olympic arena will become the home of the London Olympic institute, a new world centre of sporting excellence. Each of the venues has an agreed and clear long-term future. Each of the venues already has a 25-year business plan in place. (…) London is committed to a sporting legacy for Britain and to a far reaching legacy for the Olympic movement. • The Games guarantee this regeneration will create a community, where sport is an integral part of everyday life. This is a model for 21st century living and it is the embodiment of the philosophy of Pierre de Coubertin. The park will be an environmental showcase. The Games will dramatically improve the lives of Londoners.	• Make the UK a world-leading sporting nation. • Transform the heart of East London. • Inspire a generation of young people to take part in local volunteering, cultural and physical activity. • Make the Olympic Park a blueprint for sustainable living. • Demonstrate the UK is a creative, inclusive and welcoming place to live in, visit, and for business" (DCMS, 2007, p. 4).	• increase opportunities for Londoners to become involved in sport. • ensure Londoners benefit from new jobs, businesses and volunteering opportunities. • transform the heart of east London. • deliver a sustainable Games and developing sustainable communities. • showcase London as a diverse, creative and welcoming city. (GLA, 2009, p. 7).	• Harnessing the United Kingdom's passion for sport to increase grass roots participation, particularly by young people – and to encourage the whole population to be more physically active. • Exploiting to the full the opportunities for economic growth offered by hosting the Games. • Promoting community engagement and achieving participation across all groups in society through the Games; and • Ensuring that the Olympic Park can be developed after the Games as one of the principal drivers of regeneration in East London.

neighbourhood, making it a more attractive place to live and to invest in, reflected in raising property prices in the surrounding neighbourhoods (Hopkins and Neal, 2013: 27).[3]

The name Queen Elizabeth Olympic Park (QEOP) refers to the entire area of 560 acres (226 hectares) – equivalent to Hyde Park in London – that was used as the "Olympic Park" during the Games (see: LLDC 2014), i.e. the perimeter to which access was restricted and ticketed. It hence also includes areas, where new neighbourhood will be built over the next 30 years as part of the post-Games *Legacy Communities Scheme* (LCS).

As part of the post-Games transformation of the park temporary venues (e.g. Water-polo Stadium, Basketball Arena) were removed and other venues converted to their post-Games use (e.g. Aquatics Centre, Copper Box Arena, Velodrome, Stadium and Press and Broadcast Centre). The northern part of the park re-opened in July 2013 and most parts of the southern park opened in the summer of 2014.

West Ham United Football Club will be the anchor tenant for the Olympic Stadium, although the stadium will also be used for other events, including concerts, the Rugby World Cup 2015 and the Athletics World Championships 2017. The management of the park and the sport venues has been outsourced partly to Balfour Beatty Workplace (park and ArcelorMittal Orbit) and partly to Greenwich Leisure Limited (GLL), a social charitable enterprise (Aquatics Centre and Copperbox).

The Athletes' Village

The planning application submitted by the LLDC in 2011, before the LLDC itself subsumed planning powers for the site, included provision for up to 6,800 residential units for the post-Games phase called *Legacy Communities Scheme* (LCS), featuring a site-wide indicative quantum (i.e. not necessarily in each Planning Delivery Zone) of 42 per cent family housing (3 bedrooms and larger), a tenure split of 65 per cent market housing and 35 per cent affordable housing, whereby the "affordable housing" would be divided into 60 per cent social rented and 40 per cent intermediate housing (LLDC, 2011: 23).

Table 12.4   Family housing – Projected site wide provision

Family Housing	Total number of units	Number of family housing units	Family housing as % of tenure	Family housing as % of scheme
Market	4,432	1,778	40%	26%
Intermediate	930	149	16%	2%
Social rented	1,412	911	65%	13%
Total	6,775	2,839		

Source: Own table, based on information from LLDC 2011.

Since the end of the Olympic and Paralympic Games, the former athletes' village, which housed around 11,000 athletes and officials during the Games, has been transformed into apartments, marketed under the new name 'East Village London E20'. In 2011 the Olympic village had been sold (see also Boykoff 2014) to a consortium of Qatari Diar and Delancey (51 per cent) trading under the name *Get Living London*[4] and Triathlon Homes (49 per cent), a consortium of the development and investment company First Base and two housing associations. Chobham Manor, situated in the northern corner of the Park, will be the next neighbourhood to be completed after the Olympic village, with residents expected to start moving in from

---

3   See also Matthew Gandy's work (Gandy 2002) on real estate speculation related to the creation of Central Park in New York.

4   Their legal commitment to let the properties, rather than to sell them, will expire in eight years (House of Lords, 2013a: 63).

2015. This neighbourhood will feature a mix of building types (e.g. town and mews houses, maisonettes, and single flats) of which 70 per cent are designated as family homes, (see Plate 12.1).

In addition, four more new neighbourhoods are planned for the park: *East Wick* (near Hackney Wick, in the north-west), *Sweetwater* (south-west), *Marshgate Wharf* (in the centre of the park) and *Pudding Mill* (south-eastern tip of the park).

Governance Structures and Financial Arrangements

The key organisation in charge of the legacy of the Olympic and Paralympic Games 2012 in London is the LLDC, who inherited all land previously acquired by the London Development Agency (LDA).

The initial budget of the LLDC was provided from public funds, including national (e.g. the Public Sector Funding Package for the London 2012 Games, PSFP) and local (e.g. GLA) budgets (see LLDC, 2013: 27). Capital receipts are generated by disposing of sites for the development of new neighbourhoods by private developers. Further revenues will be generated via residential estate service charges, Crossrail licence fees, rent from the lease of the Press and Broadcast Centre, the Olympic stadium, and other venues (LLDC, 2013: 22).

While these revenues are expected to increase over time, the LLDC will be facing what has euphemistically been called a "timing issue", because, according to the LLDC's own projections, financial self-sufficiency will not be achieved until 2018, yet public funding was scheduled to expire in 2014–15 (GLA, Budget and Performance Committee, 2013: 10). The two major risks potentially jeopardising the realisation of the LLDC's legacy objectives are the lack of financial resources (GLA, Regeneration Committee, 2013a) and the fact that

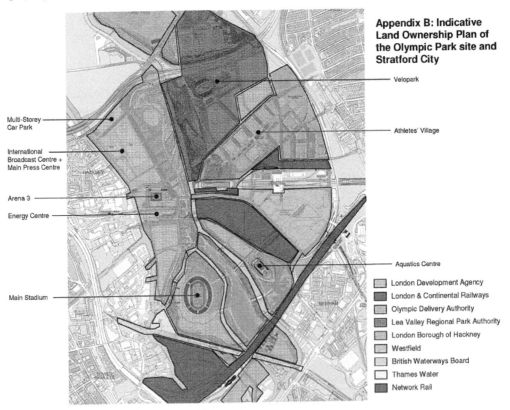

Figure 12.3   Indicative Landownership Plan of the Olympic Park and Stratford City
Source: House of Commons, Culture, Media and Sport Committee, 2010.

the LLDC will only have a limited life-span, as London's mayor Boris Johnson declared in 2013: "I do not believe that the LLDC should be immortal. (…) I would think that it will collapse like a booster rocket back to earth in about 10 years' time or perhaps even sooner" (House of Lords, 2013b: 348).

The LLDC's responsibilities would then fall to the host boroughs, while the "mayoralty" – as the "body with the democratic legitimacy in London" – would remain responsible for the co-ordination (House of Lords, 2013b: 348). Yet, neither the responsibilities nor the funding have been agreed so far. Similarly, the UK government's pledge to support the Queen Elizabeth Olympic Park financially as published in their *National Infrastructure Plan*, provides neither a timeframe nor an amount (HM Treasury, UK Government, 2013:10).

Discussion: Similarities and Differences

Despite the slightly uneven pairing – London with a population of about 8.3 million vs. Munich with 1.4 million – and, expectedly, many differences, a look back at the Olympic Games of Munich 1972 and London 2012 reveals some surprising similarities.

Similarities

The most striking similarities between Munich 1972 and London 2012 are their early and strong commitments to creating sustainable (urban) legacies, especially with regards to providing new housing and green spaces, their preference for "compact Games" in a new Olympic Park and, finally, their decision to use relatively centrally located "under-developed" brownfield sites.

The historical and social reasons may be different, yet the urgent need for more (affordable) housing was a problem that London in 2012 shared with Munich 40 years earlier. Faced with growing populations and a lack of available homes, both cities were also keen to prevent further suburbanisation and instead chose urban densification, i.e. the filling of existing "gaps" in the urban fabric.

Addressing the housing crisis via densification was one of the reasons for Munich and London to choose a former airfield (Oberweisenfeld) or an area dominated by derelict industrial sites (Stratford and the Lea Valley). The other reasons were the desire to create new green spaces and the hope that these "under-valued" spaces would generate considerable profits via subsequent land sales once the sites had been regenerated, fitted with improved infrastructure, received a new name (E20) and image and profited from some advertising by means of the Olympic Games.

Both sites are located relatively close to the heart of the city – about 4km in Munich and 8km in London – and both are, thanks to the improvements in the transport infrastructure, conveniently reached by public transport. Even though these infrastructure upgrades and new public transport networks (e.g. a new tube line and the creation of pedestrian zones in the city centre in Munich or the upgrade of Stratford Station and the completion of Crossrail in London) were planned before the Games, their completion was sped up by the tight timeframe of the Olympic preparations. Owing to the differences in size, "compact Games" in Munich meant that most venues were within 20 minutes walking distance from the athletes' accommodations, while "compact Games" in London meant that athletes could reach their competition venues in maximum 30 minutes (by car or shuttle).

Finally, both Olympic cities (initially) had the ambition to create good quality, family-friendly homes in a setting that would provide access to green spaces, foster healthy lifestyles and become a blueprint for sustainable living. If defining "family homes" as accommodation with 3 bedrooms or more, then the Olympic village in Munich delivered 70 per cent family homes. The prospected average percentage of family homes across the entire LCS in London is 41 per cent, with significantly higher levels in Chobham Manor.

In Munich, the Olympic village's radical innovation was the strict separation of car traffic from pedestrians and cyclists. By banning all cars to a below-ground network of access roads and parking, safe and child-friendly spaces were created on the pedestrianised ground level. As planning experiment and model town of Modernity (Heger, 2014) the Olympic village in Munich was a bold and contested statement; an attempt to prove that it is possible to build high-density, high quality neighbourhoods, where people will love to live. In London, the Olympic village is built to very high sustainability requirements. Yet, London 2012 seems to have lacked the ambitiousness of some other Olympics. An opportunity was missed to create a new model neighbourhood that would impress with experimental, outstanding, extraordinary or avant-garde features or innovations.

The Olympiapark in Munich and the Queen Elizabeth Olympic Park in London provide venues for sport and other events set within new urban parklands and they link previously existing, but disjointed green spaces. Munich's Olympiapark became part of a chain of parks stretching east-west and along the river Isar. London's Queen Elizabeth Olympic Park links the Lea estuary and the River Thames in the south to the existing Lea Valley Park stretching north into Hertfordshire.

The two cities also share certain problems and challenges, pronounced to different degrees, but symptomatic when it comes to creating a new residential area from scratch. Linking the new homes to existing neighbourhoods, weaving them into the urban fabric to create a continuous urban landscape, is a significant challenge. Firstly, the residents tend to form a self-selecting ghetto, socio-economically different from the surrounding neighbourhoods. The new residents of Munich's Olympic village, for instance, were over-proportionally well educated, in skilled jobs, earning above average, with young children (Heinle, Wischer and Partner, 1980), predominantly of the same age-cohort. It is likely, but due to the early stages of the process not yet empirically tested, that similar socio-economic and demographic patterns will apply in London.

Secondly, geographical barriers (e.g. the River Lea and the A12 in London, busy ring roads in Munich), cut the new neighbourhood off from the surrounding land, disconnecting them from neighbouring communities, hence reinforcing potentially existing "psychological/emotional barriers" (GLA, Regeneration Committee, 2013c: 35; House of Lords, 2013a and 2013b). While the problem of infrastructure networks providing connections between certain places for some citizens, while cutting off and segregating others, is not unique to newly built neighbourhoods (see: Marvin and Graham, 2001), the negative effects might be particularly pronounced in the case of the new Olympic neighbourhoods and could further complicate the growing together of new and old neighbourhoods.

Differences

Urban planning in 1960s Germany, then still a new discipline, was characterised by a strong belief in the welfare state (social capitalism) and in a public responsibility to provide affordable housing, sporting facilities, green spaces for recreation. It was generally expected that these services and utilities would be publicly financed and publicly accessible. The aim was to create socially just cities, reflected in the legal protection of tenants and their rights, including a cap on rent increases, which partly still persists in Germany today.

This stands in stark contrast to the situation in the UK, marked by the economic crisis of 2008, and the recent rolling out of neoliberal policies, austerity measures, cuts to public services and social benefits etc. With regards to the Olympic Park and the Olympic village these "anti-welfare state policies" of the Conservative-Liberal coalition government will, for instance, influence the socio-demographic composition of the new neighbourhoods and the question of long-term (public) funding to secure legacies.

Conclusion

London 2012 and Munich 1972 intended to create "sustainable urban legacies" by integrating the Olympic Games into long-term urban development strategies. Even though no legacy concept existed in Munich in 1972, detailed plans were in place for the post-Games use of the athletes' accommodation, the Olympic park and most sport facilities. Munich's legacy plans were grounded in the idealistic urban planning ideals of the late 1960s with their belief in technical solutions for social problems typical of modernity.

In London, the terms "legacy" and "sustainability" played an important role in the preparation of the bid. Yet, while London was extremely successful in developing and imagining positive legacies, the implementing (e.g. via urban masterplans) of the promised legacies has proven more difficult, possibly, because Great Britain has not a particularly strong tradition of creating and respecting legally binding masterplans or land-use plans, instead offering (private) developers rather flexible guidelines (Meyer and Schlaich, 2012: 51).

The experience of Munich shows that creating lasting legacies from mega-events requires long-term financial commitment and governance structures. Yet, even in Munich, where legacy plans, a legacy institution and legacy funding were in place at an early stage and for the long-term, financial problems occurred. Since the departure of Premier League football, public subsidies have been necessary. Without the city's continuous financial commitment it would not have been possible to preserve and maintain the "Olympic ensemble", which is a designated heritage site since 1998 (Bergande and Schulze, 2014).

Considering that in London the LLDC as the body responsible for delivering the promised legacies of London 2012, is only equipped with limited funding and granted a limited life-time, the challenges to delivering long-term sustainable Olympic legacies for east London are obvious. This lack of long-term financial plans and governance structures has prompted a House of Lords Select Committee on Olympic Legacies to conclude that they were "unconvinced that the Government's current oversight arrangements represent a robust way to deliver the legacy" (House of Lords, 2013a: 5).

Despite a surprisingly vast array of obvious similarities on the surface – from the site and sight of the park to the sometimes identical rhetoric of regeneration and long-term urban benefits – this chapter has also shown that the Olympic Games in Munich 1972 and London 2012 were born out of different historical, political and urban planning contexts and were grounded in almost fundamentally opposed urban models and societal ideals for our urban future.

Acknowledgement

This research was supported by a Postgraduate Grant (2013) of the Olympic Studies Centre of the IOC, Lausanne, granted jointly to Dr Holger Kretschmer, University of Cologne, and Dr Valerie Viehoff, University of East London.

References

Abad, J. M. (1996). 'Olympic village, city and organisation of the Olympic Games. The experience of Barcelona'92', in Moragas, M., M. Llines and B. Kidd (eds.). *Olympic Villages. Hundred Years of Urban Planning and Shared Experiences*. International Symposium on Olympic Villages. Lausanne: IOC, pp. 15–20.

Armstrong, G., D. Hobbs and I. Lindsay (2011). 'Calling the shots: The pre-2012 London Olympic contest'. *Urban Studies*, vol. 48, no. 15 (November 2011), pp. 3169–3184.

Bergande, B. and K. Schulze (2014). 'Der Olympiapak – Ein "Gebrauchsgegenstand" unter Denkmalschutz?', in Hennecke, S., R. Keller and J. Schneegans (eds.) (2013). *Demokratisches Grün – Olympiapark München*, pp. 201–205.

Bernstock, P. (2014). *Olympic Housing. A Critical Review of London 2012's Legacy*. Farnham: Ashgate.

Blanco, I. (2009). 'Does a "Barcelona model" really exist? Periods, territories and actors in the process of urban transformation'. *Local Government Studies*, Volume 35, Issue 3 June 2009, pp. 355–369.

Boykoff, J. (2014). *Celebration Capitalism and the Olympic Games*. London and New York: Routledge.

Brunet, F. (2009). 'The economy of the Barcelona Olympic Games', in Poynter, G. and I. MacRury (eds.) (2009). *Olympic Cities: 2012 and the Remaking of London*. London: Ashgate, pp. 97–109.

Cashman, R. (2003). 'What is "Olympic legacy"?', in M. Moragas, et al. (eds.) (2013). *The Legacy of the Olympic Games 1984–2000*. Lausanne: IOC, pp. 31–42.

Cashman, R. (2005). *The Bitter-Sweet Awakening: The Legacy of the Sydney 2000 Olympic Games*. Sydney: Walla Walla Press.

Chalkley, B. and S. Essex (1999). 'Urban development through hosting international events: A history of the Olympic Games'. *Planning Perspectives,* vol. 14, pp. 369–394.

Chappelet, J.-L. (2008). 'Olympic environmental concerns as a legacy of the Winter Games'. *International Journal of the History of Sport*, Vol. 25/14 (December 2008), pp. 1884–1902.

Cottrell, R. (2003). 'The legacy of Munich 1972: Terrorism, security and the Olympic Games', in M. de Moragas, et al. (eds.) (2003). *The Legacy of the Olympic Games 1984–2000*. Lausanne: IOC, pp. 309–313.

DCMS (2007). Our Promise for 2012. How the UK will benefit from the Olympic Games and Paralympic Games. Forword by Tessa Jowell. Available online at: https://www.gov.uk/government/publications/our-promise-for-2012-how-the-uk-will-benefit-from-the-olympic-games-and-paralympic-games

DCMS (2010). Plans for the legacy from the 2012 Olympic and Paralympic Games (December 2010), available online at: https://www.gov.uk/government/uploads/system/uploads/attachment_data/file/78105/201210_Legacy_Publication.pdf

Deutsche Olympische Gesellschaft (1962). Der Goldene Plan in den Gemeinden: ein Handbuch. Memorandum zum "Goldenen Plan" für Gesundheit, Spiel und Erholung. Wiebelsheim: Verlag Limpert.

Deutscher Bundestag (1975). Unterrichtung durch die Bundesregierung betr. Gesamtfinanzierung der Olympischen Sommerspiele 1972. Drucksache Drucksache 7/ 3066, Sachgebiet 7, (9 January 1975). Available online at: http://dipbt.bundestag.de/doc/btd/07/030/0703066.pdf (accessed January 2014)

Essex, S. and B. Chalkley (1998). 'Olympic Games: Catalyst of urban change'. *Leisure Studies*, vol. 17, pp. 187–206.

Essex, S. and B. Chalkley (2003). Urban transformation from hosting the Olympic Games. University lecture on the Olympics, Barcelona: Centre d'Estudis Olímpics (UAB). Available online at: http://olympicstudies.uab.es/lectures/web/pdf/essex.pdf

Gandy, M. (2002). *Concrete and Clay: Reworking Nature in New York City*. Cambridge, Mass./US: MIT Press

Garcia-Ramon, M.-D. and A. Albet (2000). 'Comment: Pre-Olympic and post-Olympic Barcelona, a "model" for urban regeneration today?' *Environment and Planning A*, vol. 32, pp. 1331–1334.

Geipel, R., I. Helbrecht and J. Pohl (1993). 'Die Münchner Olympischen Spiele von 1972 als Instrument der Stadtentwicklungspolitik', in H. Häußermann and W. Siebel (eds.) (1993). *Festivalisierung der Stadtpolitik. Stadtentwicklung durch große Projekte*. LEVIATHAN, Sonderheft 13 (1993), Wiesbaden: Springer Fachmedien, pp. 278–304.

GLA (2009). Towards a Lasting Legacy. A 2012 Olympic and Paralympic Games Update. Prepared by the Economic Development, Culture, Sport and Tourism Committee, July 2009.

GLA (2010). Legacy Limited? A review of the Olympic Park Legacy Company's role. Produced by University of East London for the Economic Development, Culture, Sport and Tourism Committee, available online: http://www.uel.ac.uk/londoneast/documents/FINAL_EDCST_LegacyLimitedEMBARGOED.pdf

GLA (2012). Olympic Park Legacy Corporation – Proposals by the Mayor of London. Webpage, available online at: https://www.london.gov.uk/priorities/planning/consultations/olympic-park-legacy-corporation-proposals-mayor-london (accessed August 2014).

GLA, Budget and Performance Committee (2013). Appendix 2 to the Minutes of the meeting on 22nd October 2013. Transcript of Agenda Item 6: London Legacy Development Corporation. Available online at: http://www.london.gov.uk (accessed April 2014).

GLA, Regeneration Committee (2013a). The work of the LLDC, one year on (December 2013).

GLA, Regeneration Committee (2013b). The work of the LLDC, one year on. Appendix 1: Report of the LLDC

GLA, Regeneration Committee (2013c). The work of the LLDC, one year on. Appendix 2: Transcript of Item 6: London Legacy Development Corporation.

Gold, J. and M. Gold (2013). '"Bring it under the legacy umbrella": Olympic host cities and the changing fortunes of the sustainability agenda'. *Sustainability*, Vol. 5, pp. 3526–3542.

Harrenberg, B. (1971). München '72. Bilder und Daten zu den Olympischen Spielen. München.

Haus der Bayrischen Geschichte (ed.)(2010). Edition Bayern, Sonderheft 2: München '72.

Heger, N. (2014). *Das Olympische Dorf München: Planungsexperiment und Musterstadt der Moderne*. Berlin: Reimer Verlag.

Heineberg, H. (2001). *Stadtgeographie*. Stuttgart: UTB.

Heinle, Wischer und Parter – Freie Architekten (1980). *Eine Stadt zum Leben. Das Olympische Dorf München*. Freudenstadt: Heinrich Müller Verlag.

HM Treasury, UK Government (2013). National Infrastructure Plan 2013 (Published December 2013). Available online at: https://www.gov.uk/government/uploads/system/uploads/attachment_data/file/263159/national_infrastructure_plan_2013.pdf

House of Commons, Culture, Media and Sport Committee (2010). Olympic and Paralympic Games 2012: Legacy. Written evidence submitted by the Olympic Park Legacy Company.

House of Lords, Select Committee on Olympic and Paralympic Legacy (2013a). Keeping the flame alive: the Olympic and Paralympic Legacy. Report of Session 2013–14, House of Lords Paper No. 78, published 18 November 2013.

House of Lords, Select Committee on Olympic and Paralympic Legacy (2013b). Evidence Volume; Corrected Oral and Written evidence on Olympic and Paralympic Legacy; published 16 July 2013.

International Olympic Committee (2003). The Legacy of the Olympic Games: 1984–2000, Joint Symposium, IOC Olympic Studies Centre and Olympic Studies Centre at the Autonomous University of Barcelona, November 2002, Lausanne.

International Olympic Committee (2012). Sustainability through Sport. Implementing the Olympic Movement's Agenda 21. Lausanne: IOC.

International Olympic Committee (2013a). Olympic Charter in force as from 9 September 2013. Lausanne: IOC.

International Olympic Committee (2013b). Olympic Legacy. Available online: www.olympic.org/Documents/Olympism_in_action/Legacy/2013 Booklet_Legacy.pdf (accessed Jan 2014).

International Olympic Committee (2014). Factsheet: The Environment and Sustainable Development (update January 2014), available online: http://www.olympic.org/Documents/Reference_documents_Factsheets/Environment_and_substainable_developement.pdf (accessed June 2014).

Internationaler Arbeitskreis Sportstättenbau e.V. (ed.) (1967). Wettbewerb Olympische Bauten München 1972. *Sportstättenbau und Bäderanlagen* (SB), Vol. 67 (November 1967), Sonderheft, München.

Krämer, K. (ed.) (1970). Architekturwettbewerbe (AW) Internationale Vierteljahresschrift. Olympische Bauten München 1972 (December 1970), 2. Sonderband Bestandsaufnahme Herbst 1970".

Krämer, K. (ed) (1972). Architekturwettbewerbe (AW) Internationale Vierteljahresschrift. Olympische Bauten München 1972 (July 1972), 3. Sonderband Bauabschluß Sommer 1972.

Landeshauptstadt München, Referat für Stadtplanung und Bauordnung (2008). Entwicklungsplanung Olympiapark 2018: Materialsammlung und Grundlagenworkshop. Available online at: http://www.muenchen.de/rathaus/dms/Home/Stadtverwaltung/Referat-fuer-Stadtplanung-und-Bauordnung/Projekte/Rahmenplanung-Olympiapark/Olympiapark_2018_Grundlagenworkshop.pdf

Leopkey, B. (2008). The Historical Evolution of Olympic Games Legacy. Final Report submitted as part of the IOC's Olympic Studies Centre's Postgraduate Grant. Available online at: http://doc.rero.ch/lm.php?url=1000,44,38,20090831145942-FR/Leopkey_Becca_2008.pdf

Leopkey, B. and M. Parent (2012). 'Olympic Games legacy: From general benefits to sustainable long-term legacy'. *The International Journal of the History of Sport*, vol. 29, no. 6, pp. 924–943.

LLDC (2011). Planning application, Legacy Communities Scheme: Housing and Social Infrastructure Statement September 2011 (LCS-GLB-ACC-HSIS-001). Available online at: https://planningforms.newham.gov.uk/online-applications/files/749C1A57EE7C5C161746C2FD967779BF/pdf/11_90621_OUTODA-HOUSING_AND_SOCIAL_INFRASTRUCTURE_STATEMENT_PART1–309.pdf

LLDC (2013). Three Year Business Plan 2013/14 – 2015/16 (published October 2013), available online at: http://queenelizabetholympicpark.co.uk/~/media/qeop/files/public/lldcbusinessplanv42012_132015_161.pdf (accessed August 2014)

LLDC (2014). Facts and figures. Webpage, available online at: http://queenelizabetholympicpark.co.uk/media/facts-and-figures (accessed August 2014).

MacAloon, J. (2008). '"Legacy" as managerial/magical discourse in contemporary Olympic affairs'. *International Journal of the History of Sport*, December (14), pp. 2060–2071.

Meadows, D., Meadows, H., Randers, D. L., Behrens, J. W. III (1972). *The Limits to Growth*. New York: Universe Books.

Meyer, F. and C. Schlaich (2012). Zwischenräume achten. London 2012. Interview with Stephen Taylor Architects. In *Bauwelt*, 24/2012, pp. 50–51.

Modrey, E. M. (2008). 'Architecture as a mode of self-representation at the Olympic Games in Rome (1960) and Munich (1972)'. *European Review of History: Revue européenne d'histoire*, vol. 15, no. 6, pp. 691–706.

Moragas, M. de, M. Llines and B. Kidd (eds.) (1996). Olympic Villages. Hundred Years of Urban Planning and Shared Experiences. International Symposium on Olympic Villages. Lausanne: IOC.

Nello, O. (1997). 'The Olympic village of Barcelona '92', in De Moragas, M., et. al (eds.) (1997). *Olympic Villages: Hundred Years of Urban Planning and Shared Experiences*. Lausanne: IOC, pp. 91–96.

New Economics Foundation (NEF) (2008). Fools Gold. How the 2012 Olympics is selling East London short, and a 10 point plan for a more positive local legacy (written by J. Ryan-Collins and P. Sander-Jackson); available online at: http://www.neweconomics.org/sites/neweconomics.org/files/Fools_Gold.pdf

Poynter, G. (2009). 'London 2012 and the reshaping of East London', in Imrie, R., L. Lees and M. Raco (eds), *Regenerating London*. London: Routledge, pp. 132–151.

Preuss, H. (2003). 'Rarely considered economic legacies of Olympic Games', in Moragas, M. de, C. Kennett and N. Puig (eds.) (2003). *The Legacy of the Olympic Games 1984–2000*. International Symposium, Lausanne, 14–16 November 2002. Lausanne: IOC, pp. 243–252.

Scherer, J. (2011). 'Olympic villages and large-scale urban development: Crisis of capitalism, deficits of democracy?' *Sociology*, vol. 45, no. 5, pp. 782–797.

Schiller, K. and Young, C. (2010). *The 1972 Munich Olympics and the Making of Modern Germany*. Berkley: University of California Press.

Stadt München (ed.) (1970). Olympia in München. Offizielles Sonderheft 1970 der Olympiastadt München. München: Münchener Leben GmbH.

Stadt München (ed.) (1972). Olympia in München. Offizielles Sonderheft 1972 der Olympiastadt München. München: Münchener Leben GmbH.

Süddeutsche Zeitung (12 October 2013). Wie es für den Olympiapark weitergehen könnte (by Silke Lode). Available at: http://www.sueddeutsche.de/bayern/zukunftsplaene-in-muenchen-wie-es-fuer-denolympiapark-weitergehen-koennte-1.1793359 (accessed 12 January 2014).

Toohey, K. (2008). 'The Sydney Olympics: Striving for legacies – Overcoming short-term disappointments and long-term deficiencies'. *The International Journal of the History of Sport*, 25, no.14, pp. 1953–1971.

Toohey, K. (2010). Is Sydney still an Olympic city? Report produced for the IOC's Postgraduate Grant Selection Committee.

Toohey, K. (2012). Olympic Sustainability Reporting and Sport Participation. Paper written as part of the IOC's Postgraduate Grant Selection Committee.

Traganou, J. (2012). 'Foreword: Design histories of the Olympic Games'. *Journal of Design History*, Vol. 25 No. 3, pp. 245–251.

Vogel, H. J. (1969). Speech by the Mayor of the City of Munich, Dr. Hans-Jochen Vogel, to the International Olympic Committee in Warsaw on June 8, 1969 (IOC archives, record no. 14297, JO-19723 – RAPPO, SO3: Rappel de Willi Daume à la session de Varsovie 1969, appendix 9).

Vogel, H.-J. (1972). Die Amtskette. Meine 12 Münchner Jahre. Ein Erlebnisbericht. München: Süddeutscher Verlag.

Wimmer, M. (1976). *Olympic Buildings*. (Translated from the German original by Herbert Liebscher) Leipzig: Edition Leipzig.

Chapter 13

Sport Mega-events as Catalysts for Sustainable Urban Development: The Case of Athens 2004

Constantinos Cartalis

Introduction

Mega-events provide a strong opportunity for the sustainable urban development of the host city. The larger the city, the easier it is for the mega-event to be incorporated into its long term planning. The smaller the city, the more impact a mega-event may have in terms of shaping the city's future structure.

The critical challenge is whether planning adjusts the city to the needs of the mega-event or uses the mega-event as a catalyst for sustainable urban development. In the former case the benefits from the mega-event expire shortly after its completion and legacy is limited. In the latter case, benefits are lasting and the post-event legacy is important as well as supportive for the sustainable transformation of the city. The scale of the Olympic Games makes it unique amongst mega-events both in terms of the complexity of the event and its potential to influence urban development and promote sustainability. This chapter focuses upon Athens (2004). It describes the planning guidelines, argues that a legacy of sustainable urban regeneration for the city of Athens and the region of Attica was achieved and presents a number of preconditions to be met for urban legacies to be maximized and sustained.

Urban Development as Related to the Olympic Games

Urban development impacts of the Olympic Games are usually not confined to sporting venues, but also include airport, port, road and public transport improvements, new hotels, new commercial centers, upgrading of communication and other technological systems, new parks and public spaces and recreation and leisure spaces. Such wider infrastructure investment has the potential to generate multiplier development in the post-Olympic period (Gold and Gold, 2008). At the same time, the type and scale of urban development generated by the Olympic Games, depends also on public funding, public/private partnerships, market demand, the urban governance model and the prevailing political, economic and social conditions (Preuss, 2007; Tziralis et al, 2006). It also depends on several key parameters that potentially shape the nature and scale of development in the post-Olympic period.

The first parameter is the spatial model of Olympic infrastructure. In some cities the centralized model was selected with sporting venues concentrated in a main Olympic pole (Sydney 2000, London 2012). Conversely, there have been Olympic Games where the decentralized, 'scattered model' was selected, resulting in the spread of the Olympic venues to various parts of the urban web of the host city (Barcelona 1992, Athens 2004, Beijing 2008). The second parameter is the nature of the Olympic infrastructure itself. In some cities, Olympic Games have largely used existing sports facilities (Los Angeles 1984) or, where new sports facilities were built, they were in their majority of a temporary character (Atlanta 1996). There have also been Olympic Games that invested in new sports facilities (Athens 2004, Beijing 2008, London 2012) and/or stimulated urban development beyond the Olympic sports facilities themselves (Sydney 2000, Athens 2004, London 2012).

A third parameter is the land ownership pattern in areas potentially most affected by Olympic infrastructure. The existence of publicly owned land within the urban area facilitates the integration of Olympic projects (and related capital investments) into the city; on the contrary, the lack of such areas will limit the potential for extensive urban regeneration projects and increase the financial burden of the Olympic event on the public purse, because the necessary expropriations usually incur high costs due to the elevated market demand for available areas. A fourth parameter is the urban governance context. In cities where the private sector leads major development proposals, achieving post-Olympic development might be easier due to market driven forces. In other cities with a more public sector–led development framework, post-Olympic development may be more difficult due to the need for additional public funding as well as due to political changes at the level of the central or local governments which may affect the design and/or the implementation of post-Olympic planning.

Organizing the Olympic Games 2004

In Athens, the social, economic and political conditions continued to change throughout the Olympic preparation period, as a function of a) the fiscal policy of the Government (which resulted in the decrease of the ratio of public debt to Gross Domestic Product – GDP below 100 per cent by 2003), b) the effort of the State to become a member of the Eurozone and thus shift from the vulnerable "drachma" to the stable "euro" (2001) and c) the opposition to OG2004 by a number of political parties and NGOs. Greece was the smallest country ever to organize the Olympic Games apart from Finland, where the Games were held in 1952, when the organizational needs of the Games were still significantly lower, with the number of athletes, for instance, being only 4,955. In Greece, the Olympic projects were mostly based on public funding as the experience in public-private partnerships was limited.

Furthermore, Greece was the first country to organize the Olympic Games as a member of the European Union, i.e. with European legislation fully in force. Although Spain in 1992 (Olympic Games of Barcelona) was also a member of the European Union, this was still during a period of legislative derogations, a fact which allowed the organizers to, among other matters, apply simpler procedures for the construction of the venues. Sydney also simplified the appeal procedure regarding construction works so as to expedite their construction schedules.

Planning for Legacy

In June 1999, the main Olympic Law for the 2004 Olympic Games (hereinafter referred to as OG2004) was ratified by the Parliament of Greece; it provided a detailed description of the locations of the Olympic sporting venues (and the most important non-sporting venues) and also defined the post-Olympic use of each venue. More importantly, the law linked the Olympic constructions to the revised (in 1999) masterplan of Athens, in an obvious effort to use OG2004 as a catalyst for the urban regeneration of the capital city of Greece.

The Olympic Law also introduced a new planning tool, namely the "Special Plan for the Integrated Development of Olympic Poles" (which was thereafter applied with considerable success for critical, in development terms, areas of Athens, such as the coastal zone of the city) and was also complimented by a thorough study (Ministry of Environment, 1999) for the spatial and urban planning of Athens. The usefulness of the latter was proved in the course of the Olympic preparation, when over fifty appeals against Olympic works and projects submitted by citizens or NGOs at the level of the Council of State were rejected on the basis that venues were compatible to the priorities of the revised masterplan of Athens.

One critical issue was the mix of permanent and temporary facilities. In Athens, the majority of venues were constructed as permanent ones so as to enhance the legacy of OG2004 in a city with several infrastructural needs. To avoid the problem of "*white elephants*" (i.e. Olympic facilities of large size being left under-used in the post-Olympic period), several of the Olympic sporting venues in Athens were constructed in a manner so as to allow their reduction in size and spectator capacity in the post-Games period (Olympic Aquatic Center, Olympic Boxing Center, Olympic Baseball Center, Olympic Softball Center, Olympic Hockey Center, etc.). Others were designed so they could be converted after the Games; for example the Taekwondo and Handball Olympic venue, in the coastal zone of Athens, was from the beginning designed in a way that it could be converted to a Conference Center after the Games, thus supporting the urban regeneration plan for the coastal zone of Athens as well as city tourism.

It should be mentioned that the share of new Olympic works corresponding to permanent facilities, accounted for 87 per cent of the total. This share is higher compared to other Olympic host cities, yet demonstrates the strategic choice of the Greek state to enhance Olympic legacy and improve urban infrastructure. In practical terms a set of five criteria was defined and applied for each Olympic intervention/action/project (Table 13.1).

Table 13.1   Sustainability criteria for Olympic interventions/actions/projects

Sustainability criteria for Olympic interventions/actions/projects
Criterion 1. Does the intervention/action/project result in: a) Long lasting value b) Return on investment
Criterion 2. Does the intervention/action/project create business interest and could the business interest be considered as being sustainable?
Criterion 3. Does the intervention/action/project upgrade: a) the natural environment b) the urban environment c) the cultural environment
Criterion 4. Does the project support: a) Urban regeneration b) Social needs at the local or wider scale c) Social capital
Criterion 5. Does the intervention provide: a) New employment opportunities b) Positive impact on the GDP and such economic indicators as public debt,   public deficit, economic growth rate

Sustainable Urban Development and the Olympic Games 2004

The Sites of Olympic Venues

In OG2004, the choice of sites of Olympic sporting venues followed a decentralized pattern (see Plate 13.1), with the main venue being the Olympic Sports Complex (Olympic Stadium, Olympic Velodrome, Olympic Tennis Center, Olympic Aquatic Center, Olympic Basketball Center) located in Marousi (a suburban area, 7 kms north of the center of the Athens) and the remaining sporting venues scattered in various parts of the urban agglomeration of Athens (Greece, General Secretariat for the Olympic Games – Ministry of Culture, 2002). Such a pattern was considered beneficial to the city (Beriatos and Gospodini, 2003).

In particular, the reasoning behind this planning pattern was: a) to develop a main Olympic pole, b) to establish local development poles in degraded parts of the city c) to link OG2004 to the spatial planning and development priorities as defined by the revised masterplan of Athens, especially with respect to the coastal zone of the city and d) to develop a network of sporting facilities of high standards, with the capacity to operate, after the Games in a coherent and complimentary manner. New non-sporting Olympic venues were also constructed or existing facilities were upgraded in various parts of the wider urban agglomeration of Athens with a view to supporting the requirements of OG2004 as well specific public needs.

Transport and Road Network

The decentralized pattern chosen for the locations of Olympic sporting and non-sporting venues was supported by the upgrade of the transportation and road network of Athens. The upgrade was based on public funding, not a part of the Olympic budget, and was provided by the central government and from the structural funds of the European Union. It included a brand new underground railway (metro) that now connects more than 20 municipalities in the wider urban agglomeration of Athens, the construction of the suburban rail which linked the (new) Athens International Airport to the city centre, to several suburban areas of the city and to three adjacent cities, a brand new tram railway (23 km in length) connecting the city center to the southern suburbs of Athens and 120 km of new highways in the urban agglomeration of Athens serving, to varying extent, all fifty Municipalities (Greece, General Secretariat for the Olympic Games – Ministry of Culture, 2004).

As a result, all Olympic (sporting and non-sporting) venues were fully supportive of the transportation needs of athletes, officials and spectators; areas of the city became fully accessible as compared to the period prior to OG2004 when they were poorly interlinked, a solid shift to mass transit systems was observed (resulting in more than 250,000 passengers per day) at the expense of the use of private cars and the traffic conditions were improved (a reduction, on average, of two hours of stationary traffic per day). The Olympic Games can also be credited with the successful reengineering of the public transport system and the road network as the event provided considerable thrust and a fixed deadline for the completion of all the related projects (Tziralis et al., 2006). To this end, the legacy of OG2004 in achieving the sustainable urban development of Athens was not only testified, verified and considered as highly significant.

The transport plan for OG2004 as well as for the post-Olympic period is presented in Plate 13.1 (Greece, Athens Organizing Committee, 2005). For comparison reasons, the Athens' public transport network in 1998, only consisted of the dark green line in the centre of the illustration!

Environment

The preparation for OG2004 was based on a thorough strategic environmental impact assessment (EIA) for the wider urban agglomeration of Athens as well as on EIAs by venue, in full compliance with European legislation. The preparation proved beneficial for the environment in Athens as several new open spaces (a total area of 650 hectares) were gained for Olympic and post-Olympic exploitation, whereas a number of buildings along the coastal zone were demolished to improve public access to the sea. More importantly, the improvement of the traffic conditions and the reduction in the use of private cars in favor of mass transit systems, resulted in the gradual improvement of air quality (Greece, General Secretariat for the Olympic Games-Ministry of Culture, 2004).

Special reference should also be made to the plan for the environmental protection and upgrade of the wetland (also a Natura 2000 site) in the Marathon area through the construction of the Olympic Rowing Center (Figure 13.1). In particular, the rowing canal coincided with the runway of an airport which was closed down so as to protect the wetland, with the overflow of water from the canal to the wetland secured its needs for water resources. Despite the strong environmental dimension and the innovative character of the plan, considerable dispute arose due to claims that the site of the Rowing Center coincided with the one where the battle between Athenians and Persians took place in 490 BC. The dispute was terminated following a thorough geophysical survey of the area which proved that the area was a lagoon in ancient times and thus could not have served as a battlefield. In addition, excavations by the Archaeological Service in the area of the Rowing Center confirmed the geophysical survey as no remains were found (Greece, General Secretariat for the Olympic Games-Ministry of Culture, 2003).

Urban Regeneration Projects

Three major urban regeneration projects (Greece, General Secretariat for the Olympic Games – Ministry of Culture, 2003) were promoted with the potential to support the sustainable urban development of Athens in the post-Olympic period: the unification of the archaeological sites of Athens, the regeneration of the coastal zone through the hosting of Olympic sporting and non-sporting venues and its conversion, in the post-Olympic period to a coastal park with a mix of public and private facilities, and the conversion of the area of the former airport of Athens to an Olympic pole, at a first phase, and to a metropolitan park after OG2004.

The first project for the unification of the archaeological sites around the Acropolis (including the conversion of streets around the Acropolis into pedestrian zones) was completed in 2003 and provided a "breath-taking promenade" through classical antiquities in the historic part of the city of Athens (Figure 13.2). The project reflects a major urban legacy for Athens and contributed to the sustainable urbanization of the centre of Athens in the post-Olympic period, as a new public space was developed. This new space now hosts

Figure 13.1   Olympic Rowing Center in the wetland of Marathon

Figure 13.2   Unification of archaeological sites around the Acropolis

a wide variety of social and cultural events and is frequently visited by domestic and international tourists. The operation, since 2009, of the new Acropolis Museum in the same area as well as the construction of a station of the new Metro of Athens next to the Museum, further enhanced the sustainability of the project.

The second project related to the coastal zone of Athens and was organized in two phases: a) the Olympic phase (Figure 13.3) which included the removal of the racehorse track field (it was moved to the new Olympic Equestrian Centre to the southeast of Athens), the construction of two Olympic sporting venues (Olympic Taekwondo and Handball Center and Olympic Beach Volleyball Center) and the development of two marinas for sailing and motor boats and b) the post-Olympic phase which aimed at the overall regeneration of an area of 80 hectares to become a new park for Athens at its coastal front. The project, in terms of its Olympic planning, was completed by the end of 2003 developing an important urban legacy, practically a new public space for the city by its coastal waterfront. The post-Olympic plan was, however, only partially realized. New momentum was given to the project in 2010 by the decision of the Greek State to convert part of the area to a cultural park hosting the new Library of Athens and the National Opera. The construction of both venues, on the basis of the architectural plans of Renzo Piano and with private funding as provided by the Niarchos Foundation, are to be concluded by 2015/16.

Figure 13.3   Urban regeneration plan for the coastal zone of Athens – Olympic phase

In terms of the third project for the area of the former airport of Athens (along the coastal zone of Athens, size of 530 hectares, i.e. 3.5 times as big as Hyde Park in London or 1.5 times as big as Central Park in New York), the plan was to free the area from any aviation or other uses so as to be converted to a major Olympic pole hosting several Olympic venues (Olympic Hall for basketball, Olympic Fencing Center, Olympic Baseball Center, Olympic Softball Center, Olympic Hockey Center and Olympic Canoe/Kayak Slalom Center) and linked to the Olympic Sailing Center at the adjacent coastal area (Figure 13.4) The planning was completed early; the new Olympic pole has proven highly supportive for the needs of OG2004 and demonstrated its capacity to become a major (new) metropolitan park and development area for Athens in the post-Olympic period. Despite the fact that several of the Olympic venues were used for sporting or recreational needs following OG2004, the finalization and implementation of an integrated exploitation plan experienced major delays, mostly for political reasons. The area was finally included (2011) in the Government's privatization program and in 2014 a concession agreement was signed between the State and a major private firm for the long term exploitation of the area with mixed residential, business, entertainment and recreational uses.

Figure 13.4   The Olympic Sailing Center; in the post-Olympic period it was converted to a tourist marina

Tourism

Tourism, in the wider agglomeration of Athens as well as in neighboring prefects, enjoyed a massive boost due to OG2004 (see also Table 13.4). In 2000, a long lasting ban regarding the construction of new hotels in Attica was lifted resulting in the construction of twenty-three new hotels.

In addition, the vast majority of existing hotels (categories two stars and above), were upgraded to be used for the accommodation needs of OG2004 (Olympic family and domestic and international visitors); to this end the Government supported the upgrade with a subsidy of 3,000 euros per room on the precondition that at the completion of the works, all hotels (and 10 per cent of the rooms per hotel) would be accessible to people with disabilities. Furthermore, the operation of the new airport of Athens and the expansion of the port of Piraeus to host three cruise ships during OG2004 for the needs of the Olympic accommodation supported, in a sustainable manner, the touristic needs of the city in the post-Olympic period.

Culture

An important pillar of OG2004 was the *"Cultural Olympiad 2001–2004"* which included more than one hundred high level performances and events in Greece and abroad, under the unifying theme *"for a Culture of Civilizations"*. In addition, the historic landscape of Athens was linked to OG2004 as the Olympic cycling event took place in the area around the Acropolis (Figure 13.5) and Olympic archery in the Panathinaikon Stadium (built in the third century B.C. and renovated to host the first modern Olympic Games in 1896). Furthermore, an extended program for the upgrade of the main museums in Athens was initiated, supporting plans for the post-Olympic legacy of the city.

Figure 13.5   Olympic cycling event around the Acropolis

Post-Olympic Use

Tables 13.2 and 13.3 summarize the post-Olympic use for OG2004 (Greece, General Secretariat for the Olympic Games – Ministry of Culture, 2004). In particular, Table 13.2 provides the exploitation status as divided into five main categories (Transportation, Road works, Accommodation, Sporting, Health), whereas Table 13.3 refers to the post-Olympic use of non-sporting Olympic venues.

Cost or Investment?

In most cases, the discussion following the organization of Olympic Games concentrates on the final cost of the event. The discussion is complicated by the fact that a unified approach or agreement about the projects to be included in the final cost is often missing and there is a tendency to also include the cost of capital

Table 13.2   Post-Olympic use per category

Category	Exploitation after OG2004
Olympic Transportation Projects	Full by 2004.
Olympic Road Works	Full by 2004.
Olympic Accommodation projects (Olympic Village, Media Villages, Villages for Referees and Judges)	Gradual. Full by 2007.
Olympic Sporting Venues	Gradual. Post-Olympic use still to be resolved for some venues.
Olympic Health Projects	Full by 2004.

Table 13.3   Post-Olympic use per Olympic (non-sporting) venue

Olympic venue	Post-Olympic use	Remark
Olympic Village	Social workers	In use as of 2005
Olympic Village – International zone	Institute for Geology and Mineral Exploration	In use as of 2007
Olympic Village Polyclinic	State Hospital	In use as of 2005
Olympic Media Village close to Olympic Stadium	Ministry of Education	In use as of 2007
Olympic Media Village (close to the Olympic Rowing Center – Marathon)	Summer camps for the employees of the Ministry of Health, the Ministry of National Defense and the Municipality of Athens	In use as of 2005
Olympic Media Villages (private projects)	Private housing	In use as of 2005
Olympic Media Villages in the University of Athens and the Technical University of Athens	Student dormitories	In use as of 2005
Olympic accommodation for referees and judges	Police Academy	In use as of 2005
International Broadcast Center	Shopping Mall	In use as of 2006
Main Press Center	Ministry of Health	In use as of 2007

investments, which would have been incurred even if the Olympic Games had not been awarded to the host city (Kitchin, 2007).

The Budget of OG2004

Several publications (BBC, 2004; Christian Science Monitor, 2004, 2008; Business Insider, 2010; Olympic Time, 2012) refer to the Olympic Games of Athens as being among the most expensive ones, with the cost soaring to 15 billion euros. There is no publication however which provides any justification with respect to the above mentioned cost; instead, these publications are based on speculation and disregard official documents (Greece, Ministry of National Economy, 2005; Staikouras, 2012; Stournaras, 2013) as ratified or presented at the Parliament of Greece.

It should be clarified that the cost of OG2004 does not include, *according to the international practice,* capital investments and works which although they were linked to the overall improvement of the capital city of Athens (and were also beneficial for OG2004), they were planned by the Greek State prior, and independently, to the assignment of the OG2004 to Greece. Although these capital investments (Attica highway, suburban rail, urban regeneration projects, unification of archaeological sites and upgrade of public hospitals), which improved the capacities of the host city, did gain momentum due to OG2004 and even though many believe that they would have never have been completed without the Olympic Games, the respective expenditures should not be added in the category "Olympic expenditures". Despite the above clear distinction, some analysts erroneously add the cost of these capital investments (from 1.5 to 2.5 billion euros depending on the public works chosen for inclusion) to the final cost of OG2004.

Taken that the budget of the Organizing Committee Athens 2004 (ATHOC) was balanced (expenditures-revenues) at the amount of 1.962 billion euros, the discussion focuses on expenditures from the public purse divided into four distinct categories:

a. Olympic works linked to the obligations resulting from the bid file,
b. works or projects, which despite the fact that there were not obligations of Greece for OG2004, were considered necessary for the smooth organization of the event,
c. expropriations, technical studies, personnel as hired by public entities for the needs of OG2004, etc.,
d. Olympic security, as they were formulated in the aftermath of 9/11 (from roughly 0.4 billion euros prior to 9/11 to 1.1 billion euros after 9/11).

Taking the above, the cost for OG2004, as far as public expenditure is concerned, accounts for 4.93 billion euros (including approximately 0.6 billion euros in VAT). The Olympic Games also generated approximately 1.1 billion euros in income for the Greek State reflecting: the return of VAT for Olympic projects/works as funded by the State, tax corresponding to the income of the employees in the Athens Organizing Committee and to income of hired personnel explicitly for Olympic works, tax corresponding to the profits of the companies (construction, commercial, touristic, etc.) which undertook Olympic projects and Olympic construction works and, finally, tax relating to the profits of Olympic sponsors.

Assessing the Catalytic Effect of OG2004

A number of publications (for example, BBC, 2004) claim that OG2004 were responsible for the explosion of the public debt of Greece and the economic crisis of the country. At the completion of the OG2004, the rate of public debt to Gross Domestic Product (GDP) was roughly 100 per cent. During the period 2006–2009, the respective percentage increased considerably reaching 129.7 per cent in 2009 and 148.3 per cent in 2010. In particular, in 2009 and 2010, the public debt corresponded to 299.7 and 329.5 billion euros respectively (Greece, Ministry of National Economy, 2010).

Taken that the cost of OG2004 is estimated at 4.93 billion euros, the respective share to the public debt of 2009 and 2010 is 1.64 per cent and 1.5 per cent respectively, a fact which clearly demonstrates that despite the preparation and organization of the Olympic Games required considerable funding, the Games cannot be considered responsible for the economic crisis of Greece. Even if the cost of OG2004 was assumed to be 15 billion euros, as some publications erroneously claim, the respective share to the public debt of 2010 would still only have been approximately 4.5 per cent. It should be also mentioned that the amount of 4.93 billlion euros was spread over a period of seven fiscal years (1998–2005) and corresponds to 0.4 per cent of the aggregate GDP for the period 1998–2005.

In terms of assessing the catalytic effects of Olympic Games, a special indicator is used, namely the ratio of total expenditure in context activities (i.e. activities related to the OG2004 but not being obligations from the bid file) to total expenditure on Olympic activities. The higher the ratio, the greater the catalyst effect of the Games. In the case of OG2004, 41.1 per cent of the expenditures were related directly to OG2004, whereas the remaining amount was investments contributing to the Olympic legacy. In particular, the indicator acquires the value of 1.44 which implies that the collateral activities triggered by the hosting of OG2004, were

1.44 times higher in cost as compared to the activities in direct relation to the preparation and organization of OG2004 (Tziralis et al., 2006). It is important to note that according to the OECD (2004):

'Athens is benefiting from investments for the 2004 Olympic Games but it needs clear strategic planning to take advantage of the opportunities that globalisation and eastward expansion of the European Union will bring. Organising the Olympic Games has proved to be a unique challenge not only for Greece's capital city but for the entire national administration. ... Preparations for the Olympic Games in August 2004 and financing from EU Community Support Funds have boosted investment in the hotel sector, year-round sports facilities and a modern region-wide transport network. This includes a brand new international airport, urban highways and ring roads to decrease congestion, upgraded rail links, a new metro, a non-polluting bus fleet, and tramway lines which connect the city centre and the suburbs. A programme to enhance architectural heritage and environmental assets has transformed central Athens and the area around the Acropolis. Like Barcelona, Athens now boasts easy access to a landscaped coastal zone at Faliron which offers a wide range of leisure and sports activities'.

The Impact of OG2004 on tourism

Of particular importance is the examination of the impact of OG2004 on tourism in the post-Olympic period. Table 13.4 (Greece, SETE, 2012) shows the positive impact of OG2004 to tourism in Greece, at least for a period of five years following OG2004. For comparison reasons, data is provided for the period 2000–2009 and analyzed with respect to two 5-year periods, namely 2000–2004 and 2005–2009.

Table 13.4 **Number of international visitors to Greece and revenues in billion euros for the period 2000–2009**

	Number of international visitors	Revenues (in billion euros)
2000	12.378.282	10,06
2001	13.019.202	10,58
2002	12.556.494	10,28
2003	12.468.411	9,49
2004	11.735.556	10,38
2005	14.388.182	10,73
2006	15.226.241	11,35
2007	16.165.265	11,33
2008	15.936.806	11,63
2009	14.914.537	10,40

In particular, Table 13.4 reflects the "explosive" increase in international visitors (excluding financial immigrants) to Greece following OG2004. The respective number for the period 2005–2009 increased by 14.473.086 visitors compared to the period 2000–2004. This increase is directly attributed to the wide promotion of Greece due to OG2004, the promotion of the "authenticity of places, ideas and ideals" of OG2004 as linked to the unique history of Greece and to the ancient Olympic Games, and to the impeccable organization of the sporting event despite the negative international publicity, especially in the period 2002–2003. The "explosive" growth stalled after 2009, when the economic crisis in Europe (and in particular in the European South) affected considerably the flow of tourists to the country.

Table 13.4 also provides the revenues from tourism (in billion euros) during the period 2000–2009. The aggregate revenues for the period 2005–2009 are 4.65 billion euros higher compared to the respective revenues for the period 2000–2004 (an increase of 9.1 per cent). Despite the decrease in tourism in 2004 as compared to 2003, the respective revenues in 2004 due to tourism are higher compared to 2003. This is attributed to the mix of Olympic visitors in the summer of 2004 and their increased spending capacity.

Economic Benefits for Greece due to OG2004

During the period 1997–2004, the Greek economy experienced a gradual improvement in fiscal terms. The improvement was mainly due to the fiscal policy of the Government, but was also influenced by the organization of OG2004 and its boost to the national economy (Greece, Ministry of National Economy, 2009):

1. GDP increased from 117 billion euros in 1998 to 190.7 billion euros in 2004 (an increase of 63 per cent) and to 195 billion euros in 2005 (anincrease of 67 per cent).

2. The contribution of OG2004 in terms of GDP increase for the period 2000–2004 accounts for 9 billion euros (16.9 per cent of the overall increase of the GDP).

3. The ratio of GDP per capita in Greece increased from 84.1 per cent in 2000, to 94 per cent in 2004 and 92.8 per cent in 2005.

4. The ratio of public debt to GDP declined from 108.2 per ent in 1997 to 103.2 per cent in 2000 and 98.8 per cent in 2005.

5. The growth rate (per cent of annual change) of the national economy ranged, during the course of Olympic preparation, in considerably high levels: 4.6 per cent in 2000, 3.4 per cent in 2002, 5.5 per cent in 2003 and 4.8 per cent in 2004.

6. The average annual rate of increase of investments during the period 1997–2004 increased to 12 per cent.

7. Several productive sectors were greatly enhanced (constructions, metal, cement, furniture, telecommunications technology, etc.) due to OG2004.

8. An extended repository of public assets was developed, in line with the new masterplan of Athens until 2020 (Greece, Ministry of Environment, 2013) and the privatization plans of the Government.

Conclusion

A detailed analysis of the Olympic Games 2004 in Athens demonstrates the capacity of a mega (sport) event of the size and complexity of the Olympic Games to support sustainable urban development in the post-event period and to contribute positively to the quality of life, the living and working conditions and the development prospects of the host city. However a number of preconditions need to be met for urban legacies to be maximized and sustained:

1. The choice of locations for Olympic (sporting and non-sporting) venues should support urban regeneration projects as well as the regaining of important city areas for potential exploitation in the post-Olympic period.
2. The Masterplan of the host city may need to be revised so as to take into consideration the new parameters as related to the hosting of the Olympic Games.

3. The post-Olympic use of each (sporting and non-sporting) venue should be defined from the moment the bid file is being put together and finalized in the initial phase of Olympic preparation, in line with the (revised, if necessary) masterplan of the host city.
4. If the 'decentralized or scattered model' of urban regeneration and development is chosen, i.e. with venues located not in one single Olympic Park, but dispersed across the entire host city, then a solid connection of the model to the masterplan of the host city needs to be established and carefully monitored.
5. The size of the Olympic venues needs to be adjusted to their anticipated use in the post-Olympic period. Negotiations with the International Olympic Committee (IOC) and the international sporting federations should be completed as early as possible, to guarantee the sustainability of the venues in the post-Olympic period and to avoid "white elephants".
6. Legal issues which may be critical for the exploitation of the Olympic venues or for the promotion of the urban regeneration plans, need to be resolved prior the completion of the Olympic Games. Following their completion, it may be too late.
7. It is useful to develop and operate a special public company or organization for the exploitation of the Olympic infrastructure in the post-Olympic period.
8. Considerable space should be given to the involvement of the private sector in capital investments related to the Olympic preparation as well as in the exploitations plans in the post-Olympic period.
9. A number of international (sporting and non-sporting) events should be scheduled as early as possible in the post-Olympic period to sustain the momentum and facilitate the integration of the Olympic venues into city life.
10. A post-Olympic communication policy needs to be implemented to sustain international attention to the host city and to support tourism and foreign investments.
11. Any governmental or political changes should not influence post-Olympic planning, whereas continuity in public administration should be secured.
12. The urban governance model needs to acquire a "metropolitan" dimension to maximize benefits and avoid local conflicts.

In terms of Athens 2004, a legacy of sustainable urban regeneration for the city and the region of Attica was achieved, the living and working conditions in the city were overall improved and tourism was substantially strengthened both for the city and the country. Finally, the development prospects for the city in the post-event period could have been better exploited, if the strategic plan had been carefully implemented and continuity secured.

References

BBC News (2004). *Greece's Olympic bill doubles.* [Online] Available from: newsbbc.co.uk (accessed 18 April 2014).

Beriatos, E. and A. Gospodini (2003). *Built Heritage and Innovative Design versus on-competitive morphologies – The case of Athens 2004.* 39th ISoCaRP Congress 2003, 'Glocalisation' & Urban Landscape Transformations.

Business Insider (2010). *What Bankrupt Greece? It was the Olympics.* [Online] Available from: www.business.insider.com (accessed 15 April 2014).

Christian Science Monitor (2004, 2nd October 2008). *A Post-Olympic hurdle for Greece: The whopping bill* [Online] Available from: http://www.csmonitor.com/2004/0901/p07s01-woeu.htm (accessed 15 April 2014).

Gold, J. R. and M. Gold (2008). 'Olympic cities: Regeneration, City rebranding and changing urban agendas'. *Geography Compass*, 2(1), pp. 300–318.

Greece. Association of Greek Tourism Enterprises (2014). *Greek Tourism Basic Figures*, www.sete.gr

Greece. Athens Organizing Committee (2004). *Official Report of the XXVIII Olympics.* Athens, vol I and II.

Greece. Ministry of Environment (2013). *Masterplan of Athens 2020*, Athens.

Greece. Ministry of National Economy (2005) *National Budget 2006: Parliamentary Report*, Parliament of Greece.

Greece. Ministry of National Economy (2009). *Data regarding National Economy*, www.mnec.gr

Greece. General Secretariat for The Olympic Games, Ministry of Culture. (2002, 2003, 2004). *Olympic Progress Report*, Athens.

Kitchin, P. (2007). 'Financing the Games'. *Olympic Cities: City Agendas, Planning and the World's Games, 1896–2012*, pp. 103–119.

OECD (2004). Territorial Review of Athens 2004, Paris: OECD.

Olympic Time (2012). *Was it Worth it? Debt-ridden Greek questions the cost of the 2004 Olympics* [Online] Available from: Olympic.time.com

Preuss, H. (2004). *The Economics of Staging the Olympics: A Comparison of the Games 1972–2008* Cheltenham: Edward Elgar.

Preuss, H. (2007). 'The conceptualisation and measurement of mega sport event legacies'. *Journal of Sport & Tourism*, 12(3–4), pp. 207–227.

Staikouras, C. H. (2012). *Parliamentary Hearing*, Parliament of Greece.

Stournaras, I. (2013). *Parliamentary Hearing*, Parliament of Greece.

Tziralis, G., I. Tolis, I. Tatsiopoulos and K. Aravossis (2006). 'Economic aspects and sustainability impact of the Athens 2004 Olympic games. Environmental economics and investment assessment, WIT'. *Transactions on Ecology and the Environment*, 98.

Chapter 14
Atlanta's Centennial Olympic Games – Before and After, Their Lasting Impact

Michael Dobbins, Leon S. Eplan, and Randal Roark

What Would be the Legacy of the 1996 Centennial Olympic Games in Atlanta?

To evaluate that legacy, now 18 years later, it is important to give some context. How did the City of Atlanta succeed in securing the Olympics in the first place? What did the City do in order to host the Games? What did the event do for the city? What were its lasting effects? What would have happened anyway?

The Background: How Atlanta Won the Games

In a visionary moment in 1986, an Atlanta attorney, Billy Payne, hatched the idea that Atlanta would be well-suited and capable of hosting the Olympic Games. While seemingly a farfetched idea at the time, he was able to form a team and create the Atlanta Committee for the Olympic Games (ACOG), the legal entity that would govern the event. They pursued the idea at first probingly and then in earnest. He enlisted the support of then-Atlanta Mayor Andrew Young, formerly a U. S. Congressman and later President Jimmy Carter's Ambassador to the United Nations, to make the run for the Games. Step one was gaining the endorsement of the United States Olympic Committee (USOC) to become the United States candidate for hosting the 1996 Games.

In their formal presentation to the USOC in Salt Lake City in 1988, Payne and his team stressed Atlanta's history and background in sports activities, its wealth of existing facilities suitable to host the Games, and a commitment to construct other sports facilities as needed. He extolled Atlanta's international airport, the City's exceptional capacity for accommodating large gatherings, its region's existing 56,000 hotel rooms, its modern rapid rail-centered transit system, and its network of television facilities, especially CNN. In addition, the Atlanta team promoted the idea of revitalizing Downtown, where most of the venue, hospitality, and transit facilities would be focused, as a core purpose of the bid.

The bid described Atlanta's long history and leadership in the civil rights movement, as the home of Martin Luther King, Jr. and as the first large Southern city to integrate its public facilities. Finally, the presentation noted the City's experience and ability to raise private funding from large corporations and foundations, essential in light of the federal and state governments' disinterest in underwriting the Olympics.

The bid convinced the USOC and Atlanta was chosen as the U. S. contender for the 1996 Olympics. The bid package delivered to the International Olympic Committee (IOC) in Lausanne, Switzerland, further developed and promoted Atlanta's strong points, especially the already in-place infrastructure and facilities and the experience in managing civil rights tensions. In Tokyo, in September 19, 1990 Atlanta was announced as host city for the Summer Olympic Games in 1996.

ACOG's successful initiative to secure the Games was virtually entirely a private venture, narrowly focused on the venues and the production of the Games themselves, to be funded mainly privately. Until the somewhat unexpected announcement in Tokyo, the City government and most of its citizens had not paid much attention to the Olympic selection process. Although the civic leadership was excited and galvanized by the news, many in the nearby neighborhoods and beyond raised concerns and criticisms about the potential impacts on their quality of life.

"The Twin Peaks of Mt Olympus:" Two Plans for the Games

While there was great potential for the Games to transform the core city, there were also formidable challenges to overcome. The City had experienced a population decline from about 500,000 to about 400,000 people since the 1970s, even as the Atlanta region had ballooned from one to over three million in the same period. The Downtown and Midtown areas, where the Games' activities would be concentrated, exposed their disinvestment in patches of blight. Most of the office and hotel spaces recently built in Downtown were inward-looking, turning their backs on the public realm, thus leaving uninviting street and sidewalk conditions. Conditions in the nearby neighborhoods, with mostly low-wealth and African American populations, were likewise in a deteriorating state. In short, the physical environment would require reinvestment, redevelopment, and repurposing of existing derelict buildings, forlorn streets and parking lots.

Atlanta has an often uneasy alliance between the City's business leadership, mostly white, and the City's political leadership, mostly African-American since the 1960s. For the Olympics, the former saw the Games in terms of business opportunities, and the latter saw a chance to ameliorate socio-economic inequities that had persisted over decades. The Olympics award revealed the realities of the heart and soul of a city in transition. The Olympic moment represented both a mirror to the past and the opportunity for a window into the future.

Responsive to the concerns and needs of the broad constituencies that were skeptical of realizing any benefit from the Olympics, Mayor Maynard Jackson announced his intention to use the Olympics as leverage to try to answer the skeptics. He recognized the challenge, yet believed that the lasting value of the event would be measured in how the Games delivered on the larger city policy agenda. He characterized the challenge as climbing the "Twin Peaks of Mt. Olympus", where one peak was to host a spectacular Olympics and the other was to use the Games to revitalize Downtown and inner city neighborhoods.

ACOG was mainly responsible for the first "peak", while the City and other public agencies had to deliver on the second. The management of this dual agenda fell to two distinct entities. ACOG's commitment was to build the Olympic Athletes Village, Centennial Olympic Park and any necessary sport venues not already in place, manage the volunteer programs, carry out the Cultural Olympiad, and put on the show.

The City had the job of creating a welcoming and attractive public environment to host the millions of expected visitors. Sorting out Mayor Jackson's "twin peaks" policy led to agreements between the two entities that came to be termed "inside the fence" for ACOG and "outside the fence" for the City. This arrangement was focused in what came to be called the "Olympic Ring", a three mile diameter centered in the heart of Downtown. Other smaller venues were dispersed across the larger region. (See Plates 14.1 and 14.2.)

The ACOG Plan: "Inside the Fence"

ACOG delivered on the venues, athletes' village housing, volunteer recruitment and management, Cultural Olympiad commitments, and on producing the Games.

Their centerpiece athletic venue was the Olympic Stadium. ACOG entered into creative financing agreements whereby the Stadium would be built to Olympic specifications in partnership with the Metropolitan Atlanta Olympic Games Authority (MAOGA), which was set up to formalize the public partnership for carrying out the Games. Post-Games the north quarter of the stadium would be demolished and reconfigured to house the City's major league baseball franchise, the *Atlanta Braves*. The former Braves stadium, which was to house soccer for the Games, would be demolished to provide parking for the new stadium, named Turner Field for the team's then owner, Ted Turner, founder of CNN.

Issues of how the existing low wealth neighborhoods might benefit or be impacted by these project continued to simmer as the ground breaking for the Olympic Stadium approached. Neighborhood leaders and the local unions mounted a campaign to assure that jobs and job training for jobless people in the impacted neighborhoods would be included in the development mix. Their non-violent actions included sit-ins in ACOG's offices, and as the groundbreaking ceremony approached, organizing for a major protest demonstration. These actions led to an all-night negotiating session that resulted in the provision of jobs and job training for 250 people in need of work, transforming the planned protest into a celebration. Importantly, the agreed training and employment provisions did not disqualify the consideration of people with criminal or substance abuse backgrounds, provided they stayed clean while employed.

ACOG developed many smaller venues that added to or upgraded athletic facilities for the Georgia Institute of Technology (Georgia Tech), Georgia State University, nearby neighborhoods, and for the Atlanta University Center, the largest complex of historically black colleges and universities in the nation.

The athletes' village was focused on Georgia Tech's campus, where existing dormitories were upgraded, supplemented by major new residential construction immediately to the south. The funding for the $241 million project included $47 million of private funds raised by ACOG Georgia State University and then by Georgia Tech students.

Figure 14.1   The Athletes Village reconfigured as university housing
Source: Jeb Dobbins.

This project too raised community concerns, as the site for the new housing marked the beginning of the already planned replacement of an adjacent public housing project. Ultimately, that whole project was redeveloped as a new mixed income community, the first built under the new HOPE VI program of the federal Department of Housing and Urban Development (HUD).

The volunteer and Cultural Olympiad programs were successful and well received. Altogether, close to 500,000 people and 1700 companies throughout Georgia participated in the "Atlanta Force" volunteer effort. And, while not adhering strictly to the City's procurement policies that mandated percentages of publicly sponsored work going to disadvantaged businesses, ACOG's performance was superior to most private sector developments of the time. Overall, ACOG was able to stay close to its initial $1.7 billion budget, with the final tally amounting to $2.3 billion.

In late 1993, a little belatedly since most of the venues and other projects were already under construction, Downtown leaders and Billy Payne realized that there was no provision for a major public gathering place for the Olympic multitudes. The small Downtown Woodruff Park would be simply overwhelmed by the numbers and the demands for sponsors' tents. To meet that need, ACOG identified an opportunity to achieve two goals

Figure 14.2   Mixed Income Centennial Place Housing Development
Source: Jeb Dobbins.

at once. A mostly derelict area separated the major concentration of hotels and the intense anticipated use of the Omni Arena and the Georgia Dome venues. The opportunity to bridge this gap was the construction of Centennial Olympic Park, which became the principal public gathering space during the Games.

Because of the tight timeframe and the need to acquire multiple properties, mostly parking lots and warehouse type buildings, ACOG was able to persuade the State of Georgia, through the Georgia World Congress Center (GWCC), to undertake and deliver the project. Armed with eminent domain powers, the GWCC was charged with acquiring the property and building, owning, and managing the park. Since the Games, the park has become the central venue for all manner of events and activities.

CODA PLAN: "Outside the Fence"

The City realized that on top of its ongoing service obligations it would not be able to deliver on the extraordinary demands for major new streetscapes, parks, and for the neighborhood revitalization planning required to greet the visitors and address neighborhood needs. Following the "twin peaks" theme, in early 1992 the City drew from its annual Comprehensive Development Plan a program for utilizing the Olympics as a way to meet the challenge. Called *An Olympic Development Program for Atlanta*, it set the vision for selecting actions and projects in three categories: (1) accommodate visitors, for which 49 measures were identified; (2) improve nine Olympic neighborhoods; and (3) create a lasting legacy. The actions and programs, prioritized to meet the Olympic deadline, would extend into the post-Olympic future.

To implement the plan, the City formed a public non-profit corporation to take responsibility for implementing the Mayor's vision. Accordingly, the Corporation for Olympic Development in Atlanta (CODA) came into being with a board including public, private, and neighborhood representatives, co-chaired by the Mayor and a business leader. Its charge was to produce a welcoming and attractive public environment and to launch the neighborhood revitalization process.

The streetscape program was designed to connect the venues, hotels, and transit stops in the Downtown core, where much of the activity would occur. Thus CODA oversaw the design and construction of wide sidewalks, tree planting, lighting, way-finding and historic signage, and coordinated street furniture. The immediately positive impacts of these efforts overturned the previously unthinkable notion that Atlanta could become a walkable city.

Capitol Avenue, Atlanta, 1993   Capitol Avenue, Atlanta, 1993

Figure 14.3   Before and after images of typical streetscape improvements
Source: Randal Roark.

CODA similarly managed the improvement of a number of parks in and adjacent to the Olympic Ring areas, most centrally the design and construction of Woodruff Park, establishing a well-scaled, fountain-enlivened "green room" in the core of Downtown. Accompanying and complementing all these improvements to the public environment, CODA initiated and implemented an extensive public art program, whose artifacts added a touch of grace to an otherwise somewhat gritty city.

And CODA moved aggressively to work with neighborhood leaders to develop master plans for the revitalization of five neighborhoods in the Ring. This work complemented the establishment or the enhancement of six community development corporations in low wealth neighborhoods, bringing focus to how City policy and resources should be prioritized to make progress on Mayor Jackson's goal.

In retrospect, CODA's work represented a remarkable spurt of achievement in a City not known for efficient delivery of capital projects, and all on a $65 million budget and a three year delivery ultimatum.

There were many other City initiatives that paralleled CODA's plans and actions. To accommodate a surge of tourists and visitors at Hartsfield International Airport (now Hartsfield-Jackson), during and after the Games, the airport issued bonds of $350 million to build a dedicated international concourse, opened in 1994, and a $250 million program of passenger improvements. Since 1998, a new one-and-a-half billion dollar international terminal has opened, keeping the airport as the world's busiest.

Other projects and programs accompanied the Olympic preparation energy and were very likely accelerated by the Games' timeline urgency. The long simmering community-driven struggle to convert a planned freeway right-of-way into a new park resolved in the community's favor, resulting in the new 200 acre Freedom Park and catalyzing reinvestment in Atlanta's east side neighborhoods.

After two failed earlier attempts, Atlanta citizens approved a major bond referendum in 1995 for $150,000,000 whose proceeds dealt with a raft of city-wide deferred maintenance items including roads, bridges, sewers, storm drainage, and parks improvements. Some of these projects directly complemented Olympic preparations, and altogether they signaled the will of the citizenry to tax themselves to improve conditions. It remains hard to say whether the bond election would have succeeded without the Olympics, yet the impending Olympics surely figured positively. While a few of the bond-funded projects were under construction during the Olympic run-up, most were completed over the next years and served both to remedy neglect and to upgrade infrastructure to attract new private investment.

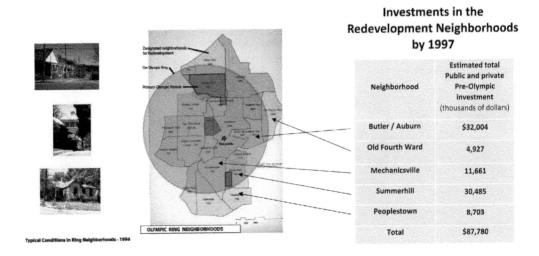

Investments in the Redevelopment Neighborhoods by 1997

Neighborhood	Estimated total Public and private Pre-Olympic Investment (thousands of dollars)
Butler / Auburn	$32,004
Old Fourth Ward	4,927
Mechanicsville	11,661
Summerhill	30,485
Peoplestown	8,703
Total	$87,780

Figure 14.4   Olympic Ring map showing CODA neighborhoods
Source: Randal Roark.

In addition, while nominally not committed to sponsoring Atlanta's and ACOG's effort, major funding and program support came from county, state, and federal sources. Thus, major upgrades to the highways and their IT systems, beautification initiatives, security combines, infrastructure upgrades, and other capital and programmatic support greatly helped the City to prepare. At the same time, Georgia Power Company, which had a significant role in entertaining and ultimately recruiting foreign business, made major upgrades in its power grid. With all of these different actors, pulling together multijurisdictional and private sector entities required collaborations that were often prickly, yet they resulted in creative and constructive improvisations to get the job done and the deadlines met.

The Aftermath

The Games proved to be an event for the everyday American, as well as for a raft of international visitors. Over 8.5 million tickets were sold, the most for any Olympics so far, with ticket prices more affordable to a broader citizenry. Unlike some Olympics, where many tickets were bought but not used, Atlanta's events mostly attracted a full house. Even now, Atlanta's Games boast the most participants and visitors of any Olympics. General feedback indicated that most visitors had a good time.

The Atlanta Olympics cost considerably less than any other in the last 50 years. Produced on a private shoestring budget, the overall costs were but a fraction of other cities' Olympics costs. The City incurred no lasting debt. ACOG gained a $10 million profit.

The City, the universities, and the neighborhoods realized significant additions to their sports-oriented facilities. The neighborhoods made headway toward sustaining improvements in the quality of life, some more than others based on the effectiveness of neighborhood leadership structures. These included hundreds of new and rehabilitated housing units, infrastructure improvements, and strengthened capacity in neighborhood-based community development corporations.

In the local and international media, the most criticized aspect was the "look" of the Games. There were no "gee whiz, wow" works of architecture. The public parks and streetscapes throughout downtown sported a dominance of in-your-face vending stalls and tents. The local media focused on the tents of the "little guy", a program of the City meant to offer benefit from the Games to small entrepreneurs. Lining many streets, these stalls were seen locally as ragtag, tacky, and contrary to the image the local business leadership wished

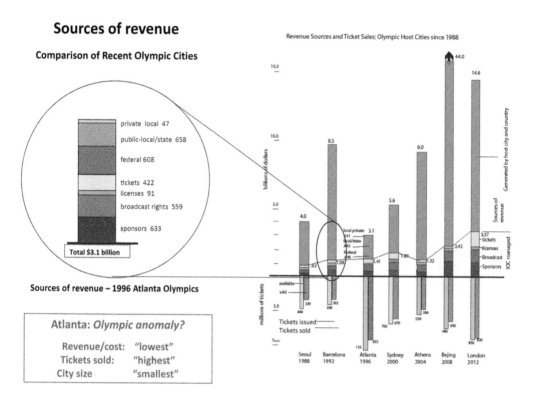

Sources of revenue

Revenue Sources and Ticket Sales; Olympic Host Cities since 1988

Comparison of Recent Olympic Cities

private local 47
public-local/state 658
federal 608
tickets 422
licenses 91
broadcast rights 559
sponsors 633

Total $3.1 billion

Sources of revenue – 1996 Atlanta Olympics

Atlanta: *Olympic anomaly?*

Revenue/cost: "lowest"
Tickets sold: "highest"
City size "smallest"

Figure 14.5   **Chart comparing the costs, revenues by source, and ticket sales for the last seven summer Olympics**

Source: Randal Roark.

to portray. The international coverage, while similarly noting the little tents, seemed more offended over the commercialization of the Games in general through the huge, corporate sponsorship tents covering most of the parks, some calling the event the "Coca-Cola Olympics".

At both ends of the spectrum, the little tents and the big tents obscured much of the underlying quality of the streetscape and park development efforts. Ironically, if all the vending stalls and sponsors' tents had been stripped away, the millions spent on streetscapes, pedestrian amenities, and Downtown parks would have been seen, but there may not have been the street life and maybe not so much city life behind them.

The main stain on the three weeks experience was the tragic bombing in Centennial Olympic Park that took the lives of two people. The terrorist, who planted the bomb, is now serving a life sentence.

The International Olympic Committee (IOC) didn't like the Atlanta Olympics. IOC president Juan Antonio Samaranch withheld his usual "Best Olympics Ever" accolade and instead dubbed Atlanta's the "Most Exceptional Olympics". There is speculation about this apparent downgrading, centering on who controls the Games. In Atlanta there was no single authority controlling the funding or the management of the process, with responsibilities instead split between ACOG, the City and other public and private players. The IOC since has required full governmental backing and a unified implementation authority as a condition for hosting the Games.

Immediately after the Olympic Games and using the same facilities, Atlanta also hosted the Paralympics. For the first time the Paralympic Games gained media and other sponsorships to pave the way for what is now a formalized component of the Olympic movement. With a budget of just over $80 million, over 3000 athletes from some 100 nations participated, marking a lasting bridge to inclusion. All in all, the Olympics left the City with pride in having accomplished a grand feat.

Figure 14.6   Street vendors' tents and corporate sponsors' tents dominated downtown parks and public ways
Source: Randal Roark.

The Legacy

When the Olympic Games concluded in August 1996, a sort of post-delivery haze enveloped the City. When Atlanta was awarded the Games in 1990, the question was "Can we pull this off?" Now the question was "Now what do we do?"

The local media reflected the malaise of business and political leaders. Keeping in mind that the City comprised only about a tenth of a booming region's population, there was a lot of skepticism about the City's future. Would the Olympics be the shrinking City's last hurrah – a bump in what had been a long, slow decline? Or was the City prepared to move forward on the momentum the Games had generated? Would the City's population stabilize following a 25 year slide? Would the City's gradual decline as a business center simply resume when the Olympics-focused hype faded? Would the city be able to translate its moment on the international stage into lasting relationships and investment? Would the buildings, monuments, streetscapes, parks, and public art be able to maintain their functional, aesthetic, and symbolic value? Finally, would the core City focus of the Olympics be sufficient to lift the quality of life in the neighborhoods around where the Games were concentrated?

For guidance on these questions, Mayor Bill Campbell, who succeeded Mayor Jackson in 1994, convened a group of civic and business leaders to advise the City on policy and program measures to take best advantage of the Olympic "bump". Called the Renaissance Policy Board and chaired by the CEO of Coca-Cola, Roberto

Goizueta, its goals included attracting the middle class into the city, stimulating population growth, attracting new business investment, reducing crime, improving public education, attacking poverty, rebuilding public housing, and improving transit access to jobs and services. Many of the Renaissance Policy Board's goals were achieved, while the plan also helped to set the course for both public and private initiatives to achieve progress that since has visibly and substantively transformed many parts of Atlanta.

The Olympics benchmarked the City's shift from shrinking to growing. The forces of suburbanization and racist reactions to civil rights gains that had drained the City of 100,000 residents from the late '60s reversed as 30,000 more people had moved into Atlanta by 2006. Investment marked significant upticks in the post-Olympic years in both Midtown and Downtown and in many neighborhoods where the urban life style attracted new demographics. Most of that growth was in the middle income category called for in the Renaissance report. Providing 25,000 new middle income housing units in ten years seemed daunting in 1997, yet that goal was achieved in just seven years. In the five years following the Olympics, 42 new companies, some of them international enterprises, located in Atlanta, adding 6000 new jobs.

The Atlanta Housing Authority followed its two Olympic era project redevelopment initiatives, Centennial Place Downtown and The Villages at East Lake (a project five miles east of the Olympic Ring) with a complete redevelopment of all its over 15,000 housing units. (While criticized, because of its temporary and permanent displacement of low income people, the Authority achieved mostly stable mixed income communities, with, on average, 40 per cent low income, 20 per cent low to middle income and 40 per cent market rates).

The Renaissance Policy Board did not reflect upon the longevity of the venue, streetscape, parks, public arts, or neighborhood planning and development achievements. Yet one of the most enduring Olympic legacies is that most of the improvements are still operational and in good use, or they have even been expanded based on increased demand.

Centennial Olympic Park is even more vibrant after receiving a major expansion and catalyzing a ring of development around it. It is the go-to venue for large outdoor events and serves daily residents, tourists, and downtown workers with more intimate and small scale play and picnic areas. Its northern edge fronts major new attractions:

- the Georgia Aquarium, one of the nation's largest and most committed to the advancement of marine science
- the Coca-Cola Museum, which attracts thousands of paying customers per year
- and the soon to be completed Center for Civil and Human Rights, that, beyond state of the art exhibitry tied to human rights efforts around the world, will house the papers and other artifacts of Atlanta native son Rev. Martin Luther King, Jr.

Soon after the Olympics, the leadership of the Downtown business organization, Central Atlanta Progress (CAP), established the Centennial Olympic Park Area (COPA) district to promote development in the Park area. The City established a tax increment financing district to induce new investment in this underdeveloped area, resulting in continuing residential, office, restaurant, and hotel development, so far totaling about $1.8 billion in new investment.

The baseball stadium is still the baseball stadium, housing the successful Atlanta Braves franchise, though the team plans to leave for a new venue northwest of the City in 2017. Most of the other venues and parks are still functioning, serving neighborhoods and athletic needs at the Atlanta University Center campuses. Many, like the Omni Arena, the Georgia Tech arena, and soon the Georgia Dome have been or will be replaced or upgraded in situ, as the success of the programs that use them has risen.

The Olympics Aquatics Center at Georgia Tech has been expanded and enclosed to house one of the nation's premier college recreation centers. The athletes' village is still student housing, having received a remodeling in 2012 as Georgia Tech students replaced Georgia State students.

The CODA streetscape and park improvements continue to provide the model for major extensions of their high quality designs, improving the public environment in an ever expanding ring around their initial installation. The idea of the Olympic Ring was picked up by a Georgia Tech graduate student and expanded into what is now known at the BeltLine, a ring of parks and trails around abandoned rail rights-of-way, the first parts of which enjoy heavy use.

Beyond the immediate Olympic development initiatives, the growth and positive change in the City have depended on a number of interacting factors, in which the Olympic period played a seminal role. The southeastern U. S. has been fast-growing for decades, with the Atlanta region growing as fast as any. Significant for the City itself, though, have been shifting demographics and markets, where the demand for "urban living" began to emerge as a trend at about the time of the Olympics. Demographic changes and market shifts, e.g. fewer families with children, empty nesters, growing numbers of seniors, and young people who grew up in suburbs but preferred the vitality and diversity of cities have, played an important role in repopulating and reinvesting in the core City.

Among the central conundrums in assessing Olympic impact in Atlanta as well as in other Olympic cities is sorting out the blend of cause, effect, and coincidence. A theme that runs through any such analysis is the focusing impact of an absolute and irrevokable deadline, which might accelerate long standing aspirational projects and also cause rethinking of old problems or reveal new opportunities. In Atlanta, the deadline for putting on the Games had a catalyzing impact. Projects that had been talked about for years suddenly got a boost. As mentioned earlier, after two failed bond elections, 1995 produced a winner, necessary to carry out the Olympic program plus long neglected infrastructure needs throughout the City for the years to come. Increased confidence in the City's ability to improve, led to the successful vote for an additional $150 million "Quality of Life" bond issue in 1999.

The realization of the need for a robust transit system awakened mostly white suburbanites to the fact that the City actually had such, the Metropolitan Atlanta Regional Transit Authority (MARTA). Its intense use during the Games began to break down race and class driven resistance to using the system, though patterns of cultural antipathy continue to hamper its ability to reach its full potential.

Two federal programs prioritized Atlanta as recipients for major grants and it can be suspected that the Olympics energy either moved them up in the priority list or accelerated their implementation: (1) The aforementioned HOPE VI program fueled the transformation of the Atlanta Housing Authority's communities over the next fifteen years. (2) The $250 million-project of Empowerment Zone designation was a mixed success. It funded framework plans for 26 low wealth neighborhoods encircling the Downtown core, which stimulated reinvestment and redevelopment activities, until they were rudely interrupted by the "Great Recession" of 2008.

In the same vein, the mandated timelines launched collaborative organizational arrangements necessary for meeting the deadlines. People within and between the private and public sectors, unaccustomed to working with each other, had to break through patterns of suspicion and exploitation that had dominated the development climate in the past.

With the mandated dissolution of CODA in 1997, its mission was subsumed into a public "superagency", the Atlanta Development Authority (ADA, now Invest Atlanta). The ADA consolidated five existing agencies into a single multi-purpose authority. As part of the ADA consolidation, the City established and co-managed a "development council" that regularly brought together the major public and community development agencies for the purpose of communicating and coordinating their information and activities. It was clear that the City needed to adjust its planning and development policies and priorities to sustain the Olympic momentum and respond positively to the new markets. While in no way a cause of the market shifts, the City's Olympics response created the kinds of walkable, well-lit street environments, the kinds of parks, mixed density, mixed use developments, and the dawning environmental consciousness that attracted prospective urban dwellers.

Through its collaborative efforts and in strengthening guidance from the City's Neighborhood Planning Unit (NPU) structure, the City undertook a number of initiatives. In partnership with various neighborhoods and with the City's three major business centers, Downtown, Midtown, and Buckhead, the City carried out major rezoning initiatives. These put the guidance in place to assure that new and retrofit development would encourage mixed use, pedestrian friendly, and parking shielding patterns to enliven formerly car-dominated cityscapes.

Using the transformational success of the CODA projects, the City prioritized its public improvement commitments and instituted new development finance tools to help pay for the requisite improvements to the public environment. These included tax increment financing (called tax allocation districts in the state of Georgia), tax deferred development incentives, and creative blending of public and private fund sources, all to meet the emerging demands for urban settings that both worked and appealed. It used regionally controlled funds under a Livable Centers Initiative (LCI) program to both plan and implement streetscape and plaza development. In addition, the City was an essential partner in large scale redevelopment projects that tapped the new markets for in-town living, working, and shopping complexes as viable options to the suburban-dominated patterns that had characterized the region's growth.

Many positives were generated during the Olympic period and carried on in the following years, some directly attributable to the Games, some more coincidental but catalyzed by the time deadlines, and still others reflecting market shifts. Some of Atlanta's most intractable challenges, however, persist. The City ranks near the top of the Gini Index, which measures inequality, and a recent study concluded that Atlanta is at the bottom of major U.S. cities in affording its low wealth citizens the opportunity to escape poverty. The goals of the Renaissance Policy Board to "attack poverty", boost public education, and improve access to jobs, particularly for those without cars, have not been met. The band of poor, mostly African-American, neighborhoods that runs from the northwest to the southeast is still afflicted with low wealth, high unemployment, poor schools, and inadequate transit. Though easing somewhat, race still plays a significant role and neither business nor political leadership has shown any sustained commitment to improve these conditions.

Despite the success of the community and union-driven attack on poverty with a jobs program associated with the construction of the Olympic stadium, this model has not been replicated for other large development initiatives. Neither private nor public City leadership has been willing to be bound to enforceable community benefit agreements, in which jobs are the most urgent need. In current times, the wealth gap, persistent poverty, and social immobility remain as the City's most daunting challenges. No Olympic "bump" or civic action has successfully addressed this problem.

Still and all, Atlanta's Olympic moment represents a major milestone in Atlanta's development progression. It either catalyzed or coincided with major shifts in the marketplace, where in-town living, environmental sustainability, choice in living and working environments, choice in travel modes, and appreciation for more compact and diverse communities all came together. Using the Olympics opportunity, the City and its private sector adopted new policies, plans, city design, and projects to respond and further stimulate the new city building partnerships. Reporting these facts is important, especially when the reporting on the Atlanta Olympic experience by others has been limited and narrowly focused.

References

Atlanta Committee for the Olympic Games (1997a). *The Official Report of the Centennial Games*, Volume I: Planning and Organizing. Atlanta: Peachtree Press.

Atlanta Committee for the Olympic Games (1997b). *The Official Report of the Centennial Games*, Volume II. The Centennial Olympic Games. Atlanta: Peachtree Press.

Atlanta Committee for the Olympic Games (1997c). *The Official Report of the Centennial Games*, Volume III. The Competition Results. Atlanta: Peachtree Press.

Atlanta Committee for the Olympic Games (1997d). *Community Outreach, ACOG's Legacy of Human Impact.* Atlanta: Peachtree Press.

Atlanta Committee for the Olympic Games (1997e). *The Cultural Olympiad.* Atlanta: Peachtree Press.

Andraovich, G, M. Burbank, J., Heying, H. Charles (2001). 'Olympic cities: Lessons learned from mega-event politics'. *Journal of Urban Affairs*, 23.

Battle, Ch. (2013 and 2014). Unpublished notes and interviews from a key ACOG official. 2013 and 2014.

Dobbins, M. and D. L. Sjoquist (1998). *Atlanta Renaissance Program*. Atlanta: City of Atlanta

Economist Intelligence Unit (2012) 'Legacy 2012: Understanding the impact of the Olympic games', *The Economist*. 2012.

Engle, S.M. (1999). The Olympic Legacy in Atlanta. *University of New South Wales Law Journal 902*.

Eplan, L. (1992). *An Olympic Development Program for the City of Atlanta: Official Report of the City of Atlanta,* City Department of Planning and Development. Atlanta, Georgia, July 22, 1992.

French S.P. and M.E. Disher (1997). 'Atlanta and the Olympics: A one year retrospective'. *Journal of the American Planning Association,* 63(3), pp. 379–392.

Frey, W. H. (2012). *Population Growth in Metro America since 1980: Putting the Volatile 2000's in Perspective.* Washington D.C.: Brookings Metropolitan Program.

Keating, L., M. Creighton and J. Abercrombie (1997). 'Community development: Building on a new foundation', in Sjoquist, D. L. et al. (1997). *The Olympic Legacy: Building on What Was Achieved.* Atlanta: Georgia State University, School of Policy Studies, Research Atlanta Inc.

Lohr, K. (2011). "The Economic Legacy of the Atlanta Olympic Games", *National Public Radio*, (August 4). Available at: http://www.npr.org/2011/08/04/138926167/the-economic-legacy-of-atlantas-olympic-games

Lomax, M. L. (1997). *The Arts: Atlanta's Missing Legacy.* Atlanta: Georgia State University, School of Policy Studies, Research Atlanta Inc.

Newman, H. K. (1999). 'Neighborhood impacts of Atlanta's Olympic games'. *Community Development Journal,* 34, no. 2, pp.151–159.

Padgett, R and J. R. Oxendine (1997). 'Economic development: Seeking common ground', in Sjoquist, D. L. et al. (1997). *The Olympic Legacy: Building on What Was Achieved.* Atlanta: Georgia State University, School of Policy Studies, Research Atlanta Inc.

Patton, C. V. (1997). 'Downtown: The heart and soul of Atlanta', in Sjoquist, David D. L. et al. 1997. *The Olympic Legacy: Building on What Was Achieved.* Atlanta: Georgia State University, School of Policy Studies, Research Atlanta Inc.

Piper, V. (2005). *Case Study Atlanta.* Washington: Brookings Institution.

Preuss, H. (2002). *Economic Dimension of the Olympic Games.* Centre d'Estudis Olympics (UAB) Barcelona.

Quesenberry, P. (1996). 'The disposable Olympics meets the city of hype'. *Southern Changes,* 18, pp. 3–14.

Roark, R. (1994). 'Atlanta: Formal Paradoxes And Political Paratactics', in *Atlanta.* Bernado, J and R. Prat, Barcelona: Actar.

Roark, R. et al. (1995). *The Civic Trust.* Atlanta: The Corporation for Olympic Development in Atlanta.

Rutheiser, C. (1996). *Imagineering Atlanta: The Politics of Place in the City of Dreams.* London: Verso.

Sjoquist, D. L. (1997). *The Olympic Legacy: Building on What Was Achieved.* Atlanta: Georgia State University, School of Policy Studies, Research Atlanta Inc.

Terrazas, M. (1996). 'Let the Arts Begin: The cultural Olympiad', *Georgia Tech Alumni Magazine Online.* Summer 1996. Available at: http://gtalumni.org/Publications/magazine/sum96/CultOly.html

U.S. General Accounting Office (2006). *Olympic Games; Federal Government Provides Significant Funding and Support. Report to Congressional Requesters.* General Accounting Office (September).

Vaeth, E. (1998). '1996 Olympics: A defining moment in Atlanta's history: was it worth it?' *Atlanta Business Chronicle* (June 15). Available at: http://www.myajc.com/gallery/lifestyles/flashback-fotos-atlantas-midtown-1970–1990/gCGx4/#4322045

Chapter 15
Mega-events in the South: Offside for the Poor? FIFA 2010 Legacy in Durban, South Africa

Brij Maharaj and Vyasha Harilal

Introduction

Neoliberal global economic restructuring has forced cities around the world to compete for investments in an international market in order to boost declining economies. Marketing the city as a mega-event destination has become a prominent urban promotion strategy (Hall, 2006). With a few exceptions, the scholarly focus has been on the western experience. Following the trend in the global north, mega sporting events are increasingly being viewed as a strategy to promote economic growth and to reduce poverty in cities in the Global South, e.g. 2010 Commonwealth Games Delhi; FIFA 2010 South Africa; FIFA 2014/ Olympic Games 2016 Brazil.

The hosting of mega-events is touted as an opportunity to fast track development in the host city or country and to create a positive international image, which will in turn attract foreign investments and tourists. However, there are some concerns about the costs of hosting a mega-event and competing priorities in terms of public spending (Baumgarten, 2010). FIFA is not an altruistic or philanthropic organization, but an international body interested in selecting locations and venues that would maximise its profits (Matheson and Baade, 2004). Bidding for, and hosting such mega-events have been equated to a gamble and a risk which:

> 'Embody a palpable aura of endless promise. They thrive on that uniquely late capitalist drug, speculation, which represents one of the biggest gambles a nation can take – betting high with the promise of limitless returns but laced with great risk since so much can go wrong' (Pieterse, 2012:83).

Other concerns are that the infrastructure that was developed for the event itself would be of little or no use to the host city or country after the event, and will be costly to maintain. Furthermore, the maintenance of the infrastructure may steer public funds away from the improvement of other services and infrastructure, such as transportation, housing and the provision of basic services (Tilley, 2006).

Hence, strategies to transform declining city zones are 'often written into bids for major sporting events, as the costs of staging such an event are so high they can only be justified when they are envisaged as leading to a major programme of urban regeneration and improvement' (Essex and Chalkley, 1998: 187). Sophisticated infrastructure development is a prerequisite for hosting sporting mega-events, and is often viewed as a catalyst for urban development. Although there are many types of infrastructural developments that take place in preparation for hosting a mega-event with huge injection of public funds, a key concern is what happens after the event, especially in developing countries.

South African cities have not been immune to these global trends. Against this background, this chapter discusses the various types of infrastructural projects that were initiated in Durban for the 2010 FIFA World Cup, and assesses their legacy implications. This includes the Moses Mabhida Stadium, the King Shaka International Airport, and the upgrade of the beachfront. There is also some critical reflection on the legacy implications for the poor. Before proceeding further, a brief reflection on mega-events in the developing world is necessary.

Mega-events in Developing Countries

Developing countries usually have many social priorities such as the provision of basic services to the poor, including water, sanitation, housing and education. Some of the advantages of hosting a mega-event in a developing country are concerned with the anticipated influx of capital and profits which could be used to address such basic needs. Also, the host city or country's image will be boosted by being linked to influential and high ranking events such as the Olympic Games or the FIFA World Cup (Andranovich, Burbank, and Heying, 2002). Furthermore, with the influx of capital that is expected into the country, together with the infrastructural investment that is required for the event, the development of targeted urban zones could be fast tracked (Bass and Pillay, 2008). Zimbalist (2010:11) has argued that 'with proper planning, hosting a large event can serve as a catalyst for the construction of modern transportation, communications, and sports infrastructure, which generally benefits less-developed areas more'.

A counter argument is that the money that is spent on infrastructure development could alternatively be used to address the social challenges in the host country which are largely related to poverty and inequality. Also, the infrastructure developments should be adapted to the priorities of the country, rather than the narrow demands of the mega-event in question. Also, the cost of developing and maintaining infrastructure is extremely high (Matheson and Baade, 2004).

Preparation for mega-events in developing countries is time consuming and costly, and is not only focused on infrastructure development, but also on formulating various protocols, committees and structures to ensure the smooth running of the event. Also, the majority of preparations in developing countries are started from scratch. The economies of developing countries are usually vulnerable and susceptible to negative influences. The high level of spending that is associated with preparations for a mega-event weighs heavily on the economy of a developing country (Matheson, 2006).

Mega-events are short term spectacles that are intended to inject large amounts of capital into the economies of the host countries, with the expectation that tourists and investors would be attracted to the country or city, which would benefit through various knock-on and multiplier effects (Andranovich, Burbank, and Heying, 2002; Matheson, 2006). These events are said to have millions of spectators at the event itself, and also attract billions of viewers through satellite transmission. These types of events are deemed to be invaluable to the host nation in terms of boosting its international image as a global destination for economic investments and tourists (Baade and Matheson, 2004). However, there are also many doubts about the benefits of mega-events for the host country. Frequently costs are underestimated and benefits are exaggerated. The costs and impact of hosting a mega-event in a developing country may have serious negative implications. The hosting of a mega-event may reap short term benefits, but there is no concrete evidence that proves that such events contribute to poverty reduction and addresses developmental challenges (Bass and Pillay, 2008).

One of the key differentiating factors between developed and developing countries is the extent of infrastructure development. Developed countries have excellent infrastructure facilities and are thus well equipped to host mega-events, with minimum cost (Ling, 2005).

Developing countries hosting mega-events spend far more than their northern counterparts e.g. the 1994 USA FIFA World Cup less than $30m; the 1998 FIFA World Cup France less than $500m; the 2002 South Korea $2bn, South Africa 2010 $4.1bn (Cottle, 2009:52). Another issue is the use of facilities after the event, especially in the developing world, 'where prudent allocation of limited resources is critical' (Goliger, 2005:176).

Durban – A Sporting Mecca?

The port city of Durban (also called eThekwini in the democratic era) located in the province of KwaZulu-Natal (KZN) (Figure 15.1[1]), often cited as 'the warmest place to be' and 'Africa's playground', is one of South Africa's premier tourist destinations. Durban is considered to be an all year-round tourist destination because

---

1   We are grateful to our colleagues Dr Njoye Ngetar and Jatin Channa who developed this map.

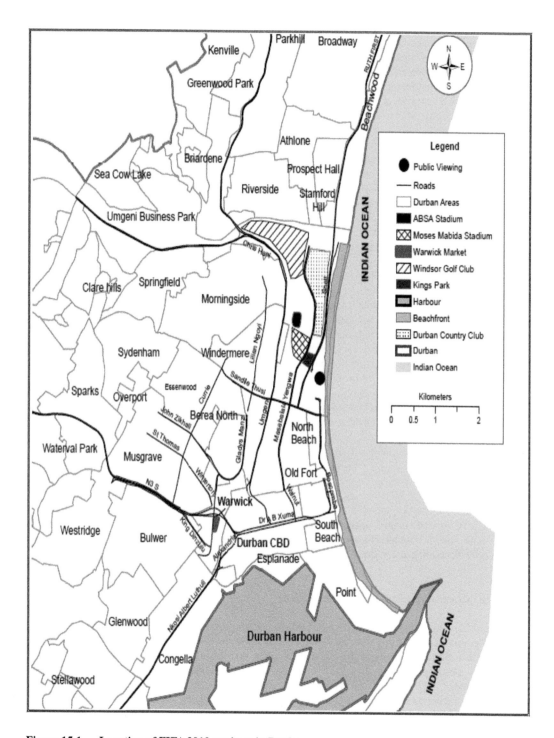

Figure 15.1   Location of FIFA 2010 projects in Durban

of its sub-tropical climate and warm beaches. The city's golden mile comprises of a '6 km long strip of beachfront lined with beaches, lawns, promenades, luxury hotels and restaurants' (Oliver and Oliver, 2001:92).

The move toward democratisation in South Africa opened cities to forces of globalisation and the pressure to restructure local economies in order to promote economic growth, and Durban was no exception. In Durban prominent local economic development strategies included the construction of the International Convention Centre and the uShaka Marine Park, public-private partnerships which ratepayers continue to subsidise (Maharaj and Ramballi, 1998).

Durban also had a reputation for important sporting activities such as the Rothman's July Handicap, the Comrades Marathon and several surfing events. Until 1990, the target market for such audiences was largely national and white. However, since then there has been an aggressive attempt to market Durban as a city with the potential to host world class sporting events. The marketing of Durban as a sporting destination started in 1991 with the staging of the first leg of the International Powerboat Racing Grand Prix in the Durban harbour. The event was viewed in 60 to 70 different countries and it was estimated that over 150 million people world-wide became aware of Durban (Dayanand, 1996). In 1991, Durban together with 15 other cities worldwide (including Cape Town and Johannesburg) also tendered a bid to stage the 2004 Olympics. In January 1994 Cape Town was chosen as the South African candidate city for the 2004 Games bid. The reality was that Durban did not have the infrastructure to host an event of such magnitude (Maharaj, 1998).

These strategies emerged as Durban grappled with rapid population growth, a slow economic growth rate, housing backlogs, an increasing number of informal settlements, escalating poverty, high unemployment rates and an inadequate supply of basic services to the majority of the population (Maharaj, 1996). Durban's focus on sport was largely influenced by the notion that an entrepreneurial approach to urban governance could help boost investment and generate economic growth (Centre for Development Enterprise, 1996).

On 15 May 2004 South Africa won the bid to host the FIFA 2010 World Cup. In terms of legacy, in its 2009 Election Manifesto the ANC (African National Congress) emphasised its intention to:

> Ensure that the 2010 FIFA World Cup leaves a proud legacy that our children and our communities will enjoy for many years to come, and contributes to the long term development of the country. The ANC government will work with all stakeholders to ensure that this world event contributes to create decent work opportunities, particularly for the youth, women and street traders.
>
> Source: www.anc.org.za/elections/2009/manifesto (accessed 10 May 2009).

As the host of the 2010 FIFA World Cup, the South African government was obliged to commit to certain types of infrastructure development in order to host the event. This included the construction of stadia in the provinces across the country, upgrading of the transport system in the country, including airports and public facilities such as beaches. Not surprisingly, as part of the FIFA 2010 strategy, Durban was promoted as 'Africa's premier sporting and events destination'.

FIFA 2010 Infrastructure Development in Durban

In 2005 the '2010 and beyond strategy for Durban' plan was adopted by the city management and the KZN provincial government, and this was 'an event-led economic development strategy ... aimed to use the World Cup to build the economy of the city and to ensure that the infrastructure developed for the World Cup provided a lasting legacy for the city'.[2] The focus was on developing the city as 'Africa's sporting and events destination'; upgrading physical and sports infrastructure in the city and surrounding region; providing attractive investment opportunities; promoting urban renewal and inner city regeneration. According to the City Manager, Dr Michael Sutcliffe:

> This strategy was supported by favourable regional climatic conditions, geographical features and the excellent infrastructure, most importantly our well-located sporting infrastructure. Currently, eThekwini has the potential

---

2   eThekwini Municipality State of Local Innovation Report, October 2011, p. 6.

to host the 42 main Olympic sporting codes in a single precinct within easy access to the city centre, beaches and accommodation (Sutcliffe, 2006:4).

In terms of this strategy the planned 2010 legacy infrastructure in Durban would include the construction of the Moses Mabhida Stadium; the addition of public transport lanes to the M4 highway in Durban; developing pedestrian and cycle paths in the city; the inner city distribution (people mover) system; the beachfront upgrade; the Warwick junction infrastructure (fly-over bridge) and urban regeneration project; and the construction of the King Shaka International Airport.[3] The focus in the next sections of this chapter will be on the Moses Mabhida Stadium, the King Shaka International Airport, the Beachfront upgrade and the Warwick Junction urban regeneration project.

Moses Mabhida Stadium

In preparation for FIFA 2010 five new stadiums were built, and five stadiums were refurbished at an initial cost of R8.4bn which doubled to R16.3bn without any public oversight (Maharaj, 2011). In Durban the initial proposal was that the ABSA Rugby Stadium could have been upgraded for R54million (venue for Rugby World Cup), and this was accepted by FIFA. However, in a less than transparent process, the Durban Municipality took a decision to build new stadium in order to demonstrate its 'world class status'. The initial projected cost was R1.8bn but this escalated to R3.4bn.[4]

The 70 000 seater Moses Mabhida Stadium (named after an ANC struggle hero) was built on the site of the Kings Park Soccer Stadium. It was designed as an iconic architectural edifice with an arch in the form of a 'Y' (similar to the South African flag), symbolically reflecting social cohesion and unity. Planned to 'ensure 365-day-a-year usage, the stadium includes retail space, an open air amphitheatre, a cable car, which takes visitors to a viewing platform on top of the 350m arch, and the 'Big Swing', the world's only stadium swing and the largest swing of its kind'.[5] The intention was to bid for more international events, and this was emphasised by the head of strategic projects, Julie-May Ellingson:

> With this kind of stadium – and the proposed sporting precinct – we are confident that we will be able to host any world- class event, including the Olympic Games and the Commonwealth Games (Makhaye, 2006:12).

International journalist Dave Zirin described the Moses Mabhida stadium as the 'most breathtaking he'd ever seen', but he called for a reality check: 'This is a country where staggering wealth and poverty already stand side by side. The World Cup, far from helping this situation, is just putting a magnifying glass on every blemish of this post-apartheid nation' (Zirin 2010: 7). Similarly, Bond, Desai and Maharaj (2010:419) pointed out the contradictions illustrated by the stadium:

> Durban's 70,000-seater Moses Mabhida stadium is delightful to view, so long as we keep out of sight and mind the city's vast backlogs of housing, water/sanitation, electricity, clinics, schools and roads, and the absurd cost escalation. Harder to keep from view is next-door neighbour Absa Stadium, home of Sharks rugby, which seats 52,000 and which easily could have been extended (see Plate 15.1).

According to the eThekwini's Head of Strategic Projects, Julie-May Ellingson, the completion of the Moses Mabhida Stadium, a 'world-class engineering and architectural feat', was a 'true example of what can be achieved by team work, the sharing of skills and the collaboration between public and private enterprise'.[6]

---

3   2010 FIFA Soccer World Cup: Durban, KZN: Business Plan, Dr. Michael Sutcliffe – City Manager: eThekwini; Ms. Julie-May Ellingson – Head : Strategic Projects and 2010 Programme, www.pmg.org.za/docs/2006/060829ethekwini1. ppt (accessed 5 November 2013).

4   See Sole (2010) and Taal (2011) for an analysis of how the politically connected elite won tenders for the construction of the stadium, and how the Competition Commission ruled that major construction cartels colluded to artificially inflate costs of stadium.

5   eThekwini Municipality State of Local Innovation Report, October 2011, p. 6.

6   http://www.property24.com/articles/moses-mabhida-stadium-completed/ (accessed 9 November 2010).

There was consensus amongst trade unions, civil society organisations and the media that the government ought to harness the skills and organisational strategies that produced the successful world cup to deliver basic needs for the poor. The editor of the Durban based *Sunday Tribune* Philani Mgwaba argued the government needed to work with the same single-mindedness of purpose in serving the needs of South Africans as it had done to meet the stringent requirements of FIFA:

> Somehow we have to get them to accord us, the people of this country, the same respect and reverence they
> have shown to the world soccer body. The urgent problem areas are well known: abysmal state hospitals,
> dysfunctional public schools and dangerous criminals having a field day while the majority of law-abiding
> citizens cower from fear of violent crime. Equally important are what should be simple and routine tasks: timeous
> and regular rubbish collection, pothole filling and keeping our traffic lights working (Mgwaba, 2010:28).

A major concern was the sustainability of the ten stadiums after the 2010 World Cup, and the possibility that they could become costly white elephants, adding to the financial burden of municipalities. The majority of host cities were 'financially incapable of footing the bills for their stadiums over the next 30 years', and the annual maintenance costs would have to be borne by ratepayers (Eybers, 2010:8). Trevor Phillips, former director of the South African Premier Soccer League, posed a critical question:

> What the hell are we going to do with a 70,000-seater football stadium in Durban once the World Cup is
> over? Durban has two football teams which attract crowds of only a few thousand. It would have been more
> sensible to have built smaller stadiums nearer the football-loving heartlands and used the surplus funds to have
> constructed training facilities in the townships (Minto, 2010:1).

By May 2010, the cost of maintaining the Moses Mabhida stadium had not been established (Eybers, 2010). In November 2010 its maintenance and operation costs was estimated to be R28m annually (Seale, 2010). In 2013 the Moses Mabhida Stadium had an operating loss of R34.6m.

In Durban popular sports like cricket and especially rugby had the capacity to fill stadiums. At the planning stage there were suggestions that the specific requirements of cricket and rugby should be incorporated into the design and configuration of the new stadium in order to contribute to its sustainability. However, the city manager Dr Michael Sutcliffe refused to consult with the cricket and rugby authorities. After the World Cup, attempts were made to persuade rugby officials to shift to the new stadium. There were two factors inhibiting the rugby move:

> The new stadium only has 150 suites, while Kings Park currently leases out 350 private suites, and the drop in
> suites would translate into a drop in revenue, as these are quite lucrative at the current stadium. The lay out of
> the new stadium also sees fans sitting further away from the action, which dampens the atmosphere somewhat
> (Fergusan, 2011:1).

South African Rugby Union President Regan Hoskins 'criticised those managing the Moses Mabhida Stadium ... for failing to implement a plan before they had built the new stadium ... it is tragic for us as a nation that we now have to act in reverse gear' (Bishop and Mgaga, 2010:1). According to the CEO of South African cricket, Gerald Majola, the playing area at the new stadium (was) too small to host cricket games (and) blamed the authorities for failing to consult cricket authorities before construction (Bishop and Mgaga, 2010:1). In August 2010 the stadium was described as follows:

> Empty retail space and low foot traffic at Durban's iconic Moses Mabhida Stadium has raised concern among
> tenants ... more than one month after the city hosted the semi-final between Spain and Germany the retail area
> at the stadium is anything but a buzzing hive of activity (Umar, 2010:3).

In a counter-view, sport journalist Luke Alfred contended that one had to adopt a broad, futuristic perspective, 'beyond debates about what they will cost and how they will be used', where the stadiums will become iconic global monuments like the 'Statue of Liberty, the Eiffel Tower or the Sydney Harbor Bridge':

With sport to some extent replacing nationalism (or being one of the ways in which the nation expresses itself in these post-nationalistic times) the stadiums for the world cup will express the best of what South Africa has to offer as the century progresses. They will become monuments by which the world recognizes this country and by which we define ourselves (Alfred, 2008:1).

A more general question was whether huge soccer stadiums would 'survive as the centerpieces of World Cup contests?' (Shaug, 2008:8). There were mixed views with some agreeing, while others suggested future stadiums would be smaller and designed for television, accommodating 30 000 elite spectators comprising FIFA officials and the affluent (Shaug, 2008). However, it is evident from planning for Brazil FIFA 2014 that the giant stadiums are still around e.g. Arena de Itaquera (Sao Paulo) 68 000 capacity; Estádio do Maracanã (Rio de Janeiro) 78,838 capacity.[7] In Durban another major infrastructural project was the King Shaka International Airport.

King Shaka International Airport

The King Shaka International Airport (KISA), located approximately thirty kilometers north from central Durban in La Mercy replaced the Durban International Airport (DIA) which was located south of the city. The airport was conceptualized in the 1970s, but was only built as part of the FIFA 2010 infrastructure projects. Construction began in 2007, and KSIA was officially opened for business in May 2010. KSIA was constructed primarily because of the inability of the DIA to handle large aircraft traffic, and there was no room for the expansion of the Durban International Airport's runway (Crosby, 2013).

However, the construction of the airport has been viewed as an extravagant, rushed development, costing approximately R7.4 billion, as the Durban International Airport was adequate for handling the volume of air traffic in Durban. Furthermore, Durban did not warrant a new airport because the DIA was upgraded in 2005 at a cost of approximately R158 million in three distinct phases – the upgrade of the passenger terminal, the parking structure upgrade and an increase in the apron capacity (Crosby, 2013). One of the major criticisms of KISA is that the volume of international air traffic at the airport is not enough to justify its costly construction. Currently, there is only one daily international incoming flight at the King Shaka Airport, from Dubai (Crosby, 2013). If the airport is to recover costs, and become sustainable, then more international incoming flights need to be secured.

Durban Beachfront Upgrade

The upgrade and remodeling of the beachfront in Durban was one of the major 2010 FIFA World Cup infrastructure developments the city. This consisted of four key elements, namely, the promenade; the development of key nodes along the promenade; roadworks; and landscaping.[8]

The long term goal of upgrading the promenade is geared toward extending it from the Port's northern breakwater to the mouth of the Umgeni River. However, a more immediate goal was extending the Promenade from the uShaka Marine World beach to the Country Club Beach. Also included in the construction plan were walkways leading to the Moses Mabhida Stadium.

However, a major concern was that upgrading the Durban beach front was used as a strategy to displace hundreds of informal traders and thousands of subsistence fishermen from the area, and depriving them of their livelihoods. The informal traders have not been offered any compensation or guarantee that they would be allowed to move back to their spaces after construction work was completed. A major issue is what is the legacy for the poor or disadvantaged when mega-events are hosted in developing countries?

---

7   http://www.telegraph.co.uk/sport/football/world-cup/10383648/World-Cup-stadiums-a-venue-guide-for-Brazil-2014 (accessed 20 April 2014).

8   Durban: Beachfront upgrade information sheet. eThekwini Municipality, Strategic Projects Unit.

Displacement: A Legacy for the Poor?

In May 2010 the Deputy Director-General of Social Development, Selwyn Jehoma revealed that about half of South Africa's population (24 million) was 'living in abject poverty. These included children, the elderly, the unemployed, and caregivers of children' (Ross, 2010:6). According to Greene (2003:163) 'under the logic of event-oriented development, the visibility of poverty becomes paramount in renewal schemes, and preparations often involve removing the poor from high-profile areas surrounding event venues, without significant attention to long-term solutions to slum problems'.

Hence, it was not surprising that FIFA 2010 was used as a smokescreen by the Durban and KZN authorities to try to remove informal traders and shack dwellers. The intention was to hide the poor and the homeless who would be invisible to international visitors. The KwaZulu-Natal Elimination and Prevention of Re-emergence of Slums Act (2006) was promulgated to eliminate shacks in and around the city, and the intention was to extend this law to other provinces. However, this law was declared to be unconstitutional by the Constitutional Court of South Africa on 14 October 2009 (Selmeczi, 2011).

In terms of street traders the FIFA by-law document required Durban and other host cities create exclusion zones where informal trading will not be allowed. This left many informal traders very vulnerable, with an inability to make a living. According to the FIFA by-law:

> no person, except with prior written permission from the city specifically granted for the competition, may carry on the business of street trading at any controlled access sites, exclusion zones, restricted areas, public garden parks, on pavements next to auto teller machines, pavements next to declared national heritage resource buildings, next to state or municipality buildings or churches or pavements next to public amenities among others during the competition.[9]

Urban regeneration projects associated with mega-events can also impact negatively on local communities in terms of displacement of livelihood opportunites, and forced relocations. A classic case in Durban would be strategies to destroy the centry-old Warwick Market.[10] The Warwick Avenue Triangle, an inner city community, and one of the oldest mixed residential areas in Durban, had defied the apartheid state's strategies to destroy it.

The public outcry against the Durban municipality's plans, to replace the historic market with a mall, displace poor traders from the Warwick Avenue area and deny them their livelihoods, reveal significant continuities between the apartheid and democratic eras. This displacement was promoted as part of the city's FIFA 2010 infrastructure projects and urban regeneration strategy. The essence of the city's case for the destruction of the market and the construction of a mall was that: the present site was a dirty, disruptive and 'illegal' blot on a modern city (very much similar to the 'sanitation syndrome' of the apartheid era which equated contact with blacks with disease and contamination); and the public-private partnership with the mall developers would improve the transport infrastructure in the area.

According to City Manager, Dr Mike Sutcliffe and Deputy Mayor, Logie Naidoo, those who were opposing the development of the mall in Warwick Avenue were variously: preventing poor people from enjoying the privileges associated with malls; pursuing narrow, ethnic, racist agendas; opposing a 'golden opportunity for investment'; opposing the democratic majority; and wanting the traders to remain 'trapped in the second economy'.

Traders, street vendors, unions, civics, NGOS, architects, planners, academics and researchers were united in condemning the destruction and displacement as this would adversely affect the livelihoods and heritage of black communities in the city. The construction of the mall has serious consequences for the livelihoods of 600 traders, as well 8000 other formal and informal traders and street vendors who depend on the Warwick Market for their supplies. In a last ditch attempt to save the market, the traders have resorted to legal action. The mall developers have abandoned their project, and the market is safe – for now.

---

9   http://www.iol.co.za/news/south-africa/fifa-dictates-to-city-on-informal-trading-1.385942 (accessed 12 November 2013).

10   This section on the Warwick Market draws generously from Maharaj (2013).

Organizations and advocates for the poor in Durban contended that large events such as the Soccer World Cup resulted in urban inequality as opposed to promoting equality and social justice: 'Durban fisherfolk have witnessed rich people fishing off expensive boats and yachts unhindered while working-class subsistence fishermen suffer police harassment and arrests'.[11]

On 16 June 2010 (the anniversary of the 1976 Soweto riots) the Durban Social Forum comprising a coalition of more than 20 civil society organisations organised a march to the city hall and handed a memorandum to officials expressing concern about the 'way in which the World Cup was implemented by FIFA, its corporate partners, politicians and bureaucrats'.[12] The issues highlighted in the memorandum are summarised in Table 15.1. The Deputy Mayor, Logie Naidoo, responded that everyone will benefit from the world cup and that the key legacy was the country being at the centre of international attention (Mdluli, 2010:7).

Table 15.1   Memorandum of grievances against FIFA 2010

Only the rich will score in 2010. The ANC has not given a World Cup for all, but again chose to deliver to the rich!
Who can watch the live games? The rich sit in R3 billion stadiums. Only the elite are allowed to profit.
It is illegal for ordinary people to make a living out of the World Cup – anyone caught will be fined and jailed. Locals, street traders, fisherfolk, vendors and artists' rights denied.
Everything free for Fifa ; uninterrupted electricity – but electricity costs go up for us. No water, electricity, or health (care) for the poor. No facilities or services in poor areas!
Building of schools stopped for stadiums.
R17 billion would house over one million people.
Corruption whistle blowers asasssinated.
Under-resourced, public hospitals forced to keep wards empty for players and tourists while treatment for the poor is deferred.
Tax payers must maintain unused stadiums.
New apartheid: forced removals of the poor, vulnerable, children, homeless, and refugees to 'transit camps'.
Dereliction of police duties – crime increases, children targeted by traffickers and molesters, drunken driving, assaults, and abuse of women and children.
Few short term benefits – huge long term costs. No country has benefited from big, international, sporting events; they are still in debt.
High environmental costs while government paints green wash picture

Source: Adapted from 'South Africa: World Cup for All – Durban Social Forum, 16 June 2010'

Conclusion

This chapter assessed the legacy implications of FIFA 2010 related infrastructure in Durban. A key issue was sustainability, especially in the case of the Moses Mabhida Stadium. A damning indictment against city officials was their failure to consult and accommodate the needs of rugby and cricket before the stadium was designed. In terms of the challenges facing KSIA there is a need for more international flights. In the context of the current economic climate, the implications of spending R7.4 billion on an extravagant airport is questionable, especially since the DIA could have been used if proper management plans were executed.

---

11   Memorandum of Grievances to: KZN Premier Zweli Mkhize, Durban Mayor Obed Mlaba, Deputy Mayor Logie Naidoo and Durban City Manager Michael Sutcliffe – RE: Grievances about World Cup 2010 management.

12   'South Africa: World Cup for All – Durban Social Forum, 16 June 2010'.

The beachfront upgrading had displaced subsistence fishermen and informal traders. Threats to destroy the Warwick market also undermined the livelihoods of street traders. Civil society organisations protested about the extravagant expenditure of public funds against the background of widespread poverty and increasing socio-economic inequalities.

There have been concerns that the different 2010 projects only paid rhetorical lip service to reducing the socio-economic inequalities and addressing the needs of the poor; while they were largely driven by elite corporate interests; and were underwritten with public funds, with limited or no public participation. This was compounded by the failure of the dominant elite alliance to consider the critical issues raised by civil society organisations with strong grassroots support. The question of who benefits needs to be considered when hosting a mega-event, especially in a developing country context.

Although events such as the FIFA World Cup 2010 do produce benefits, the international experience suggests that the privileged tend to benefit at the expense of the poor, and that socio-economic inequalities tend to be exacerbated. There are many reasons for the failure of such projects, all of which centre around the unequal relationship between the political and business elite, and the poor, which manifests itself or is translated into unequal legacies.

This was because public sector infrastructure expenditure favoured the elite and bypassed the poor, and in some cases exacerbated their condition with threats of forced removals, displacement and loss of livelihoods. Hence, notwithstanding the pro-poor rhetoric, the ultimate impact was anti-poor. Green (2003:163) has labelled such mega-event experiences in urban centres in the South as 'staged cities' in order to illustrate the contradictions 'between the mega-event as a means of constructing an image of 'development' and the actively concealed landscape of the urban poor'.

References

Alfred, L. (2010). 'The 2010 Story No one Tells'. *Sunday Times*, 5 October, p. 1.

Andranovich, G., M. J. Burbank and C. H. Heying (2002). 'Mega-events and urban development'. *The Review of Policy Research,* 19, pp. 179–202.

Baade, R. A. and V.A. Matheson (2004). 'Mega-sporting events in developing nations: Playing the way to prosperity?' *South African Journal of Economics*, 72, pp. 1085–1096.

Baumgarten, J. C. (2010). *The Role of Mega-Events in Triggering Large Scale Infrastructure Developments*. World Tourism Council, Johannesburg.

Bass, O. and U. Pillay (2008). 'Mega-events as a response to poverty reduction: The 2010 FIFA World Cup and its urban development implications'. *Urban Forum*, 19, pp. 329–346.

Bishop, J. and T. Mgaga (2010). 'Sharks say no to Moses'. *The Witness*, 20 November, p.1 http://www.witness.co.za/index.php?showcontent&global (accessed 15 February 2011).

Centre for Development Enterprise (1996). *Durban: South Africa's Global Competitor?* Johannesburg: Centre for Development Enterprise.

Cottle, E. (2010). World Cup and the Construction Sector – Campaign for Decent Work. Building and Woodworkers International, http://www.sah.ch/data/D23807E0/ImpactassessmentFinalSeptember2010EddieCottle.pdf (accessed February 18, 2011).

Crosby, M. (2013). Aerotropolis Urban Growth? A case study of the King Shaka International Airport. Unpublished Masters Thesis, University of KwaZulu-Natal.

Dayanand, S. (1996). Local Government Restructuring and Transformation: A Case Study of Durban. Unpublished Masters Thesis, University of Durban-Westville.

Desai, A., P. Bond and B. Maharaj (2010). 'Lessons from the World Cup', in Maharaj, B., A. Desai, P. Bond (eds.) (2010). *Zuma's Own Goal – Losing South Africa's 'War on Poverty'*. Trenton, NJ: African World Press, pp. 417–432.

Essex, S. and B. Chalkley (1998). 'Olympic games: Catalyst of urban change'. *Leisure Studies,* 17, pp.187–206.

Eybers, J. (2010). Stadiums will be a burden. *City Press*, 16 May, p. 8.

Fergusan, R. (2011). Moses Mabhida v Kings Park. http://www.sharksworld.co.za/2011/11/03/moses-mabhida-v-kings-park (accessed 12 April 2012).

Goliger, A.M. (2005). 'South African sports stadia – from the perspective of the 2010 FIFA World Cup'. *Bautechnik*, 82, pp. 174–178.

Greene, S.J. (2003). 'Staged cities: Mega-events, slum clearance, and global capital'. *Yale Human Rights and Development Law Journal*, 6, pp. 167–187.

Hall, C.M. (2006). 'Urban entrepreneurship, corporate interests and sports mega-events: The thin policies of competitiveness within hard outcomes', in J. Horne and W. Manzenreiter (eds.) (2006). *Sports Mega-Events: Social Scientific Analyses of a Global Phenomenon*. Oxford: Blackwell Publishing, pp. 59–70.

Ling, C. (2005). Mega-events and infrastructure improvements – The Case of the Olympic Games in Beijing 2008. Masters Thesis, Centre for East and South-East Asian Studies, Lund University.

Maharaj, B. (1998). 'The Olympic games and economic development – Hopes, myths, and realities: The Cape Town 2004 bid', in R Freestone (ed.), *Twentieth Century Urban Planning Experience*. University of New South Wales: Sydney, pp. 583–588.

Maharaj, B. (2011). '2010 FIFA World Cup: (South)"Africa's time has come"?' *South African Geographical Journal*, 93, pp. 49–62.

Maharaj, B. (2013). Contesting post-apartheid urban displacement: the struggle for the Warwick Market in Durban. Paper presented at the International Conference on Development-Induced Displacement and Resettlement, Refugee Studies Centre, University of Oxford, 22–23 March.

Maharaj, B. and K. Ramballi (1998). 'Local economic development strategies in an emerging democracy: The case of Durban in South Africa'. *Urban Studies*, 35, pp. 131–148.

Makhaye, C. (2006). 'City's hopes built on R1.6bn stadium' *Sunday Tribune*, 18 June, p. 12.

Matheson, V. (2006). *Mega-Events: The effect of the world's biggest sporting events on the local, regional, and nation economies*. College of the Holy Cross: Department of Economics.

Mdluli, A. (2010). 'Stadium proves city can deliver. "South Africa: World Cup for All – Durban Social Forum, 16 June 2014"'. *Mercury*, 17 June, 7.

Mgwaba, P. (2010). 'Our own people also deserve the best'. *Sunday Tribune*, 27 June, p. 28.

Minto, J. (2010). 'Cup thrives at poor's expense' http://www.stuff.co.nz/the-press/opinion/columnists/john-minto/3811829/Cup-thrives-at-poors-expense (accessed 11 November 2010).

Oliver, S. and W. Oliver (2001). *Touring in South Africa*. Cape Town: Struik Publishers.

Pieterse, E. (2012). 'World Cup promise and consequences for South African cities', in Z. Asmal (ed.), *Opportunities – Design, Cities and the World Cup*. Designing South Africa, pp. 82–87.

Ross, K. (2010). 'Poverty is a huge challenge'. *Daily News*, 6 May, p. 6.

Schaug, E. (2008). '2010: last of the giant stadiums'. *Daily News*, 7 February, p. 8.

Seale, L. (2010). 'Do our stadiums have a future?' *Mercury*, 3 November, p. 7.

Selmeczi, A. (2011). 'From shack to the Constitutional Court': The litigious disruption of governing global cities'. *Utrecht Law Review*, 2, pp. 60–76.

Sole, S. (2010). 'Durban's Moses Mabhida Stadium', in C.S. Herzenberg (ed.) (2010). *Player and Referee – Conflicting interest and the 2010 World Cup*. Pretoria: Institute for Security Studies, pp. 169–201.

Sutcliffe, M. (2006). 'It's not just kick-off time in Germany' *Metro/eZazegagasini*, 9 June, p. 4.

Taal, M. (2011). 'Their Cup Runneth Over: Construction companies and the 2010 World Cup', in E. Cottle (ed.), *South Africa's World Cup – A Legacy for Whom?* Pietermaritzburg: University of KwaZulu-Natal Press, pp.73–100.

Tilley, V. (2006). 'Scary economics of the mega-event'. *Business Day*, 4 May, www.businessday.co.za (accessed: 16 September 2010).

Umar, R.S. (2010). 'Stadium tenants struggling'. *Daily News*, 13 August, p. 3.

Zimbalist, A. (2010). 'Is it worth it?' *Finance and Development*, (March), pp. 8–10.

Zirin, D. (2010) 'Colossal World Cup Foul'. *Mercury*, 27 October, 2010, p. 7.

PART V
Social Dimensions: Communities and Urban Transformation

Social Dimensions: Communities and Urban Transformation

David Powell

The relationship of communities to the mega-events hosted by their cities is influenced by the political, economic, social and cultural conditions which those communities experience in normal circumstances. This is heightened by the effects – positive and negative, planned and unexpected, tangible and intangible – of living in close proximity to large scale public and commercial projects. Large scale projects such as the Olympic and Paralympic Games bring with them accelerated change requiring major investment not just in sporting infrastructure but transport, security, tourism and hospitality facilities, housing and environmental development.

Mega-events and the preparation for them are in some ways not so dissimilar – in their effects on local neighbourhoods and the communities most directly exposed to them – to large scale development and regeneration projects. The urban areas which attract the promoters of mega events to a city – offering space to plan and to play with – are often the same areas which have had serial government and commercial interest as the location for large scale projects which other more intensively developed areas cannot easily accommodate. It is the privilege and pain of communities living next to areas such as London's Lower Lea Valley – site of the London 2012 Olympic and Paralympic Games – to be the subject of decades of large scale planning and building. In such landscapes and their timelines, mega-events arrive at some speed, with their focus on the immovable starting date and their hugely ambitious, often deliberately extravagant, claims for the complete transformation of the futures of cities, the regeneration of neighbourhoods and the transformation of individual lives.

The promoters and their partners in national, city and local government focus on the outward and visible signs of sporting excellence, world city ranking, international and domestic reputation, and the wholesale improvements to people and places. People living in the places and communities nearest to the site of these events, those who live closest to the impact zones, have to seek their own ways to deal with these heightened pressures, expectations and uncertainties which come when major changes happen at speed.

One factor which many of the very large scale events and regeneration projects share is a new, often single purpose delivery organisation, put in place to represent the government, financial and promoters "stakeholders" interests and to ensure that the substantive bodies responsible for planning and delivery speak and act with a concerted voice. These tend to be superimposed onto or inserted into the existing institutional arrangements for decision making, governance and accountability. The presence of the directly elected and therefore locally accountable tiers of government (in London's case the 5 "host" local authorities and the London Mayor) is often designed to represent the local communities.

Often, people living and working in the areas of greatest impact, locally based small businesses, community organisations, local interest groups and cultural enterprises do not feel adequately represented by or cared for by their local democratic institutions. Too often, experience shows that such new development and delivery bodies represent – or at least are widely perceived as – another layer of bureaucracy and yet another place where decisions are taken without much regard for local views, voices and interests.

Neither is it clear that as a rule, such bodies have within them the institutional memory or capacity to listen to, engage with and respond respectfully to local concerns and suggestions. By respectfully here, I mean with the capacity to listen and change position as a result of measured conversations with those voices, concerns and interests which are normally excluded from or marginally engaged in decision making.

The established – the establishment – position may presume that communities of place and of interest will automatically be against major change and development, that stasis is better than progress. The "top down and centre out" presumption is often that, locally, people are hostile or at least indifferent to the aspirations and ambitions with which mega-projects are generously freighted by their promoters, and therefore to the benefits and opportunities which might be generated by them. Experience in London, with the Olympics and around other substantial development sites such as Kings Cross or the former Docklands areas in East London, shows that community responses to mega-events are far more nuanced.

The range of responses will run across the spectrum from early adopters to implacable opponents, through those who fall in or out (or in and out) of love with the big project and its big ideas. The indifferent and those for whom other matters press in with more threat or urgency – issues of poverty, poor health, security of housing or employment, and personal safety – will of course be present in large numbers in types of communities and neighbourhoods which find themselves in the penumbra of the largest scale regeneration and events. And of course people's positions will change over time, and are likely to be greatly affected by a whole host of personal and external circumstances, events as well as the wider public perceptions of hostility, indifference and support which events like the Olympic and Paralympic Games inevitably experience.

The same range of responses is also likely to be true amongst certain communities of interest. In a short research project in 2013 Debra Reay and I looked at the different strategies adopted by groups of artists, arts and cultural organisations in three distinct neighbourhoods, each on the boundary of the London Olympic park and so close that they were either within or just outside the Olympic exclusion zone (the 35-day, one-kilometre Brand *Exclusion Zone*).

In Stratford, a group of established cultural organisations set up a professionally run network in 2007 (Stratford Rising[1]) as a consortium to support the growth of Stratford's cultural sector and its contribution to the regeneration and reputation of the area. This group knew that the journey through the Olympic project would be vitally important for them, but understood that their sustainability as long term contributors to the regeneration of East London was non-negotiable. With Stratford as the main access point for all visitors to the Games, and with the intense pressures on the theatre, arts centre and other venues which could not maintain their normal programmes through the summer of 2012, it proved to be invaluable to have a strong, communicative and collaborative arrangement between these organisations during the Games period as well as providing a platform for their longer-term intentions.

In Bow, Bow Arts Trust[2], a long established social enterprise providing affordable artists' studios, took a strategic decision not to get involved directly with the Olympic cultural project, but rather to focus on its long term survival and on the development of Bow's emerging cultural scene and the wider East London visual arts market. Working with its own artists and with local partners – regeneration agencies and artists, independent studios, galleries, cafes and others – it established "London's Artist Quarter"[3] which it described as "a valuable brand for an important community that is engaged (with) and supporting the massive regeneration programme of this historic gateway to London". It did this as a deliberate move to establish an identity for the area, enabling many hundreds of creative, cultural and related social enterprises to benefit from this elective shared identity.

In the third area, Hackney Wick and Fish Island, which is immediately to the north-west of the Olympic Park, a wide range of artists, arts companies and creative micro-enterprises have since the 1980s co-located to make an informal creative micro-cluster. This consists of galleries, arts and social spaces and many studios and production units, alongside the small scale commercial businesses which still operate there, many of them the last survivors of the pre-Olympic Lower Lea Valley economy. Much of the networking and intelligence sharing in the area (in the context of the years running up to the Games) was relatively informal, and focussed on the meeting places – work and social – in the area. Some of this was focused in 2008 on the first Hackney Wicked Festival.[4] The Hackney Wick and Fish Island Cultural Interest Group[5] was founded in 2009 to "facilitate a permanent, sustainable, creative community in Hackney Wick and Fish Island and particularly to advance the arts and culture". This group established formal relationships with a range of institutional stakeholders, and provide a focal point for the concerns of a very disparate and autonomous group of artists and cultural enterprises.

In each case, these cultural communities moved from informal to more formalised ways of navigating their way towards, through and past the major disruptions and beguiling opportunities which the Olympic event offered. The genesis of each network was different, but fundamental to all three was the need to establish a platform that was bigger than the individual member of the community. This requirement was important, because one of the issues which confronted these three groupings was the difficulty of the promoters of the

Games – the delivery agencies – in dealing with the very small, the hyper-local as well as the very large. This is a characteristic likely to be present in situations where the largest scale regeneration, infrastructure or event-hosting projects is proposed for dense urban areas.

There may be many reasons for this institutional difficulty with listening, and many factors involved. An important question for people living and working in the areas most closely affected by the mega-project in question, is how to speak with authority to the existing and new government and local decision making agencies. How do these hyper-local groups gain the authority to speak with authority to authority?

The anecdotal evidence suggests that most political and organisational energy is spent integrating the major institutions and interests required to deliver very large projects, so that agencies and institutions can work together to deliver effectively, to time and to budget, in these relatively infrequent circumstances. Most effort is expended in keeping these players of scale working together: the absence of such good working relationships will have featured heavily in the mega-projects risk analysis. The absolutes of a specific, immovable deadline will keep this in the forefront of the promoters' minds.

One corollary is that much less effort is committed to getting the very large and the very small to communicate and collaborate effectively. We find that across the major projects in East London over the last 30 years or more, there has been a cumulative failure to find effective ways of engaging with communities, or to sustain effective ways of developing and support local leadership. One explanation advanced was that the sheer scale of London's mega-event was such that only the largest cultural producers and organisations would bring sufficient confidence that they had the capacity (to guarantee delivery) as well as the reputation (to assure and attract global talent and a global audience). But this reaction – to bring in the large at the expense of the local and small – is seemingly hard-wired into the way in which development and delivery bodies are required to operate. It has been apparent within (and also equally informally acknowledged by) the agencies steering the development of London Docklands and the Olympics legacy.

There is little evidence that large public or public-private agencies involved in projects like London's Olympics have the capacity to learn how to deal with the hyper-local voice. All the evidence suggests that individuals living in local communities directly affected by very large scale projects learn such lessons very effectively. They may learn with difficulty, and they may not succeed in rendering sufficient benefit out of the glorious behemoths in whose shadows they live to balance the short and longer term pain that their proximity may bring them. The analyses in the following chapters throw further helpful light on this, even though the voices of those who are required and even privileged to live so closely with mega-projects and events may not always be audible in the narrative.

Notes

1   Stratford Rising web link http://stratfordrising.com/home (retrieved 24/07/2014).

2   Bow Arts Trust web link http://www.bowarts.org/ (retrieved 24/07/2014).

3   London's Artist Quarter web link http://www.londonsartistquarter.org/ (retrieved 24/07/2014).

4   Hackney Wicked Festival web link http://www.hackneywickedfestival.co.uk/hackney-wicked-festival/ (retrieved 24/07/2014).

5   Hackney Wick and Fish Island Cultural Interest Group we blink http://hackneywick.org/about/ (retrieved 24/07/2014).

Chapter 16

The Mega-event Experience: The Formula 1 Grand Prix in Turkey

Semiha Sultan Eryilmaz and Hüseyin Cengiz

Introduction

Formula 1 (F1), with the construction of the track it required, the triggering of further investment it induced within the surrounding location and the numbers of local and foreign participants attending, is the biggest mega-event held in Istanbul to date. As a result of its strategically important location on the Bosporus, Istanbul has played an important role throughout its history. It was the capital of the eastern part of the Roman Empire and of the Byzantine and Ottoman Empires. The city brings together different religions and cultures as well as being the site of their construction and integration. In terms of socio-economic development, Istanbul is Turkey's leading city and is in the process of becoming a post-industrial international business centre, it has also become a city that hosts global events.

Events organised in Istanbul have significantly increased the number of local and foreign tourists coming to the city. In addition to cultural activities such as the International Film Festival, Theatre, Music, Jazz Festivals and the Istanbul Biennial, the city was also the EU Capital of Culture in 2010. Over recent years, the total number of international congresses and sports events held in Istanbul has increased and their character diversified.

Istanbul's great importance to the national and regional economies is matched by its importance in terms of service sector activities within the global urban network. While in 1999 Istanbul was still classified as a 'Gamma City' (Beaverstock et. al., 1999), by 2012 it had advanced to a city among the Alpha-group in the hierarchical ladder prepared by the Globalisation and World Cities Study Group (Derudder and Taylor, 2012). Alpha-cities are world cities that are associated with leading economic regions and states in the world economy.

Starting from the period during which an F1 grand prix was held in Istanbul between the years 2005 and 2011, this chapter reveals the impacts of hosting the mega-event on urban locations, an important theme in mega-event literature. The purpose of this chapter is to:

- explain the relationship between tourism and branding oriented urban development policy and mega-events,
- show how F1 might affect a city or region,
- highlight the importance of F1 among the major events held in Istanbul,
- reveal the spatial impact of the chosen location for the F1 track, on the region.

Tourism, Branding and Mega-events

The impact and importance of tourism as an industry with economic, cultural, and social dimensions is gradually increasing. Tourism is mobilizing economic activities, providing currency exchange in the country/region, raising employment, increasing personal income levels and providing an increase in public revenues through, for example, taxation. Additionally, it contributes to the renewal of urban infrastructure and the development of the urban image. It is these aspects that make the tourist industry play such an important part in the development of local and regional space and causes policy makers to produce plans for the development of tourism in their locations.

According to Harding and Le Gales (1998), cities, since the 1980s, have emerged as potential centres of governance generating and arranging the interests of groups and institutions (Harding and Le Gales, 1998 cited in Martins, 2004: 1). Sassen (1991) argues that with the integration of European countries, the political and economic significance of cities is growing and mayors have recognised their increasing importance. Local administrative authorities, in charge of directing the economic development of their city, have developed policies targeting enhanced competitiveness by creating, (1) consumption centres where prestige and culture is developed to attract tourists, (2) international company management centres, (3) prestigious public offices and (4) different areas of public and private investment (Sassen, 1991 cited in Martins, 2004: 1).

In times of industrial decentralisation, economic restructuring and globalisation, cities have recognised mega-events as important vehicles for urban renewal and regeneration, urban status promotion and as a way to attract new investments and modernise the economy. In this context, mega-events can be a popular choice in urban politics. The impact of mega-events on the hosting city are often more important than the profile and scale of the event being organised (Essex and Chalkley, 2004: 201–2). Roche states (2000) that economic, artistic or sporting mega-events attract worldwide public interest via mass media and cause numerous participants and audiences to travel. Roche also argues that mega-events create 'global concentration moments' and that they have a position that establishes, shows and reflects the process of globalisation (Roche, 2000: 1–30).

In the literature related to events a variety of concepts and definitions is used, some of which are interchangeable. Events can, for instance, be classified by the time when they are organised, the region where they are marketed and their theme. Scale and theme based classifications of events in the literature are often referred to in different ways such as 'hallmark, mega, major and special' (see Table 16.1).

Hallmark and mega-events have been an important part of the tourism literature since the 1980s with an increasing number of cities and countries competing to host sport mega-events like the Olympic Games, FIFA World Cup and Formula 1 (F1), and many countries adopting this strategy as a component of their national sports policies. Competition has also increased as a result of the adoption of the legacy narrative which promises urban renewal, new sports facilities, increased tourism and positive image effects as well as wider social and cultural benefits (Gratton et. al., 2005: 233). Effects usually taken into account during the planning, development and marketing stages of events are discussed under economic, tourism/commercial, spatial/environmental, political/administrative and psychological headings. However spatial/environmental, social/cultural and psychological effects are often not fully evaluated.

Formula 1 as a Mega-event

The F1 event consists of races generally held on tracks that are organised in different countries with each placed in the racing calendar and referred to as a Grand Prix (GP). Europe, as its most important market, is the traditional centre of F1. Every season new Grand Prix races are included or existing GPs are removed from the list. F1 includes competitions between car designers, engine and tire manufacturers and drivers (Bessit, 2006). The F1 event season usually begins in March and ends in November with an average of 19 races held over the course of the racing season. F1 events span different continents and require the construction of high cost tracks with the exception of the Monaco, Valencia and Albert Park Grand Prix, where no tracks were constructed because 'street tracks' in the respective city centres are being used instead. Depending on their location, tracks may feature river banks, woodland, desert, artificial islands etc. Tracks can be chosen to be placed in small settlements such as Spa, Hockenheim or in large metropolitan areas such as Shanghai, Istanbul and Sao Paulo. Each F1 GP is actively viewed by an average of 120,000 people at the racing venues and followed by at least 300 million people on television; making it clearly a mega-event.

F1 is organised in each of the hosting cities once per year on the dates announced in the event calendar for as long as the contract between the FIA (Fédération Internationale de l'Automobile) and each host city continues. Each Grand Prix has important economic effects and generates considerable media interest. According to the classification developed by McArdle (1998) in terms of time, location and theme, F1 is deemed to be in the motorised sports events category with a definite starting date and takes place in a specially designed area (McArdle, 1998: 97).

Table 16.1   Mega-events classification on the basis of scale and theme

Author	Scale	Theme Type / Subject
Ritchie, 1984	National International	Hallmark events • World fairs/expositions • Unique carnivals and festivals • Major spor events • Significant cultural and religious events • Historical milestones • Classical commercial and agricultural events • Major political personage events
Getz, 1997	Mega Hallmark Regional Local	Planned events • Cultural celebrations • Political and state • Arts and entertainment • Business and trade • Educational and scientific • Sport Competition • Recreational events • Private events
Hall, 1992	Internatiomal National Regional Local	Religious and sacred events Cultural events • Carnivals and festivals • Historical milestones Commercial events Sports events Poltical events
Roche, 2000	Global World Regional National Local	Mega Major Hallmark Community
Jago, 1997	National International	Routine or common events Special events • Minor special events • Festivals • Major special events – Hallmark events – Mega-events
Masterman, 2004	Ordinary (unplanned) Special (planned) • Major – Hallmark – Mega • Minor	Winter and Summer Olympics FIFA World Cup UEFA European Championship Formula One motor racing Grand Prix …

Source: Based on Getz 2008: 407; Ritchie, 1984: 2; Jago, 1997: 42–9; Roche, 2000: 4; Masterman, 2004: 16.

Effects of Formula 1 Events

Mega-event literature usually deals with the following phases: (1) prior to the event: the planning process including the construction of the infrastructure and the facilities in the local area, the urban development analysis; (2) during the event, the economic benefits and costs in relation to such themes as tourism and trade; and (3) after the event, the evaluation of the 'urban legacy'.

The first ever study of Formula 1 as a mega-event explored the impact of the Adelaide Grand Prix (Burns, Hatch and Mules, 1986), especially its effect on tourism, transportation, accommodation and restaurants (Gratton et.al., 2006: 43). In their study of 11 different regions during F1 season 1997–1998, Franco and Lilley (1999) researched the number of spectators attending from within and outside the region, the length of overnight stay periods, the average expenditure by types and amount, the occupations and industrial sectors most affected and the size of the geographic area in which impacts occurred. The study revealed that F1 impacted on local economies and that the hosting economies tried to draw foreign audiences to the area and meet their needs. The F1 sponsors, especially 'continuous sponsors', took part in successful pre-race sponsorship activities. Lilley and Franco also stated that repeated F1 events in specific locations attract a higher proportion of the 'foreign audience' who spend the most in the local economy and consequently, 'spectator expenditures' increased further (Lilley and Franco, 1999: 3–9).

Istanbul's Urban Space and Formula 1

The study on which this chapter is based examined the investment in the F1 Istanbul Park Track, its impacts upon the surrounding area and the spatial transformations that have taken place.

In order to ascertain the impacts of the Istanbul Park track as a mega-event investment on the urban space it was necessary to examine and evaluate the region's location before and after the investment, the importance of the dynamics of the population change, land values, investment in housing, housing values, planning decisions, and how the implementation of these decisions was undertaken. Data was obtained and evaluated within the scope of the area study in Istanbul from primary and secondary sources, including the Municipal and Tourism Documentary, statistics studies of Istanbul Provincial Directorate of Tourism and Culture and Tourism Ministry, information obtained verbatim from Istanbul's city authority web page, land usage and planning data from Akfirat Municipality of the time, the Municipality of Tuzla district and the Istanbul Metropolitan Municipality, interviews with real estate agents, data from Tuzla Land Registry Directorate and the Directorate of Land Registry and Cadastre of Ankara, data from the Turkey Statistical Institute, interviews with district headmen and interviews with companies, sales offices and real estate agents in the region.

Istanbul Metropolitan Area: Demographic Characteristics

Turkey's population in 2013 was 76,667,864 people, 18 per cent of whom (14,160,467 people) lived in Istanbul. In general, while the average population density in Turkey is 100 people/km², Istanbul has the highest population density with more than 2,725/km² people (TSI, 2014).

Istanbul Metropolitan Area: Spatial Features

The fact that service areas are concentrated in one area on the European side of Istanbul has been the biggest problem faced by the city's planning department. Therefore, as a result of the spatial decisions taken in the 1/100,000 Environment Plan, enacted in June 2009, future urban development is proposed to take place along an east-west axis; new sub-centres will be established (see Figure 16.1) to reduce the burden on the Istanbul Metropolitan Area (MIA) and to encourage, instead, a linear development of the city within a series of sub-regions (IMM, 2009).

On the Anatolian side, the plan proposes the development of a new sub-region Kartal-Pendik-Tuzla. Transport infrastructure, including the east-west railway system, will be improved with the aim of strengthening the region as an attractive centre and ensuring that it may serve metropolitan level needs (PM, 2008).

Istanbul's Tourism Potential

Istanbul, as a tourist centre, is of great importance for Turkey, particularly given the increasing competition between cities to secure the biggest possible share of the growing global tourism market. The number of

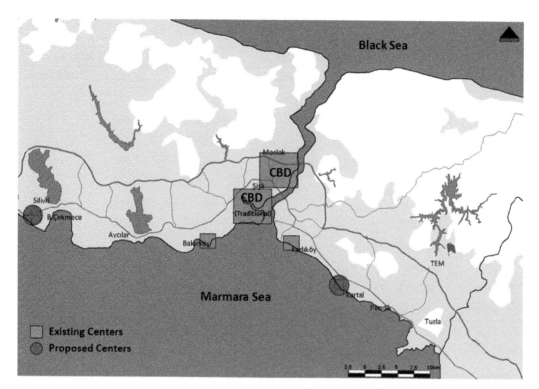

Figure 16.1   Map of existing and proposed Business Districts in Istanbul
Source: Eryılmaz, 2012.

tourists arriving in Istanbul in 2012 was 9,381,670. With increasing numbers of tourists coming to the city, the income gained from tourism is also increasing. While the tourism revenue in Istanbul in 2000 was $1.9 billion, it reached $5.8 billion in 2008, 27 per cent of the total tourism income of Turkey in that year (see Table 16.2).

The Turkey Tourism Strategy (2023), published in 2007, proposed long-term strategies for planning, investment, organisation, research, transportation and infrastructure improvement as well as the promotion and marketing, education, urban-scale branding, and diversification of tourism, including improvements in and the rehabilitation of existing tourism areas. Under the promotion and marketing strategies, branding and image are emphasised. To achieve the diversification of tourism in Istanbul, projects such as the Formula 1 track construction, improvement of the historic peninsula, increasing the number of convention centres, planning of the Kilyos tourist district, golf course construction, the creation of a platform for foreign filmmakers and the establishment of a marina on the Black Sea coast were proposed in 2007 and continue to be carried out by the Culture and Tourism Ministry (MCT, 2007).

There were 21 Tourism Centres in Istanbul in 2014. These centres are mainly located in the Historic Peninsula Beyoglu and the Bosphorus area, which offer a high tourism potential. The city's historical and cultural sites on the western side and especially the historical peninsula, are equipped with accommodation, catering, entertainment and convention centres. The eastern side of the city has a weaker structure in terms of tourist attractions and supporting infrastructure. Therefore, in addition to the proposed new centres, additional attraction points are also being created on the eastern side of the city. For this purpose, projects are being developed by the Istanbul Metropolitan Municipality (IMM), county municipalities and the Ministry of Culture and Tourism. In 2014, four of the developments planned as part of the Sport and Recreation Areas Development Project were completed, three were under construction, and 10 were in the draft stage. In addition, the Formula 1 Tuzla Akfirat Tepeoren Tourism Center was created in 2004 at the Formula 1 Istanbul Park Track (see Figure 16.2). Formula 1 races were performed on this race course between 2005 and 2011 and it was declared a tourism centre in 2009 (IMM, 2009).

Table 16.2 Growth of Tourism in Turkey and Istanbul (1996–2008)

Years	Number of Tourist Arrivals (1000 people)			Tourism Revenue ($ Million)		
	Turkey	Istanbul	Ratio (%)	Turkey	Istanbul	Ratio (%)
2000	10,428	2,350	22	7,636	1,909	25
2001	11,620	2,511	22	10,067	1,868	19
2002	13,248	2,699	20	11,900	2,118	18
2003	13,956	3,152	23	13,203	2,905	22
2004	17,546	3,473	20	15,888	3,177	20
2005	21,125	4,636	22	18,153	4,168	23
2006	19,818	5,347	27	16,831	4,550	27
2007	27,214	6,453	24	18,487	5,111	28
2008	30,980	7,050	23	21,910	5,864	27
2009	27,077	7,509	28	21,249	*	*
2010	28,632	6,960	24	20,807	*	*

Note: * not available
Source: ICOC, 2011:215.

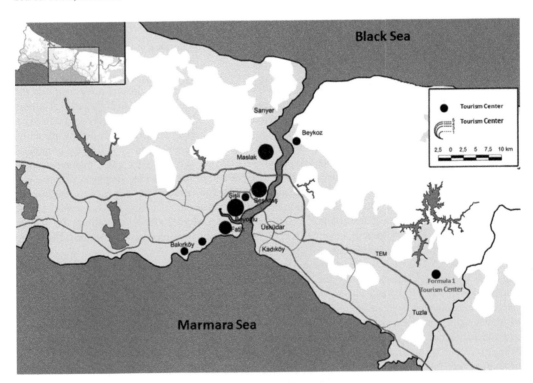

Figure 16.2 **Map of the Istanbul Province, showing existing tourism hotspots and the site of the new Formula 1 course (diameter of dots representative of the number of tourism centres in each district)**

Source: Eryılmaz, 2012.

Major Events Held in the City

In 2010, Istanbul was selected European Capital of Culture and hosted a large number of international events (see Table 16.3). In the International Congress statistics of 2012 Istanbul ranked 9th among the 128 Congresses (ICVB, 2013).

Table 16.3   International events held in Istanbul

Sporting events	Events	Conferences & Scientific Meetings
UEFA Champions League Final	Istanbul International Film Festival	NATO Istanbul Summit
FIBA Basketball World Cup	Istanbul International Theatre Festival	METREX (Network of European Metropolitan Regions and Areas)
IAAF World Indoor Championships in Athletics	Istanbul International Music Festival	OECD Summit
İstanbul Cup Tennis Tournament	Istanbul International Jazz Festival	Informal Meeting of OECD Ministers of Education
Intercontinental Istanbul Eurasia Marathon	Istanbul Biennial	International Union of Architects Congress
The Presidential Cycling Tour of Turkey	Istanbul Design Biennial	Black Sea Economic Cooperation Summit
Formula 1	Istanbul International 1001 Documentary Film Festival	World Water Forum
Moto GP	International Istanbul Short Film Festival	IMF World Bank Congress
Red Bull Air Race	Istanbul International Ballet Competition	International Istanbul Smart Grid Congress and Fair

Source: Authors' own work.

Istanbul is still nurturing plans to be a future Olympic host city and many large-scale sport facilities already exist in Istanbul and the region.

International events held in the city are situated mostly in the Beyoğlu, Şişli and Beşiktaş districts on the European side and in the Kadikoy district on the Asian side. In other words, there is a spatial concentration of events in the city's central business district and in the area, where cultural and artistic venues are located (Figure 16.3). With its historical and cultural heritage and its social diversity, the historical peninsula of Istanbul plays a unique role within the metropolitan area. Also, in terms of accessibility, being located in the centre of the city, it is an important focus for corporate offices, traditional trading venues and trade areas. With all these features being the focus for tourism, this area not only hosts accommodation, food and beverage outlets and the entertainment industry, it is also an important location for events.

The Istanbul Park Story as a Mega-event Investment in Istanbul

The Formula 1 races that took place between the years 2005–2011, were held at the Istanbul Park Circuit, situated in the Tuzla area on the Anatolian side of the city. Tuzla is located on the eastern border of the Istanbul metropolitan area, on the coast of the Marmara Sea. To the east lies Gebze, and to the west and north lies Pendik.

The Akfirat County, where the Istanbul Park track is located, is 30 km away from the city centre and 19 km north from the town of Tuzla. It is built on 5,237.6 ha land of which 1,800 hectares are scrub and

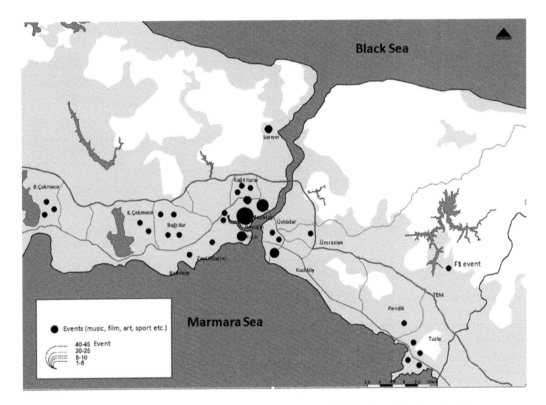

Figure 16.3   Map of the Istanbul region showing the spatial distribution of national and international cultural and sporting events, with the diameter of the dot reflecting the number of events

Source: Eryılmaz, 2012.

woodland areas. The total area of existing settlements and development consists of 3,547.4 hectares. With the Metropolitan Municipality Law No. 5216, Akfirat gained the status of first stage municipality in 2004. However, in 2008, with the decision to undertake the 'complete removal of first-tier municipalities within the Metropolitan Municipality organisation', Akfirat's legal recognition as an independent municipality ended and it was divided into two neighbourhoods in the district of Tuzla[1] (Official Gazette, 2008). The Istanbul Park race course is located in the Akfirat area, about 19 km from the centre of Tuzla. Being located only 8 km from the Sabiha Gokcen Airport, Istanbul Park is located in a prominent and accessible position (e.g. via the TEM (E-80) and E-5).

Istanbul Park track and some of the settlements in the borough are located in the Omerli Drinking Water Basin, with its 'long-range watershed protection limits'. They are situated 3 km away from the border of the inner 'absolute protection zone' (Figure 16.4).

In the 1:50,000 Scale Istanbul Metropolitan Area Plan approved in July 1980 the site of the Formula 1 course is designated as a forest and an area to be afforested. Parcels of land on which the F1 race track is located belong to the General Directorate of Foundations and have been contracted to the FIYAS Formula One Istanbul Investments Inc. for 20 years. The Metropolitan Area Master Plan Amendment, the 1:5,000 scale Master Plan and the 1:1,000 scale Implementation Plan were approved by the Ministry of Public Works and Settlement on 9 September 2003. The foundations for 'The Sport and Competition Area For Automobile and Other Motor Vehicles' at Istanbul Park were laid on 10 September 2003 in an area of 221.5 hectares and the

---

1   This study on which this chapter is based was prepared during the period when Istanbul Park Circuit's planning and F1 races' implementation stages were extant, thus taking the whole settlement of Akfirat municipality into account.

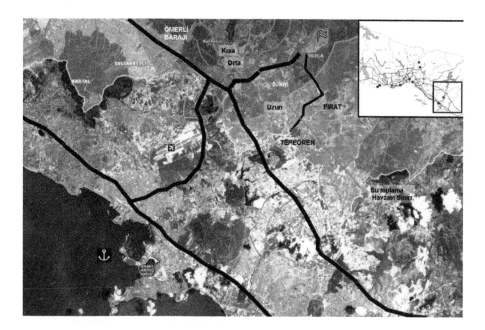

Figure 16.4   F1 Istanbul Park track and the nearby water protection zone
Source: Eryılmaz, 2012.

decision was subsequently published in the Official Gazette (14 March 2004). On 29 April 2004 the Master Plan and the Firat Neighborhood Improvement Plans were approved.

On 27 June 2005 the Omerli Drinking Water Basin Tepeoren 1st, 2nd, 3rd Stage Implementation Plan Changes and Additions Plan were approved. In the 1:100,000 scale Environmental Master Plan adopted in 2006, Akfirat County was announced as a location for 'cultural tourism' and as a tourism centre.

While the construction of the F1 Istanbul Park track developed accordingly, professional associations and civil society organisations protested, pointing out the disadvantages and risks of the project and informing the public about it. As early as January 2003, a meeting was organised by the Istanbul branch of the Chamber of Mechanical Engineers with a range of different stakeholders, including environmental NGOs, architects, engineers and urban planners where various data was presented to decide on whether or not the land was suitable for the construction of the F1 track. During the meeting the following was reported:

> In relation to increasing Turkey's promotion and tourism activities, while it was stated that F1 should be respected positively, the 2250 acres of land in Kurtkoy being located within the Omerli Dam catchment, the area being within forest boundaries and under control of the Forest Law … and that even so the parcels where the track would be built were announced as a tourism area, the Istanbul Water and Sewerage Administration General Directorate (ISKI) regulations and the watershed plan pointed [to the conclusion] that the area was not sufficient for the construction of facilities with the purpose of promoting tourism' (Gazette of Turizmde Bu Sabah, 30 January 2003).

In July 2004, the Chamber of City Planners and the Chamber of Agricultural Engineers filed a lawsuit for annulment of the plans approved by the Public Works and Housing Ministry and the decision of the Council of Ministers declaring this region as a Tourism Area. The Chambers stated that the F1 circuit, located in the medium and long-range watershed protection zone, would build up a new focus for tourism on the eastern side of the city. They claimed the cancellation of the 1:5,000 scale North Akfirat Master Development Plan was contrary to the Constitution, urban science, the upper scale plans, and the ISKI Watershed Protection and

Table 16.4   Istanbul Park Circuit and range of events (2005–2011)

Istanbul Park: 69th track in F1 history **The area covered by Track:** 222 hectares **Total length:** 5,378 meters **Seating capacity:** 130,000 people **Cost:** $ 220 Million **Owner:** Formula Istanbul Investment Joint Stock Company (FIYAS)	**Partners of FIYAS:** Istanbul Chamber of Commerce (ICC) Union of Chambers and Commodity Exchanges of Turkey (TOBB) Istanbul Metropolitan Municipality (IMM) Istanbul Special Provincial Administration
Property Rights:	**2005–2011:** Istanbul Park Organization Company (The company owner FOA President Bernie Ecclestone) **2011–2012:** FIYAS **2012–2023:** InterCity

Source: Authors' own work.

Control Regulations. However, the Administrative Court stated that this lawsuit taken out by the Chambers was not filed during the legal process period and thus the appeal was rejected.

In its first year of operation, in addition to the F1 grand prix, Le Mans, Moto GP, the Renault World Series, the FIA World Touring Car Championship and other car and motor racing events took place at the Istanbul Park circuit. Thus, the efficient use of the track in different months of the year was ensured.

For the period 2005 to 2021, Istanbul Park Circuit was leased to the Istanbul Park Organisation Company, the owner of which is Formula 1 boss Bernie Ecclestone. In 2009, Ecclestone requested from Turkey an increase of the price of the annual F1 race payment from $13.5 million to $26 million. The request was rejected by the Turkey Ministry of Youth and Sports, so the Istanbul Park Organisation Company applied to FIYAS (Formula Istanbul Investment Joint Stock Company) to terminate the operating contract (ntvspor.net, 22 April 2011). From 2005 to 2011, the Istanbul Park Organisation Company having the operation rights, progressively removed events other than the F1 racing from the competitions calendar. The stated cause being the decrease in the number of on-site spectators. After the alteration of the business management of the track in 2011, and due to lower spectator participation in the event, F1 Organisation was cancelled in Turkey from 2012. The Istanbul Park Circuit was taken over by FIYAS in 2011 and 2012. After a bidding process was completed in November 2012, the right to operate the track from 2012 to 2023 was awarded to the company InterCity (TRf1.net, 14 November 2012).

Formula 1's Impact on Istanbul's Urban Spaces

Using high-resolution satellite imagery, Demir et al (2007) monitored the impacts of the Istanbul F1 by assessing changes in land usage patterns in Akfirat and Tepeoren (2002–2005). They demonstrated that the construction of the F1 circuit had resulted in newly opened/expanded roads around the track and in spatial changes and the development of new residential areas.

With the construction of the Istanbul Park circuit, transportation investment strengthened connections with the nearby settlements. The Sabiha Gokcen Airport, located 8 km southwest of the track, is connected by a three-lane highway. As a result of the FI development and the increased popularity of the area, the number of international flights and the passenger numbers have increased.

With the planning decisions on Akfirat, where the track is located, it was aimed to increase the region's urban attractiveness by investment in low density residential areas and urban scaled investments. Factors, such as the closeness to the F1 track, the existence of the Okan University, the closeness to the Sabiha Gokcen Airport, Koc and Sabanci Universities, Tuzla and Gebze Organised Industrial Zone, played a major part in the region becoming a preferred destination for higher income groups of residents.

In the Akfirat area where the F1 track is located, 3,200 hectares of area was opened to settlement building between 2002 and 2005 (Demir et.al., 2007:7).

Table 16.5   Events held at Istanbul Park (2005–2011)

Event	2005	2006	2007	2008	2009	2010	2011
Le Mans	-	7–9 Apr.	-	-	-	-	-
Moto Gp	20–22 Apr.	28–29 Apr.	21–22 Apr.	-	-	-	-
WSRenault	23–24 Jun.	16–18 Jun.	-	-	-	-	-
Formula 1	19–21 Aug.	25–27 Aug.	24–26 Aug.	9–11 May.	5–7 Jun.	28–30 May.	6–8 May.
GP2	20–21 Aug.	26–27 Aug.	25–26 Aug.	10–11 May.	6–7 Jun.	-	-
VW Polo Ladies Cup Seat Cup	21 Aug.	-	-	-	-	-	-
LG Super Racing	16–18 Sep.	-	-	-	-	-	-
WTCC	21–23 Sep.	22–24 Sep.	-	-	-	-	-

Source: www.istanbulparkcircuit.com (2011).

Development of the Population in the Region

In the town of Tuzla, where the Istanbul Park track is located, the urban population was 91,230 in 1990. By 2000 it had increased by 18 per cent (107,883 people) and by 2007 the total population numbered 148,792. While the Firat and Tepeoren areas were previously considered to be rural settlements, now, due to the rapid urban development, they are amongst the most prestigious districts, the sites of villas and luxurious residential areas. The population increased alongside the expansion of the residential development projects, which began construction in 2003. In 1990, the population of Tepeoren increased from 928 in 1990 to 2,381 in 2007, while the population of Firat increased by 309 per cent from 1,257 in 1990 to 5,140 in 2000. Between 2000 and 2007, the population of Firat suddenly decreased by 57 per cent to 2,221, because many residents declared their property in Firat as second homes.[2]

Developing Housing Projects in the Region

For the period from the date of ratification of the development plans and the realisation date of the F1 Istanbul Park track, the main investments in the region are set down in Table 16.6.

In the brochures and advertisements for residential development projects in the area prepared by estate agents and development companies, the following key selling points are highlighted:

- only a few minutes to the F1 Istanbul Park track, Sabiha Gokcen Airport and the highway;
- close proximity to modern educational institutions like Sabanci University, Okan University, Koç High School;

---

2   The Statistics Institute of Turkey has started to implement an 'Address-Based Registration System' since 2007 in Turkey. According to this system, identification numbers of individuals are matched with their residence addresses. Only one residential address is being recorded to avoid duplications and repeated recording. Thus, the conflict of population decrease and increase in housing construction in Firat quarter arises from the people who have stated the homes in Firat quarter as their second homes.

Table 16.6   The distribution of investment in the region, Timeline

Year	Name of investment	Year	Name of investment
1988	Koc High School		Okan University
1997	Sabancı University		Bautek Kugu Houses*
			F2 Houses*
1998	Sabiha Gokcen Airport		Tepe Park Villas*
2003	Approved the Master Plans & Application Plans	**2006**	Kanarya Konakları*
	The foundation of Istanbul Park was laid.		
2003–4	Newly opened/expanded roads around the track		
2003	Tepeoren Villas*		Via/port Shopping Center
			İspanyol Houses*
2004	Arkeon Houses*	**2007**	Royal Park*
	Kırklar Valley*		Sultan Houses*
	Akfirat Houses*		
2005	Millenium Park*	**2008**	Erguvan Houses*
	Tepeoren Koru Houses*		
	Akfirat Villas*		

Note: * Residential Development Projects
Source: Eryılmaz, 2012.

- located in the Akfirat region, an untouched, natural area, rapidly gaining value;
- protection from 'urban sprawl' and illegal squatters, because the land, in terms of zoning permits, is open only to the construction of villas.

To attract high income groups to the area, investors and developers have stressed the location as being conveniently close to Istanbul, e.g. with the slogan 'to escape from Istanbul, you do not need to go to distant places'. The mega-event investment in the F1 Istanbul Park is presented as an important 'prestige' project and new housing projects in the vicinity are, by association, promoted as prestigious residential projects due to their location, resulting in rising land values.

In these residential development areas entire new (urban) landscapes have emerged, including facilities like swimming pools, walking and jogging trails, tennis courts, basketball and volleyball courts, fitness centres, hairdressers, kindergartens, and restaurants, and also some shopping malls. One of the investors choosing a location in the region is the Via/Port Shopping Centre that opened in 2008. Turkey's largest outlet centre, the Via/Port includes 189 stores, a five-star hotel, a convention centre and 617 residential flats, houses and suites.

Change in Regional Land Value

Along with the decision to construct the Istanbul Park Track in the region, the value of land, buildings and housing in the neighbourhoods of Tepeoren and Firat has increased (see Table 16.7).

When evaluating the increase in fair value of all neighbourhoods of Tuzla, Akfirat and Orhanli, the average increase of value of lands in Orhanli and Akfirat settlements in 1994 was behind the rates of the town of Tuzla. However, between 2002 and 2006, Tepeoren with a growth rate of 5,199 per cent and Firat with a growth rate of 1,059 per cent are particularly noteworthy (TM, 2009; RA, 2009).

Table 16.7   Tuzla district and neighborhood land and land value percentage growth rates/m²

Quarter		1994–1998	1998–2002	2002–2006
Tuzla district (10 quarter average)		2,160	903	86
Akfirat	Tepeoren	905	146	**5,199**
	Firat	0	170	**1,059**
The average Akfirat County (Tepeoren and Firat neighborhoods)		453	158	3,129
The average Orhanli County (5 neighborhoods)		631	136	891

Note: The data groups obtained from different institutions, are respectively; 'Minimum m2 piece of land and urban land values' published every four years by the Revenue Administration Directorate and supported by current values obtained from the Department of Real Estate & Expropriation (1994–1998–2002–2006).
§   The sales values realised in residential projects and condominium of Firat and Tepeoren obtained from the Tuzla Land Registry title deeds; (2004–2010).
Source: Eryılmaz, 2012 (Based on TM, 2009; RA, 2009).

Table 16.8   Residential development projects in the region (2010)

Location	Project Name	Land area (m²)	Number of house		House sizes(m²)		Sale price($/m²)	
			apt. House	villa	min	max	min	max
Akfirat	Akfirat Houses	140,000	-	143	175	245	1,452	-
	Akfirat Villas	17,181	-	20	280	-	985	-
	Bautek Kuğu Houses	90,000	-	98	354	400	1,941	1,998
	Royal Park	115,000	64	72	136	514	1,422	2,285
	Tepe Park Villas	48,000	-	42	330	440	1,400	1,434
Tepeoren	Arkeon Houses	370,000	-	345	194	314	1,952	2,166
	İspanyol Houses	13,500	-	10	388	-	2,577	-
	Kanarya Residence	20,000	-	15	450	-	1,402	-
	Millenium Park	282,000	-	219	360	479	1,947	2,402
	Park Ville	12,500	-	13	300	-	1,700	2,200
	Sultan Residence	33,500	-	23	320	430	1,250	1,395
	Tepeoren Koru Houses	22,000	-	22	270	-	-	-
	Tepeoren Villas	112,000	-	101	235	415	1,900	2,400

Source: Eryılmaz, 2012.

The Average m² Price of Residential Projects

Between the years 2003 and 2010, almost 20 housing projects were started in the region and 1,630 houses were built, including 64 apartments and 1,565 villas. 20–30 small unit dwellings projects or more large-scale projects with 100–350 housing units were built on the site. Most of the houses are designed as duplex / triplex villas in gardens. (See Plate 16.1.)

Housing demand in the region largely consists of people living on the Asian side with high income levels working in Tuzla-Gebze industries and Bostancı. However, there are also persons working on the European side of the city and residing there.

Housing size in the new residential projects varies between 128 and 514m² and the average housing price per m² in 2010 ranged from $ 985 to $ 2,577.

Sales Value of Property Ownership

Condominium sale values, as well as residential projects, have shown an increase in the region. In 2004, the average home sales price in the region was $11,529. It increased to $71,848 in 2005 and $386,383 in 2010, with the highest price of $465,735 paid in 2007 for the Bautek Kugu Houses. The average residential property values in the region have increased by 3,351 per cent in the years 2004 to 2010 (Table 16.9).

Table 16.9   Average housing projects condominium sales values ($), 2003–2010

	2004	2005	2006	2007	2008	2009	2010
Arkeon houses			11,524	23,178	230,531	117,233	
Kırklar valley		40,339	52,755	75,741			
Tepeoren villas	11,529	10,297	149,005	250,253	262,066	249,446	
Akfirat houses				35,329	215,571	140,715	
Millenium park		75,585	114,075	163,618	370,101	425,957	
Koru houses		161,172	141,381	171,217	201,683		
Sultan houses			138,308	154,104	159,849		
Parkville				339,405	296,443	308,115	
Akfirat villas				119,873	115,051		
Tepepark					41,303	192,249	394,766
Bautek kugu houses				465,735	391,679	385,799	377,999
Kanarya houses				41,365	49,802	38,936	
Regional Average	11,529	71,848	101,175	167,256	212,189	232,306	386,383

Source: Eryilmaz, 2012 (Adopted from TLRD Condominium values 2004–2010).

The spatial distribution of the last condominium in each project and the sales value (m² unit value) is shown in Figure 16.6.

The m² unit price of housing projects which have been developed together with the F1 Istanbul Park, may be classified into four sub regions.

1. Region: Istanbul Boulevard being positioned between the box offices-Via/Port and the F1; 246 $/m²
2. Region: Ataturk Boulevard as the F1 – Tepeoren axle; 218 $/m²
3. Region: Fatih Sultan Mehmet Boulevard as the F1- Firat village axle: 168 $/m²
4. Region: The former Ankara Highway as the Box office -Orhanli axle; 134 $/m²

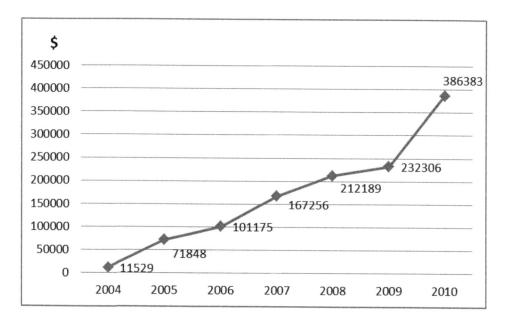

Figure 16.5   Average of sales value of property ownership in the region (2003–2010), ($)
Source: Eryılmaz, 2012.

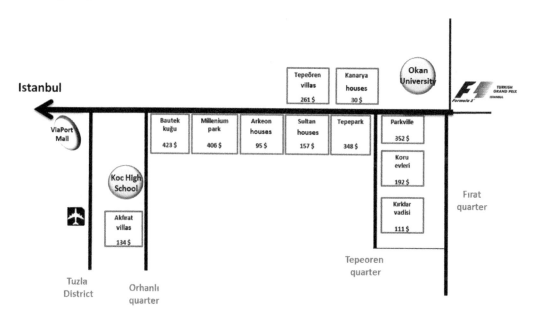

Figure 16.6   Average unit values for condominium housing projects/ m²
Source: Eryılmaz, 2012.

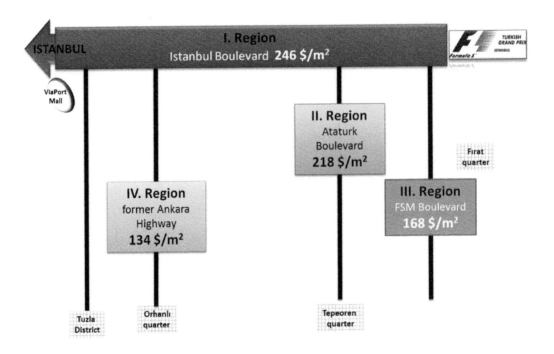

Figure 16.7   Unit values per m² in relation to location
Source: Eryılmaz, 2012.

Conclusion

Along with the processes of globalisation and decentralisation, cities have become dependent on consumer-based services industries and global capital investment. Local government and other agencies and actors, trying to adapt to changes in the local-global balance of power, have undertaken an increasingly entrepreneurial role. One of the strategies followed by local governments to provide urban economic development and enhance competitiveness has been to bid to become the host of 'mega-events'. Mega-events such as the Olympics and Formula 1 have been used as urban marketing tools to facilitate urban renewal and revival, develop new images and attract investors and tourists.

F1, as a mega infrastructure investment that requires a large and fixed facility, affects local and regional space. In the literature, effects of mega-events on urban location are categorised as *direct* (infrastructure of activities, infrastructure of transportation, the investment in urban spaces triggering further developments, etc.) and *indirect* (image, branding, urban recognition, revenues from tourism) or *measurable* and *non-measurable* effects.

Studies involving pre-event predictions often mention the positive benefits of an event; they highlight its contribution to tourism, employment, urban recognition, the development of urban infrastructure and the renewal of the urban image. However, in these studies, the costs incurred by the urban areas that will be transformed by the investments that take place alongside these events and subsequently, how the event facilities and the accompanying infrastructures will be used, and what the effects will be on the natural environment are often not considered. These types of effects are generally discussed in post-event studies. Differences are emerging, therefore, in the literature between post-event studies and those that focus on the pre- and event phases.

Today, mega-events, artistic-cultural festivals and scientific conferences have the function to create spaces of consumption, re-shaping the spatial forms of cities that seek a higher place in the global urban hierarchy. In cities with different urban forms and socio-economic structures, the impact of mega-events will vary. Mega-events that are not integrated into the strategies and actions of the city, but are accommodated through the revision of existing plans, often negatively affect the urban space.

The biggest mega-event experienced in Istanbul so far has been the F1. Despite the arguments of opponents that the F1 track was not, in fact, in line with the constitution, the urbanism principles, the upper scale plans and watershed protection regulations, the lawsuit opposing its development was rejected. Akfirat became, in a short time, an attractive region preferred by upper income groups. Luxurious housing projects have been realised and the area/land values have increased rapidly since 2003. In the case of the Istanbul Park located in the medium and long-range watershed protection area, the increase in real estate values and the local infrastructure improvement shows how a mega-event affects the texture of urban development. Therefore, prior to bidding for and hosting a mega-event, potential threats should be foreseen and considered besides the opportunities it will bring. Also, in the selection of the place for the event investments, the observance of existing local space and a thorough cost/benefit analysis must be undertaken.

The adventure of F1 Istanbul started with a group of entrepreneurs getting support from local management and the government. During the investment process for the F1 track, public-private partnerships and new company formations were actualised. Over the period during which the event was held in Istanbul, the public progressively withdrew its support for the F1 business model. When the company (the Istanbul Park Organisation Company) operating the track terminated the agreement, and the FIA removed Istanbul from the F1 calendar, the race was hosted for the last time in 2011. Today, Istanbul Park Circuit, is being revitalised with different events. It is not only important to become a mega-event host city, but it is equally important to ensure the sustainability of the event/venue. All these developments have shown that the actors involved in this mega-event – including the local government, central government, agencies and companies – did not work in harmony in organising the F1 and that there was no effective event management. However, city managers will draw lessons from their experiences and will turn these into an opportunity for other events to be organised in the future.

The F1 Istanbul example reveals that mega-events must be planned strategically. A good event strategy must be followed in order to minimise problems experienced in event nomination, the choice of location for event investment, the realisation of the event, and to ensure the event facilities do not become a burden to the city in terms of operation and maintenance. The development of the urban space should not cause environmental damage and many other problems that may be experienced in socio-economic and spatial senses. In this context, the sustainability of an urban area and its social structure must be protected and representatives of the interests of global capital should not be allowed to use the vehicle of the mega-event to influence, manage and (re-)direct urban plans.

References

Beaverstock, J.V., R.G Smith, and P.J. Taylor (1999). 'A roster of World Cities'. *Cities*, 16(6): 445–58. Available at: <http://www.lboro.com/gawc/rb/rb5.html>, [Accessed 10 May 2007].

Bessit, C.S., (2006). South African Formula One Grand Prix: A Dream or Nightmare, University of Johannesburg, Faculty of Management, Department of Business Management, p. 4. Available at: <http://ujdigispace.uj.ac.za:8080/dspace/bitstream/10210/261/1/909310525.pdf>, [Accessed 24 May 2008].

Celen Corporate Property Valuation & Counseling Inc., (2009). Property Valuation Report, Istanbul: Celen.

Chen, N., (2008). 'What economic effect do mega-events have on host cities and their surroundings? An investigation into the literature surrounding mega-events and the impacts felt by holders of the tournaments', Masters Thesis, University of Nottingham, Finance and Investment, UK. Available at: <http://edissertations.nottingham.ac.uk/2358/1/08MAlixnc3.pdf>, [Accessed 12 June 2009].

Demir H., F. Şanlı, M. Gür, C. Goksel, (2007). Sustainable and Economically Recycling Real Estate Development Project: A Case Study For Istanbul Park. International Conference on Environment: Survival and Sustainability. 19–24 Şubat 2007, Lefkoşe: Kıbrıs.

Derudder,B. and P. Taylor, (2012). Project 97: World City Network 2012, Available at: <http://www.lboro.com/gawc/world2012.html>, [Accessed 14 March 2013].

Eryilmaz, S.S., 2012. Reflections of Mega-events on the City Space Formula 1 Turkey Grand Prix Istanbul Park Example. Ph. D. Yildiz Technical University.

Essex, S. And B. Chalkley (2004). 'Mega-sporting events in urban and regional policy: A history of the Winter Olympics'. *Planning Perspectives*, 19(2), pp. 201–32.

Gazette of Turizmde Bu Sabah, 30 January 2003. "Lawsuit against Formula 1", Available at: <http://www. turizmdebusabah.com/haber_detay.aspx?haberNo=7386>, [Accessed 20 April 2005]

Getz, D., (2008). 'Event tourism: Definition, evolution, and research'. *Tourism Management*, 29(3), pp. 403–28.

Gratton, C., S. Shibli, R. Coleman, (2005). 'The economics of sport tourism at major sports events', in J. Higham, (ed.) *Sport Tourism Destinations: Issues, opportunities and analysis*. Oxford: Butterworth-Heinemann. pp. 233–47.

Gratton, C., S. Shibl and R. Coleman (2006). 'The economic impact of major sports events: A review of ten events in the UK'. *The Sociological Review*, 54(S2), pp. 41–58.

ICOC – Istanbul Chamber of Commerce (2011). Economic Report 2010, Available at: <http://www.ito.org.tr/ itoyayin/0024623.pdf>, [Accessed 20 March 2013].

ICVB – Istanbul Convention & Visitors Bureau, (2013). Istanbul strengthens its position in Top 10. Available at: <http://icvb.org.tr/wp-content/uploads/2013/10/Istanbul-strengthen-its-position-in-Top-10. pdf>, [Accessed 14 January 2014].

IMM – Istanbul Metropoliten Municipality (2008). 1:5,000 Scaled Omerli Drinking Water Basin Scaled Plan, Istanbul Metropolitan Planning Office.

IMM – Istanbul Metropoliten Municipality, (2009). 1:100,000 Scaled Istanbul Environmental Order Plan, Istanbul Metropolitan Planning Office.

Jago, L. K., (1997). Special Events and Tourism Behaviour: A Conceptualisation and an Empirical Analysis from a Values Perspective, Victoria University, Faculty of Business, Department of Hospitality, Tourism and Marketing, PhD thesis, Available at: <http://eprints.vu.edu.au/1501/1/Jago.pdf>, [Accessed 4 February 2009].

Lilley, W. And L. De Franco (1999). *The Economic Impact of the European Grands Prix*. Federation Internationale de l'Automobile/ Brussels, Belgium.

Martins, L., (2004). Bidding for the Olympics: A local affair? Lessons Learned from the Paris and Madrid 2012 Olympic Bids, paper presented at the conference entitled City Futures, 8–10 July 2004, Chicago. Available at: <http://www.uic.edu/cuppa/cityfutures/papers/webpapers/cityfuturespapers/ session7_1/7_1biddingolympics.pdf>, [Accessed 12 April 2010].

Masterman, G., (2004) *Strategic Sports Event Management: Olympic Edition*. Elsevier Science & Technology Books.

McArdle, K., (1998). Temporal, Spatial and Thematic Analysis of Special Events in Victoria (1997), Master Thesis, Victoria University of Technology, Faculty of Business, Available at: <http://eprints.vu.edu. au/15260/1/McArdle_1998_compressed.pdf>, [Accessed 9 February 2009].

MCT – T.R. Ministry of Culture and Tourism Publications – 3090 (2007). Tourism Strategy of Turkey 2023. Available at: <http://www.kulturturizm.gov.tr/genel/text/eng/TST2023.pdf>, [Accessed 4 May 2010].

MCT – T.R. Ministry of Culture and Tourism (2013). Accommodation Statistic. Available at: http://www. ktbyatirimisletmeler.gov.tr/TR,9859/tesis-istatistikleri.html, [Accessed 11 July 2013].

ntvspor.net, 22 April 2011. "Seneye yarış yapılmayacak gibi görünüyor" Available at: <http://www.ntvspor. net/haber/formula-1/39080/seneye-yaris-yapilmayacak-gibi-gorunuyor>, [Accessed 6 July 2011]

Official Gazette of the Republic of Turkey (2004). No.7214 of 28.04.2004.

Official Gazette of the Republic of Turkey, (2008). No.5747 of 06.03.2008.

PM – Pendik Municipality, (2008). Pendik City Action Plan, Available at: <www.pendik.bel.tr/UserFiles/ kenteylem.pdf>, [Accessed 2 January 2009].

RA – Revenue Administration, (2009). Land and Urban Land values/ m^2 1994–1998–2002–2006, Available at: <http://www.gib.gov.tr/index.php?id=63>, [Accessed 16 March 2009].

Ritchie, J.R.B (1984). 'Assessing the impact of hallmark events: Conceptual and research issues'. *Journal of Travel Research*, 23(1), pp. 2–11.

Roche, M., (2000). *Mega-Events and Modernity: Olympics and Expos in the Growth of Global Culture.* Routledge, London, Available at:<http://site.ebrary.com>, [Accessed 15 September 2008].

TLRD – Tuzla Land Registry Directorate, (2010). Condominium Values of Houisng Projects in Firat & Tepeoren neighborhoods, [Accessed 11 February 2010].

TM – Tuzla Municipality (2009). Directorate of Real Estate & Expropriation, Land Values 1994–2008, [Accessed 7 February 2009]

TRfl.net, 14 November 2012. İstanbul Park'ta Sürpriz Gelişme! Available at: <http://www.trfl.net/formula1_haberler/29324-istanbul-park-ta-surpriz-gelisme.html>, [Accessed 15 December 2012]

TSI – Turkish Statistical Institute, Address Based Population Registration System Statistics (ABPRS) Available at: <http://www.turkstat.gov.tr/PreTablo.do?alt_id=1059>, [Accessed 14 January 2014].

www.istanbulparkcircuit.com (2011). Events of İstanbul Park, Available at: <www.istanbulparkcircuit.com>, [Accessed, 10 May 2011].

Chapter 17

The Urban Impacts of Rio's Mega-events: The View From Two 'Unspectacular' Favelas

Matthew Richmond

By 2009, after several years of economic growth and falling violence, and the emergence of a new political alliance, there was much talk in Rio de Janeiro of a 'turnaround' (Urani and Gambiagi 2011). The city had long become accustomed to decline, having lost its capital city status to Brasília and its business pre-eminence to São Paulo in the 1960s, and then suffering more than most during the economic crises of the 1980s and 90s. The 'turnaround' cemented itself as a hegemonic narrative in October of that year when the city won its bid to host the 2016 Olympic Games. These favourable economic and political conditions, combined with the impetus of the Olympics,[1] subsequently paved the way for a range of ambitious and controversial urban reforms spanning the spheres of housing, infrastructure, transport and public security.

This chapter offers an overview of the key reforms and outlines two competing interpretations of their aims and impacts. It then goes on to look in detail at the way the reforms have affected two 'unspectacular' favelas (informal settlements) – neighbourhoods where impacts have been partial and uneven, rather than dramatic and transformative. This analysis provides the basis for some broader reflections in the conclusion on the achievements and limitations of the 'city project'.

The Policies of the 'City Project'

In its recent critical report, the *Comitê Popular da Copa e Olímpiadas* ('Popular Committee of the World Cup and Olympics'), describe the new urban policies as constituting an overarching 'city project' (Comitê Popular 2013). Although encompassing a diverse array of policies carried out by a complex mix of actors, the term provides a useful shorthand and is adopted here. This city project is being implemented across much of the urban territory. However, many of the new policies are being specifically targeted at Rio's historically excluded favelas, where around 22 per cent of the population lives (IBGE 2010).[2] As the next section will discuss, the focus on favelas is central to the rhetorical goals of the project as a whole.

One favela policy is 'pacification', whereby specialist policing units, *Unidades de Polícia Pacificadora* (Police Pacification Units, UPPs) enter and establish a permanent presence in favela territories where drug trafficking gangs had previously dominated (Cano 2012). This programme is overseen by the Rio de Janeiro state government, which runs the police force and was first introduced in late-2008. At the time of writing this has expanded to 38 UPPs covering most of Rio's inner-city favelas. Once favelas have been pacified, the '*UPP Social*' programme, run by the Rio *Prefeitura's* (city council's) research centre, has been charged with co-ordinating the development and integration of local services (UPP Social). Another favela-specific intervention is '*Morar Carioca*', a participatory upgrading programme designed and managed by the *Prefeitura* in collaboration with the Institute of Brazilian Architects (*Instituto de Arquitetos Brasileiros*, IAB), and with funding from the Inter-American Development Bank (IDB). Although at the time of writing this programme has only been implemented in a handful of locations, its stated aim is to reach all of Rio's favelas by 2020 (Prefeitura do Rio de Janeiro).

---

1 As well as the 2014 Brazil World Cup, for which Rio is one of twelve host cities.
2 Favelas are officially defined by the Instituto Brasileiro de Geografia e Estadística as "Collections of at least 51 housing units, most of which lack essential public services, which occupy or have until recently occupied publicly or privately owned land, and are characterised by disordered and dense occupation" (IBGE 2010).

There are also several programmes operating over the larger territory of the city or in specific regions. The federal government's *Programa de Aceleração do Crescimento* (Growth Acceleration Programme, PAC) is carrying out major infrastructure works in key zones across the city, including in several large favelas. There have been a number of major reforms to the transport system, including the construction of a bus rapid transit (BRT) network focussed on the west of the city, a new light rail system in the city centre, and an extension of the metro. Several major location-specific projects relating to the mega-events are being carried out by public-private partnerships led by the *Prefeitura*. The most notable of these are the port regeneration scheme *Porto Maravilha*, the redevelopment of the Maracanã football stadium and surrounding area, and the construction of the Olympic Park in the western suburb of Barra da Tijuca (see Figure 17.1 – Rio de Janeiro).

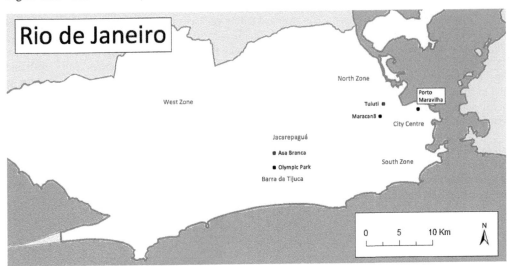

Figure 17.1   Map of Rio de Janeiro
Source: Author's own design.

A 'Post-Third World City' or Neoliberal 'City of Exception'?

Between the new favela policies and those pursuing broader urban development aims, it seems certain that the Rio de Janeiro that emerges from the city project will be a very different place. However, in the absence of an accumulated body of evidence, it is too early to say conclusively what the urban and social impacts of the city project will be. Nonetheless, two competing accounts have already taken shape.

The first of these, promoted by the authorities and much of Brazil's mainstream media, is defined here as a 'post-Third World city' narrative. This proposes that the city project is belatedly confronting the negative legacies of Rio's historical process of urban development, which at key points has been seen as paradigmatic of cities of the 'Third World' (as it was known in the language of the time). These processes include the *rapid urbanisation* of the 1950s and 60s, which left an urban landscape characterised by widespread informal housing, and weak and uneven provision of key urban services, including transport and policing. The second was the era of *urban fragmentation* during the 1980s and 1990s, when endemic economic crisis aggravated existing urban inequalities at the same time as rising gang and police violence led to a militarisation of some territorial boundaries in the city.

The post-Third World city narrative argues that the city project will reverse the segregating dynamics of Rio's historical development by focussing on measures that bring about physical and social integration, particularly between favelas and formal areas. The UPPs are conceived as a means of universalising the coverage of public security to territories where it has been historically absent.[3] *UPP Social*'s stated objectives are to "promote urban, social and economic development in the (pacified) territories; and to execute the full integration of these areas with the city as a whole" (UPP Social, Programa). Similarly, *Morar Carioca* aims for the "complete and definitive urban and social integration of Rio's favelas" (Prefeitura do Rio de Janeiro, Conheça o Programa 2011). The major transport interventions, meanwhile, are conceived as means for integrating the different regions of the city as a whole through enhanced provision and co-ordination. The *Cidade Olímpica* ('Olympic City') website states: "The (BRT) express corridors ... together with the light rail and other already existing means of transport will compose a new transport fabric interconnecting all the regions of the Olympic City" (Cidade Olímpica, Transportes 2014).

However, a growing body of academic literature has begun to challenge these claims. Although covering a broad range of themes and perspectives,[4] this literature broadly conforms to Vainer's (2011) theory of Rio as a 'city of exception'. Vainer argues that during the 1990s Rio shifted towards a neoliberal model of urban governance and began to compete more aggressively for international business and tourism flows, with a particular strategic emphasis on mega-events. To facilitate this the *Prefeitura* enshrined its own right to suspend democratic processes in the 'collective interest', creating a permanent 'state of exception' and thus making Rio a 'city of exception'. As has been demonstrated in other contexts, global mega-events like the World Cup and Olympics themselves tend to catalyse the generation of new, 'exceptional' forms of urban governance (for example, Raco 2012). For example, they typically necessitate the creation of local organising committees with special powers and public-private partnerships of various kinds. These can then execute major, long-term transformations of the urban environment while largely bypassing mainstream democratic institutions.

A wider critical literature building on this analysis argues that the long-term impacts of the city project will be the opposite of those claimed by the 'post-Third World city' narrative. Instead of producing universal public security and urban integration these authors claim that the primary impacts will instead be the securitisation and social cleansing of valuable areas of the city in the interests of powerful private actors. Many have noted the geographically uneven investments in transport, infrastructure and the location of new sporting infrastructure, arguing that these are overwhelmingly pro-rich rather than pro-poor (Comitê Popular 2013). Where new projects are focussed on historically working-class areas like the port zone or the western suburb of Jacarepaguá this has been interpreted as a gentrification strategy aimed at removing low-income populations, whether quickly or more gradually (Brum 2013).

On the implications of the city project for favelas, Freeman (2012) argues that these territories have found themselves the targets of new elite accumulation strategies based around mega-events, tourism and land speculation. Favelas stand in the way of such plans by providing a base for drug-trafficking gangs (who are portrayed as solely responsible for Rio's problems of urban violence) and otherwise impeding the development of potentially valuable areas. Consequently their populations have been the subject of diverse interventions, including repression by *UPP*s and, in some instances, outright removal. Where new infrastructure, like cable cars, museums and other amenities accessible to tourists have been installed in or around favelas, Freeman believes these constitute a kind of 'symbolic taming', which seeks to cash in on the exotic 'spectacle' of Rio's favelas while minimising their perceived threat (Freeman 2012: 108). Freeman argues that between these different interventions the city project is making the city safe for neoliberalism, rather than improving the situation of the favela population.

The View from Two 'Unspectacular' Favelas

As the above discussion suggests, the 'post-Third World city' and 'city of exception' narratives offer very different accounts of what impacts the city project is having. However, to date, both claims have tended to rest either on observation of broad trends (e.g. falling levels of violence in pacified areas, net numbers of

---

3 The official promotional video of the UPP programme 'UPP came to stay' offers a clear example of this historical narrative. See http://www.upprj.com/index.php/as_upps_us [accessed 11/03/2014].

4 See, for example, Freeman (2012), Comitê Popular (2013), Ribeiro and Santos Junior (2013), Sanchez and Broudehoux (2013), Gaffney (2010).

people threatened with removal), or of high-profile cases like big infrastructure investments or large and/ or egregious examples of favela removals. The remainder of this chapter takes a different approach. Instead of looking at a single policy or array of policies across the city or in a high-profile case study, it assesses the range of impacts in two lesser-known favelas.[5] Although located in different parts of the city and with very different histories and social conditions, the two case studies are similar in that they both lie close to, but not inside zones targeted for redevelopment by key policies. In both cases the impacts offer a useful contrast to favelas that have been spectacularly transformed, forcibly removed, or largely overlooked by the city project.

Tuiuti: Peace Without Progress

Tuiuti (population 5,718) (IBGE 2010) is a hillside favela in the neighbourhood of São Cristóvão, north of central Rio (Figure 17.1 – Rio de Janeiro). Settlement on the hill began in the early twentieth century and continued haphazardly over subsequent decades, producing the patterns of dense occupation and irregular construction that now characterise the favela (Figure 17.2 – Tuiuti). Nonetheless, in the 1990s Tuiuti benefitted from public upgrading works carried out under the *Favela Bairro* programme, which brought paved roads, sewerage, drains and new public spaces (SABREN). With the hill almost entirely built over, a high level of home-ownership (compared to renting),[6] and low resident turnover, the population of Tuiuti today is remarkably stable. This stability has meant that despite property values rising significantly in São Cristóvão in recent years Tuiuti itself has not seen major demographic or social change.

Physical Upgrading

Given its high level of consolidation, physical upgrading has not materialised as a priority of the city project in Tuiuti. Nonetheless, residents highlight some key infrastructure that is still lacking in the area, including unpaved paths and steep hills without steps that make access difficult for elderly and disabled residents. There is also a dearth of community spaces where key services could be provided. There are no health clinics and just one active crèche. Such holes in infrastructure provision would have been identified by *Morar Carioca*, which, according to its stated timetable, should eventually reach Tuiuti (although no date has ever been specified) (Phillips 2010). However, as will be discussed below, a great deal of uncertainty now surrounds the programme.

It is interesting to compare Tuiuti's experience of local 'regeneration' processes with those of two nearby favelas. The small Favela do Metrô sits next to Rio's suburban railway line across from the recently redeveloped Maracanã stadium. After the space was marked for redevelopment as part of the new Maracanã complex, residents were subjected to a painful and arbitrary eviction process (Clarke 2014). The failure to adequately resettle residents or even determine a new use for the space (despite the climate of urgency with which removal was carried out) led to the abandonment of the favela and the arrival of new squatters, who at the time of writing are yet to be removed (*ibid.*). By contrast, the much larger favela of Mangueira, which rises up the steep hill overlooking the Maracanã, has received major infrastructure investments through *PAC*. These include new housing lots, recreational spaces and a redevelopment of its famous samba school building. The city project's different treatment of the three favelas seems to reflect a kind of implicit local hierarchy. Whereas Mangueira's size, fame and visibility mean it is prioritised for improvements, Favela do Metrô is too close to the Maracanã and so must be removed, while Tuiuti is too far and therefore largely ignored.

Public Security

In the absence of upgrading works the main impact of the city project in Tuiuti has been the presence of a *UPP* since November 2011, when it was pacified along with Mangueira. Rio's first large drug-trafficking faction the

---

5   This analysis is based on ethnographic research and interviews with residents (n=35 per case study) and selected local experts carried out between January and August 2013.

6   As in the majority of favelas ownership refers to the property rather than the land, for which residents do not have legal title.

Figure 17.2   View of Tuiuti
Source: Copyright Matthew Richmond.

Comando Velho ('Red Command', CV) came to dominate the two favelas in the early 1980s and established a dynamic of territorialised conflict with rival factions and police. Due to its relative insignificance in the drug supply chain and long-standing alliances with neighbouring gangs, Tuiuti experienced comparative stability and low levels of violence in the years prior to pacification unlike Mangueira, which was the site of frequent gun battles. Nonetheless, there were still sporadic police incursions, deaths of young gang members remained a common occurrence, and traffickers continued to openly flaunt guns and sell drugs around the favela.

A majority of residents interviewed expressed qualified support for pacification, mainly because it had made firearms and drugs less visible in the community and reduced the likelihood of gang violence and other types of crime. However, residents invariably acknowledge that drug dealing continues in spite of police presence, often quite conspicuously, and that the influence of traffickers over life in the community persists. This has created confusion for some, who, in many cases, feel unable to report crimes to police or even interact with them publicly. Such influence was laid bare in February 2013 when two gang murders were committed in the locality after armed men had robbed the gun of a UPP officer earlier the same night (Werneck and Ramalho 2013). The following day small businesses closed their doors on the orders of the traffickers – a practice common prior to pacification. As this suggests, it is difficult to argue that pacification has made progress towards its stated long-term goal of providing meaningful public security and equal citizenship for favela residents.

Service Development

Compared to their muted support for pacification, residents are generally unimpressed by what has followed in its wake. Worryingly, most are entirely unaware of the *UPP Social* programme, which to date has had a very limited impact. This seems to be the result of the effective downgrading of the programme in 2011, just a year after it was unveiled. At this point it was moved from state to municipal oversight and lost its expansive service development objectives and participatory model and became, in effect, a data-collection agency (Amado 2013). Now it can only identify and refer resident demands to existing service providers, but not respond to them directly. At the time of writing, its main achievement in Tuiuti has been to produce a map of services around the UPP territory on its website. The programme's failure is particularly evident in light of the programme's aim of supporting 'local projects and organisations' and 'sports, cultural and leisure activities' for young people, which are almost non-existent in the community.

Tellingly, another, quite different, post-pacification intervention *has* been efficiently realised. Light, the state electricity company, has entered to register new customers who, as in most favelas, had previously connected to the network illegally. While some discounts are offered for households with very low incomes and most residents accept the principle of paying for electricity, many feel the company has treated the community aggressively. By placing an instant squeeze on the budgets of poorer residents, it will likely leave some households without power and could eventually even force some to leave the area. The contrast between the lack of meaningful service improvements and the punctual arrival of utility companies has led some to conclude that the purpose of pacification was not to improve conditions for residents, but to allow the state to more effectively control and profit from them.

Jobs, Education and Training

One final impact of the city project in Tuiuti has been in the area of employment. The favela's location in Rio's industrial heartland has made it particularly sensitive to deindustrialisation, as is visible in the many derelict factories and warehouses that mark the surrounding landscape. The career histories of many older residents reveal transitions from low-paid, but relatively stable employment in factories and workshops to casual employment in domestic service or the informal sector. The difficult economic environment has also complicated youth transitions into the labour market, adding to the appeal of the drugs trade.

The main impact of the city project on employment has been the creation of a considerable number of construction jobs by the large-scale regeneration of the nearby port zone and the Maracanã football stadium, although most of these are likely to be temporary. Some free training sessions have been offered locally by SENAI, the industrial training service run by Rio's Industrial Federation FIRJAN. However, these have mainly been targeted in over-crowded, low-paid sectors like hairdressing, food production and mechanics. More long-term employment solutions in this part of the city will require improvement to core education and training provision and better quality and more stable jobs. However, these have not emerged as priorities for public intervention in Tuiuti, whether as part of the city project or otherwise.

Asa Branca: Two-tier Urbanisation

Favela Asa Branca (population 3,295) is located in the suburb of Jacarepaguá in Rio's West Zone (Censo 2010) (Figure 17.1 – Rio de Janeiro). It was established through a co-ordinated invasion in 1986, and, like many of Rio's newer favelas, was built on flat terrain and has an orderly layout, with equally sized plots distributed along streets wide enough to access by car (Figure 17.3 – Asa Branca). Since the 1990s and especially 2000 Asa Branca has grown significantly through expansion and verticalisation. This increase is partly the result of the arrival of large numbers of migrants from other parts of Brazil, attracted by job opportunities in the region (see below). This has given rise to a dynamic rental market in Asa Branca. The proportion of renters increased from 10 per cent to almost one third between 2000 and 2010 (IBGE 2000; IBGE 2010).

Figure 17.3   Street in Asa Branca
Source: Copyright granted by Catalytic Communities, catcomm.org.

Infrastructure and Services

After being almost totally overlooked by *Favela Bairro*, Jacarepaguá's favelas have finally begun to attract the attention of the state. This has been particularly evident in Asa Branca. At the end of 2012 the *Prefeitura* carried out comprehensive upgrading works, paving the roads and installing drains and street lighting through the *Bairro Maravilha* ('Marvellous Neighbourhood') programme. Residents are unanimous in their praise for the transformation. It has improved the quality of everyday life, reduced the risk of flooding, and created a sense of symbolic inclusion previously denied to them (Richmond 2013). However, the upgrading process has stopped short of meeting the fundamental needs of the area. Even more than Tuiuti, Asa Branca lacks community spaces and activities and has no public squares, parks or sports facilities. The empty area at the edge of the favela serves as a makeshift football pitch and playground and there have been suggestions that it will be developed for community use, but at time of writing no confirmation has been forthcoming.

The doubts surrounding further upgrading in Asa Branca reflect the doubts surrounding the *Morar Carioca* programme as a whole. Asa Branca was to be included in the first phase of the programme and initial scoping exercises were carried out (Osborn 2013). However, they subsequently stalled and the residents' association has received no further information. Elsewhere it seems that upgrading works previously being carried out under separate programmes and without resident participation were being relabelled as *Morar Carioca* projects (*Ibid.*). Although the programme has not been officially withdrawn, the implication seems to be that the ambitious budget, scale and participatory processes outlined in the original plan will not be honoured. Some community leaders now believe that the upgrading of Asa Branca through the far less ambitious *Bairro Maravilha* was a means of avoiding the need for *Morar Carioca* and that the community will now receive no further investment.

Public Security

Unlike Tuiuti, Asa Branca lies in a region with relatively low levels of violence and without a history of organised drug trafficking. Historically, this situation has been preserved by residents acting collectively to prevent gang activity, in a context of the almost total absence of public security in the area. However, the peace is precarious, particularly with the growing influence of militias in the Jacarepaguá region. Typically formed of off-duty police officers, these organisations provide, or – depending on their degree of local legitimacy – *impose*, private security for low-income communities, often at an exacting price. The violence that accompanies such control varies from place to place (and at present seems to be low in the area around Asa Branca), but the threat of violence is implicit and ever-present. By monopolising local services and utilities some militias have become lucrative mafia-like structures whose power reaches far into political and criminal justice institutions, undermining the rule of law. Their lower visibility and greater political influence as compared to drug traffickers may help to explain why, at the time of writing, they (and Jacarepaguá) have been largely overlooked by the UPP programme.[7] This absence of public security, whether via pacification or the city's mainstream military police, leaves Asa Branca vulnerable to abuse by militias.

Elite Urbanisation and the Threat of Removal in Jacarepaguá

Jacarepaguá is a predominantly low-income region populated by favelas, housing projects and formal working-class neighbourhoods. However, it borders on the elite residential and commercial neighbourhood of Barra da Tijuca (Figure 17.1 – Rio de Janeiro). Both areas have undergone rapid processes of urbanisation since the 1970s, and these have accelerated in recent years, due in part to the decision to locate the Olympic Park on the border between them. Settlement of the region has traditionally unfolded according to a *laissez faire* model, with few public spaces or services (Herzog 2013). The city project proposes to alter this situation. An extension of the metro and three new Bus Rapid Transit (BRT) lines are being constructed to integrate the region with the rest of the city, connecting the key Olympics venues and the international airport. Post-2016 the Olympic Park itself is also set to become a 'new neighbourhood', containing public parks, schools and health clinics (Cidade Olímpica, Parque Olímpico).[8] These investments should, in theory, be of benefit to low-income residents who rely on public transport and lack community leisure spaces.

---

7   To date only one of the 38 UPPs, Batán, is in a former militia area. In Jacarepaguá only Cidade de Deus – an area dominated by drug traffickers – has been pacified.

8   However, given that the land has been conceded to a private consortium, there is doubt as to whether the new services and public spaces will be accessible to Jacarepaguá's low-income population (Wainwright 2013).

However, these public interventions are dwarfed by the scale and pace of speculative private development in the region. Immediately to the south of Asa Branca dozens of high-rise apartment blocks within gated condominiums were built in 2013 alone. A short distance to the east work has begun on the Centro Metropolitano da Barra, an empty area of five square kilometres, which will contain dozens of new housing blocks, a Hilton Hotel and a large shopping centre. This pattern of development is providing a steady flow of low-paid construction and service jobs for Asa Branca's residents and also underpins the favela's own vibrant commercial economy and housing market. However, with these economic benefits also comes a sense of foreboding.

Despite the recent urbanisation works in Asa Branca, many residents tend to see the real estate speculation and elite colonisation of the region as a threat to their future there. Although there does not appear to be any plans for the favela's removal, these concerns are not unfounded. Vila Autódromo, which lies within the perimeter of the Olympic Park, is fighting a drawn out eviction process. Meanwhile, other smaller favelas around Asa Branca have now been marked for removal to make way for BRT roads. In both cases original plans avoided the need for removals, but were altered by the *Prefeitura* (Comitê Popular 2013). Some residents now believe that freeing up valuable land for speculation and 'cleansing' the region of its favelas are the true objectives of urban policy in Jacarepaguá. Like Mangueira, in the region surrounding the Maracanã, Asa Branca's upgrading has made it a relative 'winner' from the city project so far. However, its residents continue to see the community's future as precarious.

Tuiuti, Asa Branca and the city project

The cases of Tuiuti and Asa Branca seem to reveal some important features of the city project that are neither captured by the official post-Third World city narrative nor by the city of exception critique. Indeed, analysing local processes and impacts in favelas that are less visible to policy-makers, academics and the media is essential for obtaining a rounded view of the city project and its direction of travel. Furthermore, by comparing the cases of Tuiuti and Asa Branca we can identify a number of important trends.

First of all it should be pointed out that unlike in favelas threatened by outright removal, the city project has not been experienced as entirely negative in either community. Indeed, some of the core needs of residents have been at least partially satisfied by recent interventions. Asa Branca's residents roundly praise the upgrading of their neighbourhood, and police pacification in Tuiuti is generally viewed as a qualified improvement. Beyond these successes, however, the benefits are less clear. Asa Branca remains weakly served by key institutions including the police, which leaves the community vulnerable to abuse by militias. In Tuiuti post-pacification interventions have so far failed to respond to the fundamental challenges faced by residents in the labour market and have placed a further squeeze on household budgets through rising utility bills. In both areas community services and activities to support the integration and development of different parts of the population remain non-existent.

These shortcomings should not, in themselves, be taken as evidence that the city project is failing in its stated objectives. There are over a thousand favelas in Rio de Janeiro, each with unique circumstances (Censo 2010). No citywide policy with a centralised design and inevitable resource limitations could be expected to exactly meet local demands, even if participatory principles were consistently applied. Furthermore the two flagship policies aimed at favelas – the *UPP* programme and *Morar Carioca* – were designed to be implemented over a relatively drawn-out timeframe and will, of necessity, affect different areas at different times.[9] It may be that current weaknesses and inconsistencies are bumps in the road, rather than inherent design flaws. Nonetheless, there are certain patterns in the implementation of the city project in Tuiuti and Asa Branca that suggest there is more to it than this.

Firstly, the geographical coverage and impacts of policies are unevenly spread at both a citywide and local scale. As argued by the 'city of exception' literature, the questions of 'which' policies have been implemented in Tuiuti and Asa Branca and 'how' seem mainly to have been determined by their relative locations and the way these intersect with the strategic priorities of policy-makers and elites. For example, the *UPPs*

---

9   The original plans of both programmes were to reach all of Rio's favelas by the year 2020.

have been overwhelmingly targeted at favelas dominated by traffickers, usually in central or 'strategically important' favelas. Meanwhile, security needs of peripheral favelas and those dominated by militias have been overlooked. This selective application of policies also occurs at a more local scale, with some favelas that are more 'visible' to policy-makers receiving large investments and others largely ignored – or worse, removed. Institutional 'visibility' to policy-makers may partly be the effect of something as simple as a favela's actual visibility to tourists, as suggested by Freeman. However, it also has to do with the deeper institutional processes that shape these communities' access to the state, including political clientelism (Ribeiro and Santos Junior 2013).

A second pattern is a general lack of imagination about what favela residents need and what would constitute meaningful 'urban integration', beyond the realms of security and infrastructure. The pacification approach may be a break with the past in terms of its *modus operandi*, but in its emphasis on security as a solution to the 'favela question' it shows a high degree of continuity with past approaches. The exclusive focus on physical upgrading in Asa Branca is equally one-dimensional. Meanwhile, expanded service provision, whether at the community level or in core public services like health and education, do not seem to have been conceived as a core element of urban integration. A proper application of the participatory principles of *Morar Carioca* (if it is indeed to continue), or a far more ambitious *UPP Social* programme, may have brought these issues to the fore. However, these policies have been starved of resources in ways that the UPPs and higher-profile infrastructure projects have not.

This raises the third main trend: the tendency for narrow private and political pressures to undermine stated policy objectives. In instances where the aims of integration and development have been at odds with the interests of powerful actors, it seems the city project has invariably come down in favour of the latter. The valuable land surrounding Asa Branca has almost entirely been enclosed for use as gated condominiums, despite the total lack of public spaces and services in the area. The charging for electricity in Tuiuti has placed major difficulties on the household budgets of its poorest residents. When vested interests take precedence over social outcomes – especially when the latter were so strongly emphasised by the post-Third World city narrative – it is inevitable that residents of communities like Tuiuti and Asa Branca will view the city project cynically.

This points to what appears to be the main flaw at the heart of the city project: its democratic deficit. Whether or not the policies themselves were appropriate or sufficient for fulfilling the claims of the post-Third World city narrative, they have neither had the necessary participation at the local level, nor the institutional solidity at delivery level, to be able to survive, adapt and improve as necessary. Whatever the long-term legacies of Rio's city project are, it is hard to believe that they will be the ones that a majority of residents would have wanted.

Acknowledgement

I would like to thank Theresa Williamson, Catalytic Communities, Carlos Alberto Costa Bezerra, Damiana Diniz, Gleice Valadares and Sônia Ferreira Marques, without whom the research would not have been possible.

References

Amado, G. (2013). 'Paes vai encerrar UPP Social e criar ouvidorias em favelas; moradores vão receber queixas e denúncias', *Globo Extra* http://extra.globo.com/noticias/rio/paes-vai-encerrar-upp-social-criar-ouvidorias-em-favelas-moradores-vao-receber-queixas-denuncias-7935522.html#ixzz2a4VuFIZR [Accessed 25/07/13].

Brum, M. (2013). 'Favelas e remocionismo ontem e hoje: da Ditadura de 1964 aos Grandes Eventos', *O Social em Questão*, 29, pp. 197–208.

Cano, I. (ed.) (2012). *'Os Donos do Morro': Uma Avaliação Exploratória do Impato das Unidades de Polícia Pacificadora (UPPs) no Rio de Janeiro*, Rio de Janeiro: Fórum Brasileiro de Segurança Pública.

Cidade Olímpica (official website), Transportes, http://www.cidadeolimpica.com.br/areas/transportes/ [Accessed 25/07/13].

Cidade Olímpica (official website), Parque Olímpico http://www.cidadeolimpica.com.br/projetos/parque-olimpico/ [Accessed 08/05/14].

Clarke, F. (2014). 'The Never-Ending Eviction: Demolition, Protest and Police Violence Once Again Rock Favela do Metrô', *Rio On Watch*, http://rioonwatch.org/?p=12978 [Accessed 27/03/14].

Comitê Popular da Copa e Olimpíadas do Rio de Janeiro (2013). *Megaeventos e Violações dos Direitos Humanos no Rio de Janeiro*, Rio de Janeiro: Comitê Popular.

Freeman, J. (2012). 'Neoliberal accumulation strategies and the visible hand of police pacification in Rio de Janeiro, Revista de de Estudos Universitários, 38(1), pp. 95–126.

Gaffney, C. (2010). 'Mega-events and socio-spatial dynamics in Rio de Janeiro, 1919–2016', *Journal of Latin American Geography*, 9(1), pp. 7–29.

Herzog, L. A. (2013). 'Barra da Tijuca: The Political Economy of a Global Suburb in Rio de Janeiro, Brazil', *Latin American Perspectives*, 40(2), pp. 118–134.

IBGE, Censo Demográfico, 2000.

IBGE, Censo Demográfico, 2010.

Osborn, C. (2013). 'A História das Urbanizações nas Favelas Parte III: Morar Carioca na Visão e na Prática (2008 – Presente)', *Rio On Watch*, http://rioonwatch.org/?p=8136 [Accessed 25/07/13].

Phillips, T. (2010). 'Rio de Janeiro favelas to get facelift as Brazil invests billions in redesign', *The Guardian*, http://www.guardian.co.uk/world/2010/dec/05/rio-de-janeiro-favelas-brazil [Accessed 25/07/13].

Prefeitura do Rio de Janeiro (official website) (2011), Conheça o Programa: Morar Carioca, http://www.rio.rj.gov.br/web/smh/exibeconteudo?article-id=1451251 [Accessed 27/03/14].

Raco, M. (2012). 'The Privatisation of Urban Development and the London Olympics 2012', *City*, 16(4), pp. 452–60.

Ribeiro, L. C. Q. and Santos Junior, O. A. (2013). 'Governaça empreendorista e megaeventos esportivos: Reflexões em torno da experiência brasileira, *O Social em Questão*, 29, pp. 23–42.

Richmond, M. (2013). 'Resident perceptions of urbanisation and elite encroachment in a Jacarepaguá favela', *Paper presented at the International RC21 Conference, Berlin.*

SABREN – Sistema de Assentamento de Baixa Renda, Instituto Pereira Passos, Prefeitura do Rio de Janeiro, http://portalgeo.rio.rj.gov.br/sabren/ [Accessed 25/07/13].

Sanchez, F. and A. Broudehoux (2013). 'Megaevents and urban regeneration in Rio de Janeiro: planning in a state of emergency', *International Journal of Urban Sustainable Development,* 5(2), pp. 132–153.

UPP Social (official website), Programa, http://uppsocial.org/programa [Accessed 25/07/13].

Urani, A. And F. Gambiagi (2011). *Rio: A Hora da Virada*, Rio de Janeiro: Elsevier Editora.

Vainer, C. (2011). 'Cidade de exceção: Reflexões a partir do Rio de Janeiro', Papel apresentado ao XIV Encontro Nacional da ANPUR.

Wainwright, O. (2013). 'The Rio 2016 Olympics: sun, sea and absolutely no swimming', *The Guardian* http://www.theguardian.com/artanddesign/architecture-design-blog/2013/aug/01/rio-2016-olympic-urban-plan-legacy [Accessed 27/03/14].

Werneck, A. and S. Ramalho (2013). 'UPP de fantasia: tráfico impõe terror no Morro da Mangueira', *Globo* http://oglobo.globo.com/rio/upp-de-fantasia-trafico-impoe-terror-no-morro-da-mangueira-7635299 [Accessed 25/07/13].

Chapter 18

A 'City of Exception'? Rio de Janeiro and the Disputed Social Legacy of the 2014 and 2016 Sports Mega-events

Einar Braathen, Gilmar Mascarenhas, Celina Myrann Sørbøe

Introduction

In today's globalized world, there is growing interurban competition over international flows of capital and visitors. In order to produce an image of a city that can compete for these resources on the international market, publicity strategies of "branding" the urban space gain importance (Mascarenhas, 2012). Hosting international mega sports events has recently been adapted as a branding strategy by cities in the Global South. Over the last decade, all the BRIC countries have invested enormous financial resources and political prestige in hosting mega sports events; such as the 2008 Beijing Olympics, Brazils 'double' approach of the 2014 FIFA World Cup and 2016 Olympic Games, followed by Russia's 2014 Winter Olympics and 2018 FIFA World Cup. These events place the host city and country in the world's spotlight and represent opportunities for increased trade, investments and economic growth.

To be able to pull off international events of this scale, massive public spending is required. As a way of legitimizing these expenses, leaving a positive legacy has gained importance. There are ample opportunities to use the capital and investments coming with the events as a concrete tool for social change. On the other hand, abiding to the demands of international organizing committees such as the Fédération Internationale de Football Association (FIFA) and the International Olympic Committee (IOC) often means overruling or sidestepping existing institutional frameworks and Master Plans. Building legacy thereby involves facing an essential contradiction: how can one meet the demands of the city and its inhabitants while mega sports events are increasingly aligned with large private interests, and candidate cities themselves tend to be managed in terms of urban entrepreneurship? (Mascarenhas, 2012).

There are two main views of how this contradiction has been met in Rio de Janeiro before the 2016 Olympic Games: (i) the conventional view of creating legacies through infrastructures and improvements in the built environment of the host city (physical legacy); there was a hope that Rio could become a 'positive exception' in this regard; (ii) the critical view that the main legacy is observed in the urban governance system in terms of a state of exception – or 'city of exception' (political-institutional legacy). Before elaborating and discussing these main views, we identify the global context of mega sports and then the specific context of Rio de Janeiro, characterized by a history of socio-spatial exclusion.

Mega Sports Events and the "Global City"

Sport once viewed as a form of entertainment, has emerged as an important political, social and economic force (Hiller, 2000). Sports also play an important cultural or cultural-hegemonic role. It can be used, or abused, to strengthen national identities. These different roles of sports are played out in powerful ways in the hosting of mega sports events (Tomlinson and Young, 2006). Rio de Janeiro's successful bids for the 2007 Pan-American Games, 2014 World Cup and the 2016 Olympics crowns the country's remarkable rise after decades of underachievement to becoming an economic and diplomatic heavyweight. Just as the Beijing Olympics of 2008 marked China's revival as a world power (Broudehoux, 2007), the 2016 Rio Games may be seen as a stamp of approval on the South American giant's coming of age. Over the last decade, the trend has been for mega sports events and so-called 'emerging economies' to grow closer. These economies combine three crucial elements: availability of resources; an ambition to strengthen their image as an emerging power worldwide; and relative weakness of institutions which protect the environment and human rights. The combination of these elements enables host cities to abide by the 'package' of interventions that international organizing committees require.

Rio de Janeiro's winning of the bids for these mega sports events has been attributed to a fundamental shift in the municipal leadership's strategy during the 1990s. With the re-democratization of Brazil in the 1980s, the 1988 Constitution granted more power to Brazilian states and municipalities. Matters of urban development were placed under the control of the municipal governments. Rio de Janeiro developed and implemented a Master Plan in 1992. This Master Plan is based on a prioritization of public over private interests, and contains guidelines when it comes to a democratization of access to land, infrastructure, and urban services. It also includes norms for a democratic management of the city. Ironically, the decentralization process also became instrumental in supporting a turn towards new public managerialism in Rio starting in the mid-1990s. Local scholars point at the rule of the populist-turned-neoliberal mayor, Cesar Maia, as the turning point (Vainer, 2000). In 1993, Maia invited the Rio de Janeiro Trading Association (ACRJ) and the Federation of Industries of the State of Rio de Janeiro (FIRJAN) to join the municipality in elaborating a strategic plan for the city. A key urban planner from Barcelona, Dr. Jordi Borja, was the main consultant. Inspired by the 1992 Barcelona Olympic Games, the plan emphasized the big potential of large projects and mega-events in branding Rio de Janeiro as a destination for tourists and foreign investors and transforming Rio de Janeiro into a 'global city'. In 1994, the municipality, private companies and business associations came together and created a strategic plan of Rio de Janeiro, which was approved without democratic channels of participation (Vainer, 2000: 106). In 1996 the city sent its first bid to host the Olympic Games and in 2009 it won the bid for the 2016 games (Curi, Knijnik and Mascarenhas, 2011). As a contrast to the Master Plan, the Strategic Plan is steered by business demands and interests with the goal to make the city more 'attractive' on the international market (Braathen et al., 2013: 9).

The close cooperation between the municipality and private sector leaders has been depicted by David Harvey as an international trend of transformation of urban governance towards 'urban entrepreneurship' (Harvey, 1989). Others have termed this new strategic planning either 'ad hoc urbanism' (Ascher, 2001) or *cidade empresa* – 'company city' (Vainer, 2011). In business, efficient management relies on the ability to take advantage of opportunities faster than the competitors. Thus, in the view of strategic planning, the city itself should function as a company. Political control and bureaucracy, such as responding to the institutional rights and guidelines of the Constitution or the Master Plan, erodes a city's capacity to take advantage of business opportunities, and, consequently, come across as efficient and competitive (Vainer, 2011: 5). The hosting of mega sports events intensifies these processes. As existing institutional frameworks are overruled to respond to the needs of international sponsors and private interests, the Olympic bid books become the *de-facto* urban planning documents.

Drawing on Giorgio Agamben's (1995) theories of the state of exception, local scholar Carlos Vainer (2010, 2011) claims the preparations for the mega-events have authorized, consolidated and legalized practices of legal exception in Rio. According to Vainer, two elements have been instrumental in legitimizing the transformation of Rio into an Olympic city. On the one hand, the city's patriotism led to a profound sense of pride among the inhabitants at the prospect of hosting a global mega-event. On the other, the generalized sense of an urban crisis that has characterized the city since the 1990s authorized and demanded a new form of power constitution in the city.

Rio de Janeiro: A History of Socio-spatial Exclusion

Rio de Janeiro is one of the most unequal cities in the world (UN-Habitat, 2008: 70). The socio-political space is divided between the "formal" city and the urban informal settlements known as *favelas*. The favelas date back to the end of the 19th century, and have long been a headache for the city's elites. According to Broudehoux (2001), image has always been important in the production of Rio as the "marvelous city". In that aspect, the favelas have always been central. The elite has always been preoccupied with the favela's place in the city and impact on the attractiveness of Rio. Throughout their history, the favelas have been rejected by the formal city and have continually been threatened by destruction (Perlman 2010: 26). During the period of the military dictatorship (1964–1985) favelas located in the noble areas of Rio de Janeiro were targets of public removal policies. The residents were moved to new housing estates in areas distant from the city center – both to "re-civilize" these populations and to beautify the city. As authorities proved incapable of solving the housing deficit in the city, the favelas however continued to grow uncontrollably.

In the 1980s the international drug trade came to Rio de Janeiro, and drug traffickers found a stronghold in the favelas where the state presence was weak. The violence associated with the drug trafficking grew in frequency and intensity throughout the late 1980s and 90s, and assaults, robberies, kidnappings, shoot-outs and *balas perdidas* ("lost bullets" striking innocents caught in a cross-fire) became everyday security issues. The *favelados*, pejorative for the people inhabiting these territories, have been perceived as intimately linked to all the problems associated with the favelas. Regular residents within these territories were considered as accomplices of the drug traffickers because of neighborhood relations, kinship or economic and political ties.

Giorgio Agamben (1995) has written extensively about the state of exception. He portrays how the effects of the decisions made by the state (or whoever has the sovereign power) can lead to the exclusion of somebody from the political community and the protection provided by its laws and rights. This permits the physical elimination of not only political opponents, but also of entire categories of citizens that are perceived as external and non-integral to society (Foucault, 2003; Agamben, 2005). In the public imaginary, there were no innocents in the favelas. In the 1990s, the social conflict in Rio de Janeiro became formulated as a "war", and the police took a militarized approach to combating the drug trafficking.

The war on crime produced an image of the city as notoriously dangerous and irredeemably divided, which tarnished the national and international identity of Rio de Janeiro. Public policies in the favelas had to be revised. As they had proved incapable of solving the housing deficit in the city, the removal policies were put to an end in the 1990s, and the public debate shifted to concentrating on the necessity of integrating the favelas in the city (Oliveira, 2012: 47). Programs such as *Favela-Bairro,* launched by the municipality in 1993 to upgrade all of the city's favelas, led to the notion that "the favela has won!" in the late 1990s (Zaluar and Alvito, 1998). One of their arguments was that favelas were no longer at risk of removal and most people defended their urbanization instead. When Inácio Lula da Silva from the Workers' Party (PT) was elected president in 2002 on a pro-poor platform, many were optimistic that the old divides between the 'favela' and the 'asphalt' were slowly being erased. Was Rio about to become an 'exception from the exception' – a city for the dispossessed masses?

The 'Physical' Legacy: Olympic Promises

Mechanisms of urban image construction have always shaped public policies towards the favelas, and urban renewal programs such as Favela Bairro significantly transformed the image of Rio de Janeiro in the 1990s (Brodehoux, 2001). In the process of "branding" Rio de Janeiro as an Olympic city today, the favelas once again play a central part. On the one hand, Rio de Janeiro's reputation as one of the world's most unequal cities implied a demand to implement pro-poor strategies to address the legacy of social and spatial inequalities. On the other, the desire to reach global city status in terms of attracting international investment, economic growth and tourism in order to demonstrate (Western) goals of urban achievement demanded that the city would deal with the notorious insecurity that has given the city a reputation for being a dangerous place to visit. When Rio de Janeiro first bid for the Olympic Games, the official discourse was to use the Olympic Games both to improve living conditions of the poor and improve security in the city. Lula himself headed the lobby for Brazil to receive the Olympics.

Leaving a positive legacy is one of the recent concerns of the "Olympic system" as a way of legitimizing itself (Horne and Whannel, 2012). Cities in the bidding process present a legacy plan on how they will use the event to address the city's social, economic, infrastructural, and planning challenges as a central aspect of their candidacy, often going beyond what is strictly necessary in order to stage the Games. The 2012 Olympic Games hosted by London were applauded for a legacy plan that emphasized the urban regeneration of selected underprivileged neighborhoods. Rio de Janeiro also adopted an ambitious plan to use the Olympics for citywide transformation. The federal government and the city of Rio were transcending the conventional policies for 'built-in-legacies'. In the Sustainability Management Plan of the Olympics (SMP) developed by the Municipality of Rio de Janeiro, the city government states that one of the strategic objectives of the municipal planning department is to 'organize an all-inclusive Games, leaving the city's population with a positive social balance'(Prefeitura_do_Rio_de_Janeiro.2010). Legacy plans ranged from housing improvements to crime reduction, social inclusion, and urban regeneration combined with an attempt to revive the city's national and international image (Girginov 2013: 301). The main references were the federal 'Program for Accelerated Growth' (PAC) and 'My House My Life' (MCMV) as well as the municipal 'Morar Carioca' program, which has pledged a social legacy from the Olympic Games in terms of comprehensive upgrading of all the favelas in Rio de Janeiro by 2020 (Prefeitura do Rio de Janeiro, 2010; Bittar, 2011). In terms of security, a new policing program called UPP (Units of Pacifying Police) was developed in 2007. As a contrast to former military interventions in the favelas, the program relied on the permanent placement of UPP Units in strategically located favelas. Through combining security with urban upgrading interventions and increased access to social services, the program aimed to bridge the gaps between the favelas and the rest of the city. The UPPs depended on PAC for their budget, which also was the principal fund for infrastructure associated with the World Cup and Olympics. The link to the mega-events was thus evident.

The PAC, Morar Carioca, and UPP programs were supposed to leave a "lasting legacy" and better public services in the favelas. The question is, however, to what degree these clean-up operations have done more to refurbish and secure the tourist areas, to make Rio once again attractive on the international market than to improve quality of life within the targeted communities.

Broken Promises

Vila Autódromo, a fishing village which developed into a working class neighbourhood in the upper middle class boomtown Barra de Tijuca in the Western zone, serves as an example of the conflictual relationship between local residents and the government because of the upcoming mega-events. It was threatened by collective relocation because of the construction of the main sports arenas and accommodation centres for the 2016 Olympic Games (see Braathen et al., 2013). While the government, after massive pressure from residents and civil society organizations, promised to consider an alternative plan where part of the community could remain, civil society actors claimed the municipality was still using extra juridical measures to force residents to accept relocation.

Another community impacted by the ongoing city "improvements" was the favela complex Manguinhos, located in abandoned factory areas. Since being selected for the PAC program in 2008, a brutal, drawn-out eviction process has affected the community. The authorities have strategically employed an expulsion tactic where they demolish some houses and leave the ruins, and thereby garbage, rats and hazardous conditions, behind. This makes life unbearable for those residents who remain, while sending a strong message that their eviction is imminent (Braathen et al., 2013). The Favela do Metrô Mangueira has suffered the same violent process since 2010 (Rolnik, 2010).

These cases exemplify how the tangible (physical) "benefits" and "legacy" of the mega-events are often imposed at the expense of poor communities and residents that are located near the sports facilities and the main access roads. Replacement housing being constructed through the federal housing program MCMV has overwhelmingly been located in the distant northwest of the city, where land values are cheap, employment opportunities are limited, environmental conditions are terrible, and transport connections are poor. Studies by urban planners indicate that the MCMV program reproduces the logic of older "housing estates" where the poor end up being pushed to locations far from job opportunities and without a system of transportation (Braathen et al., 2013). These people's institutional rights have had to give way to the prosperity of society in general, defined within a neoliberal discourse of economic development. In the words of Agamben, this "bare" or "naked life" represents persons or groups of persons that others, with impunity, can treat without regard for their psychological and physical well-being (Agamben, 1995).

The 'Political-institutional' Legacy: A City of Exception

Massive investments have been made in Rio in order to prepare the city for the upcoming mega-events. The short time frame, as well the demands of international organizing committees and private sponsors, resulted in market-oriented municipal governance. The consequences of this 'urban entrepreneurship' (Harvey, 1989) were quick to materialize. *Ad hoc* decisions have prevailed over abiding by existing legal frameworks such as the Master Plan. Basic democratic rights have been put on hold, and the demands and rights of ordinary citizens suppressed. These developments are especially serious for the vulnerable populations in favelas located near Olympic venues or other strategic places. The PAC and Morar Carioca interventions in the city's favelas have continuously been denounced for lacking participatory planning and for violating the human and civil rights of the families and populations "benefiting" from these programs. The word *remoção* ('removal') which was broadly used during military dictatorship is once again back on the agenda. 40,000 people are threatened by removal because of large-scale construction projects connected to the mega-events in Rio de Janeiro alone. The majority are poor favela residents. Previously acquired rights enshrined in the Constitution, such as the right to housing, were progressively being eroded on the grounds that they impede the freedom of the market and therefore restrict economic development and modernization (Dagnino, 2010).

Securitization

As mentioned earlier, increased deployment of security forces and policing is also part of this picture. According to Tony Roshan Samara (2010; 2011), urban governance in a neoliberal environment is often driven by security concerns over protecting public order and economic growth, especially in highly unequal cities. In order to produce an image of a "sellable" city in front of the mega-events; a city that can be a recipient of resources, investments, tourism and economic gains, security is essential. Security is related to liberty – in the possibility for circulation and consumption. Rio de Janeiro's reputation as a violent city was hindering its competiveness. Improving the security in the city could therefore be seen as a "branding" strategy to increase investments and economic growth.

The new UPP Police Pacification Units were reclaiming the monopoly of power in favelas that have "threatened" the sense of security in the city. While promoted as a programme to spur an approximation process between the favela and the asphalt, the UPP police practice in the favelas can be seen as a "differentiated policing of space" (Samara, 2010; 2011). The pacification represented a police mechanism that was exercised according to the spatial configuration of the city. The UPPs were only stationed in favelas. While less violent than earlier police interventions, the UPPs established a permanent militarized regime in the pacified favelas that went beyond combating the drug traffickers. The UPP security regime controlled and managed not just the behaviors of suspected drug traffickers, but the life of *all* favela residents (Sørbøe, 2013). In the run-up to the 2014 World Cup, military troops together with the military police occupied a large favela area in Rio de Janeiro – the Maré complex – strategically located next to the main access roads of the city. The troops remained until after the FIFA World Cup. In practice, a military state of emergency was introduced, justified by the hosting of mega-events. One of the most controversial aspects was a decree signed by the president, which permitted the police to break into any home without a warrant. This would be unthinkable in the city's upper and middle-class neighborhoods.

Securing these territories has therefore not been merely about improving security in the immediate locality. The improved sense of security in the city as a whole had in addition the intended benefit of attracting and securing investments.

'Rebel Cities': Towards a "Political-Cultural Legacy"?

When Lula won the elections in 2002, many civil society organizations and activists believed it represented an historic opportunity for significant change in Brazil. In response to the demands from the social movements,

Lula's administration created institutions such as the Ministry of Cities and Council of Cities, stimulating public participation in local, state and national housing and sanitation projects (Rolnik, 2011). These structures were seen as spaces where the state and civil society were expected to work together to ensure that priority-setting matched the public interest and to secure accountability in the definition and delivery of social policies (Heller and Evans, 2010). The last years have, however, seen a procedural and substantive disillusionment with the existing spaces and mechanisms to institutionalize citizens' participation in public decision-making (Baiocchi et al., 2013). The initial approval has given way to a growing sense of disappointment with how Lula and PT manage the challenges of governing Brazil (Hochstetler, 2008; Rolnik, 2011). In spite of a decade of economic growth and poverty reduction under the PT administrations, people's increased income has not automatically resulted in improved quality of life. Indicators on crime, violence, and levels of education and health remain poor. Neither has it addressed the socio-spatial segregation that characterizes Rio de Janeiro. A range of socio-economic, political, racial and cultural markers still work to exclude favela residents from many of the citizen rights enjoyed by residents of the formal city – what Holston (2008) has termed 'differentiated citizenship'. With the preparations for the mega-events, these differences became more evident.

The June 2013 Protests

In Brazil, the popular culture around sports (soccer) and festivals (carnival) has always been linked to politics or possible political abuse (DaMatta, 1991; Wisnik, 2006). As the forms of illegality and exceptions to the institutional order have multiplied with the preparations for the mega-events, growing numbers of people are questioning the true intentions of the authorities and the ultimate consequences for ordinary residents. People react with contempt at corrupt politicians who exploit public investments in the events for private gains.

In June 2013, millions of Brazilians took to the streets in what became the largest street demonstrations in recent history. What started as a protest against a price hike on the public transportation in São Paulo quickly escalated to mass mobilizations against the massive public spending on stadiums and infrastructure related to the mega-events while the quality of public services remain precarious. They also revolted against the violence used by the police forces to quell the demonstrations (Maricato, 2013).

The June demonstrations raised issues in the public debate in Brazil regarding citizenship – how to listen to the 'voice of the street', take grievances of ordinary people seriously, and improve the quality of democracy. These manifestations did not come from nowhere: they represent the culmination of years of formation of a new generation of urban movements. Organizations such as the *Movimento Passe Livre* ('movement for free transport'), student movements, urban resistance movements, favela residents' associations and movements *sem-teto* (for those without a 'roof'/house) have through occupations and demonstrations articulated in broader networks challenging the existing emptied out top- down spaces of participation. This new generation of urban movements and civic networks augurs an 'insurgent citizenship' (Holston 2007). As opposed to a statist citizenship that assumes the state as 'the only legitimate source of citizenship rights, meanings and practices' (Holston, 1998: 39), this alternative conceptualization of citizenship is active, engaged, and 'grounded in civil society' (Friedmann, 2002: 76). It moves beyond formal citizenship to a substantive one, that concerns an array of civil, political, social, and economic rights. These demands include the right to housing, shelter, education and basic health. As such, it incorporates the notion of the "right to the city" (Lefebvre, 1967) which recognizes all city residents as "right holders" in the city, defending the needs and desires of the majority and affirming the city as a site for social conflict.

Demands for 'the Right to the City'

While 'the right to the city' has been recognized by the institutional framework in Brazil, the hollowing out of the functions of the institutional spaces for citizen participation with the neoliberal reforms have left many of the promises unfulfilled (Santos Junior, Christovão et al., 2011). The new generation of civil society movements has claimed the concept of the right to the city for their own, and it was frequently seen on banners and posters during the June protests. As emphasized by David Harvey, the right to the city is 'far more than a right of individual or group access to the resources the city embodies: it is the right to change and reinvent the city more after our hearts' desire' (Harvey, 2012:4).

The right to the city has become a slogan for movements worldwide who fight against the manifestations of many modern cities in which public processes and utilities have been privatized and where development is driven primarily if not solely by corporations and markets (Santos, 2007). In protesting these tendencies, practices of insurgent citizenship have become the means through which the urban margins negotiate and contest their right to universal inclusion (Holston, 2007: 22). The June protests sparked a new wave of social mobilization, allowing local communities' struggles over localized issues to connect to a wider discourse of urban development conflicts in Rio de Janeiro and globally. All of a sudden, President Dilma Rousseff, and all the leaders of the political parties, promised to listen to 'the voice of the streets'. Several speedy reform initiatives were taken. However, while the civil society gained some small victories with the protests, there were no profound changes in the urban regime. The "Olympic project" continued to dominate city governance.

Securitization (ii): The Authoritarian Legacy

2014 was not only the year of the World Cup. It also marked the 50th anniversary of the military coup d'état. One of the principal demands of the June 2013 protests was a de-militarization of the police and of society. The politicians have however all proven to be incapable of following up on the promises they gave in the previous year to "listen to the voice of the streets". On the contrary, they prioritized the enactment of repressive measures – such as more police, new laws, and harsher penalties – with the hope of quelling the tensions and isolating demonstrators.

With enthusiastic support from right wing forces, a new "law against terrorism" was proposed by a leading senator in the government block. The law explicitly targeted street protesters suspected of using violence. Terrorism was thus defined as the act of "provoking or spreading generalized fear or panic through violating human life or the physical integrity, health or freedom of persons".[1] The prescribed penalties were up to 30 years imprisonment. The proposal was rushed through the two chambers of the Congress in order to take effect before the World Cup. At the same time, a massive police corps of 150,000 officers from the military police and 20,000 private security guards were organized in special World Cup battalions, with a budget of around 1 billion USD. Ralf Mutschke, FIFA's Director of Security, was part of the chain of command and in charge of the private security guards. In Rio de Janeiro, the World Cup battalion even developed its own elite troop, with body armor that closely resembled the "Robocop".[2] There is no doubt that the Olympics will see similar security efforts.

Conclusion

This chapter has looked at the arena of conflict between actors associated with the state, the market and civil society in the prospect of branding Rio de Janeiro as a "Global City" of international mega-events. In the process of political and urban change that has accompanied the construction of Rio de Janeiro as an Olympic city, Rio has been turned into a space for business, and no longer a space for political and democratic debate. Tendencies of social cleansing and increased socio-spatial segregation of the city were observed.

Two main views of how Rio de Janeiro has faced the 2016 Olympic Games were presented. We do not share the optimistic-conventional view of legacies created through infrastructures and improvements in the built environment, making Rio a 'positive exception' in this regard (physical legacy). We lean more towards the critical view, that the main legacy is observed in the urban governance system in terms of a state of exception – or 'city of exception' (authoritarian legacy). Carlos Vainer (2011) warned that the preparations for the mega-events had made Rio a "city of permanent exception". What was observed in South Africa in connection with the FIFA 2010 World Cup – dubious priorities, overspending, loss of sovereignty, and broken trickle down promises (Bond et al., 2011) – was also present in Brazil's hosting of this event. It can be expected on an even larger scale for the Olympics.

---

1   'Projeto Lei no. 499' (Lei Anti-Terrorismo), promoted by Romero Juca (PMDB). *O Globo*, 22.02.2014.
2   *O Globo*, 27.02.2014.

Nevertheless, this critical view easily becomes too pessimistic and one-dimensional. A synthesis of the two main views, constructed in a dialectic way, might open up a new and interesting perspective: mega sports events also create a more politicized and engaged citizenship as people respond with protest and various forms of bottom-up political engagement to the perceived negative cost-benefit (high costs and little benefits) of the built legacies. The transformation of Rio de Janeiro into an Olympic city has catalyzed social mobilization and growing politicization. This way, the main legacy of the mega events in Rio is a political-cultural one. Consequently, as the run-up to the 2014 FIFA Cup suggests, the 2016 Games in Rio may turn out to be the most contested in Olympic history.

References

Agamben, G. (1995). *Homo Sacer: Sovereign Power and Bare Life*. Stanford, CA: Stanford University Press.

Agamben, G. (2005). *State of Exception*. Chicago, IL, University of Chicago Press.

Ascher, F. (2001). Les nouveaux principes de l'urbanisme. La fin des villes n'est pas à l'ordre du jour. Éditions de l'Aube.

Baiocchi, G., E. Braathen, et al. (2013). 'Transformation institutionalized? Making sense of participatory democracy in the Lula Era', in K. Stokke & O. Törnquist (eds.), *Democratization in the Global South: The Importance of Transformative Politics*. Palgrave Macmillan: Houndmills, Basingstoke, Hampshire (UK), pp. 217–239.

Bienenstein, G. et al. (2012). 'The 2016 Olympiad in Rio de Janeiro: Who Can/Could/Will Beat Whom?' *Esporte e Sociedade*, 7(19), pp. 42–61.

Bittar, J. (2011). Morar Carioca. Rio de Janeiro, Speech delivered November 11th 2011 at a public meeting organized by Clube de Engenharia.

Bond, P., A. Desai and B. Maharaj (2011). 'World Cup profits defeat the poor', in B. Maharaj, A. Desai and P. Bond (eds.), *Zuma's Own Goal: Losing South Africa's 'War on Poverty*. Trenton: Africa World Press, pp. 417–432.

Braathen, E., T. Bartholl, A.C. Christovão and V. Pinheiro (2013). 'Rio de Janeiro', in E. Braathen (ed.) (2013). *Addressing Sub-Standard Settlements*. Bonn: European Association of Development Institutes, pp. 134-163.

Broudehoux, A. (2001). 'Image making, city marketing, and the aesthetization of social inequality in Rio de Janeiro', in Alsayyad, N. (ed.) (2001). *Consuming Tradition, Manufacturing Heritage*: *Global Norms and Urban Forms in the Age of Tourism*. London: Routledge, pp. 273–297.

Broudehoux, A. (2007). 'Spectacular Beijing: the conspicuous construction of an Olympic metropolis'. *Journal of Urban Affairs,* 29(4), pp. 383–399.

Curi, M., G.Mascarenhas and J. Knijnik (2011). "The Pan American Games in Rio de Janeiro 2007: Consequences of a sport mega-event on a BRIC country". *International Review for the Sociology of Sport*, 46(2), pp. 140–156.

DaMatta, R. (1991). *Carnivals, Rogues, and Heroes: An Interpretation of the Brazilian Dilemma*. Notre Dama, IN: University of Notre Dame Press.

Dagnino, E. (2010). 'Citizenship: A perverse confluence', in Cornwall, A. and D. Eade (eds.) (2010). *Deconstructing Development Discourse. Buzzwords and Fuzzwords*. Oxford: Practical Action Publishing Ltd.

Foucault, M. and J. Miskowiec (1986 [1967]). 'Of other spaces'. *Diacritics*. 16 (1), pp. 22–27.

Foucault, M. (2003). *Society Must Be Defended: Lectures at the Collège de France 1975–1976*. New York: Picador.

Friedman, J. (2002). *The Prospect of Cities*. Minneapolis: University of Minnesota Press.

Girginov, V. (2013). *Handbook of the 2012 Olympic and Paralympic Games. Volume One: Making the Games*. New York, NY: Routledge.

Harvey, D. (1989). "From managerialism to entrepreneurialism: The transformation in urban governance in late capitalism". *Geografiska Annaler B*, 71(1), pp. 3–17.

Harvey, D. (2012). *Rebel Cities, From the Right to the City to the Urban Revolution*. London/New York: Verso.

Harvey, D. (2011). Le capitalisme contre Le droit à la ville: néoliberalisme, urbanisation, résistances. Paris: editions Amsterdam.

Heller, P. and P. Evans (2010). 'Taking Tilly south: durable inequalities, democratic contestation, and citizenship in the Southern Metropolis'. *Theory and Society*, 39, pp. 433–450.

Hiller, H. (2000). "Mega-events, urban boosterism and growth strategies: An analysis of the objectives and legitimations of the Cape Town 2004 Olympic Bid". *International Journal of Urban and Regional Research*, 24(2), pp. 449–458.

Hochstetler, K. (2008). 'Organized civil society in Lula's Brazil', in P. Kingstone and T. Power (eds.) (2008). *Democratic Brazil Revisited*. University of Pittsburgh Press, Pittsburgh PA, pp. 33–53.

Holston, J. (1998). 'Spaces of insurgent citizenship', in Sandercock, L. *Making the Invisible Visible: A multicultural planning history*. California: University of California Press.

Holston, J. (2007). *Insurgent Citizenship*. Princeton, NJ, Princeton University Press.

Horne, J. and G. Whannel (2012). *Understanding the Olympics*. London: Routledge.

Jaguaribe, B. (2011). "Imaginando a "cidade maravilhosa": modernidade, espetáculo e espaços urbanos". *Porto Alegre*, 18(2).

Klauser, F. R. (2011). 'Interpretative Flexibility of the Event-City: Security, Branding and Urban Entrepreneurialism at the European Football Championships 2008'. *International Journal of Urban and Regional Research*, 36(5), pp. 1039–1052.

Lefebvre, H. (1967). *Le Droit à la ville*. Paris: Anthropos.

Leite, M.P. (2000). "Entre o individualismo ea solidariedade: dilemas da política e da cidadania no Rio de Janeiro". *Revista Brasileira de Ciências Sociais*, 14 (44), pp. 73–90.

Maricato, E. et. al. (2013). Cidades Rebeldes: Passe Livre e as Manifestações que Tomaram as Ruas do Brasil. São Paulo, Boitempo Editoral.

Mascarenhas, G. (2012). Globalização e políticas territoriais: os megaeventos esportivos na cidade do Rio de Janeiro. Globalização, políticas públicas e reestruturação territorial. S. M. M. Pacheco and M. S. Machado. Rio de Janeiro, Letras. 1.

Mascarenhas, G. (2014). 'O Brasil urbano na era dos megaeventos: uma nova geografia para 2014', in Melo, Victor; Peres, Fabio; Drummond, Mauricio. (Org.). Esporte, cultura, nação, estado: Brasil e Portugal. 1ed. rio de janeiro: 7 letras, 2014, v. 1, p. 188–200.

Prefeitura_do_Rio_de_Janeiro (2010). Morar Carioca – Plano Municipal de Integração de Assentamentos Precários Informais. Rio de Janeiro, Rio2016 and Prefeitura do Rio de Janeiro.

Oliveira, F. de et. al. (2012). Grandes Projetos Metropolitanos. Rio de Janeiro e Belo Horizonte. Rio de Janeiro, Letra Capital.

Perlman, J. (2010). *Favela: Four Decades of Living on the Edge in Rio de Janeiro*. Oxford: Oxford University Press.

Rolnik, R. (2010). Olimpíada e Copa trazem prejuízo social. On line, available: http://raquelrolnik.wordpress.com

Rolnik, R. (2011). 'Democracy on the edge: Limits and possibilities in the implementation of an urban reform agenda'. *International Journal of Urban and Regional Research*, 35(2), pp. 239–55.

Samara, T. R. (2010). 'Policing development: Urban renewal as neo-liberal security strategy'. *Urban Studies*, 47(1), pp. 197–214.

Samara, T. R. (2011). *Cape Town After Apartheid: Crime and Governance in the Divided City*. Minneapolis: University of Minnesota Press.

Santos, M. (2007). O espaço do cidadão. São Paulo: EDUSP, 7a edição.

Santos Junior, O., A. C. Christovão, et al., Eds. (2011). Políticas Públicas e Direito à Cidade: Programa Interdisciplinar de Formação de Agentes Sociais e Conselheiros Municipais. Rio de Janeiro, Letra Capital.

Sørbøe, C. (2013). Security and Inclusive Citizenship in the Mega-City. The pacification of Rocinha, Rio de Janeiro. Master Thesis, University of Oslo May 2013.

Tomlinson, A. and C. Young (2006). *National Identity and Global Sports Events: Culture, Politics, and Spectacle in the Olympics and the Football World Cup*. New York: SUNY Press.

Vainer, C. (2000). Pátria, empresa e mercadoria. Notas sobre a estratégia discursiva do planejamento estratégico urbano. A cidade do pensamento único. Demanchando consensos. O. Arantes, C. Vainer and E. Maricato. Petrópolis, Editora Vozes.

Vainer, C. (2011). Cidade de Exceção: reflexões a partir do Rio de Janeiro. XIV Encontro Nacional da Anpur. Rio de Janeiro, Anpur.

Wisnik, J. M. (2006). 'The riddle of Brazilian soccer: Reflections on the emancipatory dimensions of culture' *Review: Literature and Arts of Americas*, 39(2), pp. 198–209.

Zaluar, A. and M. Alvito (1998). Um seculo de favela. Rio de Janeiro: Fundação Getulio Vargas Editora.

Epilogue:
The Meaning of 'Legacy'

Michael Rustin

How are we to understand the 'legacies' of the Olympics, and indeed of other mega-events (Roche 2000) like them, such as the FIFA World Cup in Brazil which ended just as this book was going to press. Are these legacies to be understood as the main justification of these events, because of the long-term difference they make to the cities and nations in which they are staged, compared with the ephemeral life of even the grandest festivals of sport? Or are 'legacies' mere window-dressing, gestures to obscure the fact that immense resources have been consumed while giving rise to relatively little public benefit?

Concern about the legacies of the Olympics Games has grown over recent decades, as public realisation has grown of their enormous costs, and sometimes also of the large public debts and unwanted and useless edifices they leave behind them. Plans and promises for a lasting legacy have in recent years become a part of the 'offer' which national and city bidders must now provide in order to give themselves a good chance of success.

Certainly the regenerative benefits to east London, a relatively deprived part of this city-region – were a crucial 'selling point' for the London Olympic bid, and seem to have been one of the reasons for its success. Everyone remembers the multi-ethnic delegation of east London citizens which the London Olympic Committee took to Singapore to advocate its bid, and its socially-inclusive nature was given a memorable symbolisation in the Olympic opening ceremony which Danny Boyle had been commissioned to direct soon after the successful outcome had been announced. How far this attractive rhetoric corresponded to the material and political reality and legacy of the 2012 Olympics is however a more complicated question.

The London Olympic Games had certainly been closely assimilated into a long-term development strategy for the east of London, historically much poorer than its western side, and the location for much of its heavy industry and its attendant pollution. Indeed, it had been the decline of 'traditional' industries, in particular the London Docks, which had created both the physical space and the economic will for a major programme of urban regeneration. This began with the development of the business district of Canary Wharf on the site of the disused docks on the Isle of Dogs in the 1980s. Along with this came the construction of London City Airport in between the disused Royal Albert and King George V Docks, and huge investments in public transport to improve the 'connectivity' of this eastern zone with the rest of London. These have included the Docklands Light Railway; the Jubilee Line extension along the South Bank of the Thames from Waterloo to Stratford; the upgrading and development of the London Overground into an orbital railway which now circumvents inner London[1]; the siting at Stratford of a stop on the High Speed trainline to the Continent; and finally Crossrail, the new rapid transit railway which will traverse London, joining Reading in the west to Shenfield in the east, via Heathrow airport, the West End, the City and Canary Wharf. The location of London's biggest shopping mall, 'Westfield Stratford City', which lies on the path from the huge Stratford transport complex to the Olympic Park, has been another large piece of this enormous investment programme, as have other significant additions to the built environment, such as the Excel Exhibition Centre, the Millennium Dome and the University of East London's Docklands Campus.

The Olympic Games certainly brought an additional inspiration, popular appeal and glamour to this immense project to rebalance London to the east. However, virtually all of the components of this regional development predated the successful Olympic bid whose function was more to be its decoration and celebration than its strategic foundation. Had Paris or Madrid won this bid in 2005 instead of London – and the bid's

---

1   A case for the revitalisation of the 'North London Line' was made by Jerram and Wells (1995).

outcome was entirely uncertain – all of these major strategic developments would surely have remained in place or continued without interruption.

The impact of the London Olympics was in fact mainly to add one important but 'local' impact to this much broader scheme of urban and regional renewal. The construction of the Olympic Park (now the Queen Elizabeth Olympic Park) did provide the justification and funding for the development of a significant area of land (2.4 square miles) in Stratford, which might well otherwise have been left in its previous use or non-use for several more years. Some of this was empty and polluted railway land, now the site of the Olympic Park and the various enterprises and entities (for example, the Victoria and Albert Museum, and universities such as University College London) which are moving on to the Olympic site after the Games. Another part of it was an area of mixed low-level industrial use, in a district south of Stratford which had been previously disqualified from more modern investments by an unfavourable terrain prone to flooding and encumberment by electricity pylons. A major transformation is now taking place in these locations, with new residential, commercial and hi-tech industrial buildings, and the provision of the Queen Elizabeth Park, albeit sited considerably further from the main transport interchange than commercial facilities such as the Westfield shopping mall. This 'legacy' is real, and any visitor familiar with Stratford and its environs over the past decades cannot fail to have noticed how much change is taking place there.

But it is very difficult if not impossible to calculate the specific impact of the London Olympics within the much broader pattern of strategic investments which are being undertaken in this region, with Stratford, the central location of the Games, as its major communications hub. The total cost of the Games was £8.92 billion. It seems likely that the 'legacy effects' which are attributed to the Olympics, as the social and economic justification for their costs, are often confused with the effects of the transformation which is being achieved by these much larger and more strategic investments.

This somewhat equivocal and mystifying role of 'legacy' in relation to the London Olympics invites consideration of the broader meaning of legacy in the context of 'mega-events'. We need to understand their distinctive political economy in order to grasp the particular place of 'legacy' and its attributed social benefits within them. The reality is that Sporting Spectacles, like the Olympic Games, Formula One Motor Racing and the FIFA World Cup are a new mode of production and consumption, emblematic of capitalism in the 21st century (Rustin 2008, 2009). They are global phenomena, in several senses of that term. They, apart from the head offices of the entities which control them, are located in no particular geographical place, but instead migrate around the world, following contracts which have been negotiated with the particular nations and cities that compete with each other to host a particular year's event in their periodic cycles (Urry 2008). Their popular attraction and impact depends largely on global media coverage and especially on television – a billion people (21 million in the UK) are estimated to have watched the FIFA World Cup Final in July 2014 between Germany and Argentina. One of the Olympics' largest income streams derives from contracts with corporations who buy exclusive branding and selling rights in the closed sites where the spectacles take place, where there is fierce suppression of unlicensed commercial competitors, forbidden to engage in 'ambush marketing'. But they also depend on substantial funding provided by governments, on behalf of their citizens, who through various ad hoc entities, and national Olympic or other sporting associations, undertake the design, construction and management of the costly physical and organisational infrastructure needed to stage a mega-event. Being short-lived, the organisation of these events largely by-passes the normal procedures and routines of democratic decision-making; instead adopting the modality of public-private partnerships. Whereas a particular 'spectacle' may come only once in several decades to a nation or a city, which therefore does not necessarily develop long-term institutions and capacities for managing them, for the corporations who service and profit from them these structures and systems are enduring, even though their geographical locations change and although the sports events are transient. This is a new kind of compact between corporations and states, in this neoliberal world, in which it seems corporations and their hybrid equivalents like the International Olympics Committee and FIFA, [2] have most of the advantage and seem able to bend governments to their will.

---

2 Formula One motor racing is a much less ambiguous case, whose owners make no pretence of belonging in the public realm in which profit is not a goal.

What are the benefits to governments and citizens of such mega-events? One considerable benefit belongs in the domain of national and municipal prestige. Holding an Olympic Games, or a FIFA World Cup Final in one's territory gives the opportunity to make a city into a global showcase, giving a demonstration of its modernity or leading place in the world. Many cities have used sporting events like the Olympics to display themselves to the world in these terms. The Games in Sydney, Barcelona and Beijing all had this conspicuous purpose, sometimes allied to the hope of kick-starting their economic development to another level, for example through the attraction of inward investment and tourism. Norbert Elias and Eric Dunning (1986) argued that sport can be regarded, in Freudian terms, as a 'sublimation' of aggressive impulses. International sporting competitions may therefore be regarded as to some degree a functional substitute for war, arousing strong nationalistic sentiments of love for one's country and hatred of its rivals which can however find a non-violent and satisfying expression on the playing field. (The antagonism between the supporters of Brazil and Argentina in the World Cup is a recent example of this.)

These events also become a means of building or reinforcing social bonds, sentiments of belonging and pride in own community. The celebration of footballers' or athletes' successes on the field and the identification with them by fans, the involvement of volunteers who are able to become active contributors to a festival and do much to create its welcoming atmosphere, and the pleasures of collective spectatorship, whether in stadiums or in streets or pubs, all create a much more lively sense of belonging than exists in normal times. These experiences among audiences are undoubtedly augmented by the presence of a great sporting occasion in one's own country or cities, as one could see from the World Cup in Brazil. Nor of course should one neglect the intrinsic value of physical sports themselves, and the deep pleasure their performance gives to so many people (MacRury 2008, 2009). This enjoyment is in part an aesthetic one, comparable to that taken by audiences for other performance arts. Just as the ancient Greeks held Olympic Games *and* festivals in which tragedies were performed in competition with one another, so post-modern societies hold global sports events, rock concerts, and film and cultural festivals of many kinds. In so far as these are values not derived from material production and consumption (although they are often contaminated by it, through sponsorship, branding and the like) one should not be unappreciative of them.

But then there is the question of legacy, and how one weighs its importance against these other goods. Its recent importance in debates about mega-events may be an effect of nagging doubts about whether, for all of the above satisfactions, these huge and indeed by ordinary standards quite disproportional expenditures can really be 'worth it'. The claim to 'legacy' asserts that they are; that there are indeed lasting benefits from staging such events. 'Legacy', we might say, is this form of global production's gesture to social democracy. It is the acknowledgement that public expenditures on this scale need more justification than that of merely having funded a month-long party. This is especially because citizens know that these expenditures, funded by taxation, generate substantial incomes and profits to those who own, promote and sponsor these events. There may also be an element of moral puritanism, as well as a sense of egalitarian justice, embodied in the idea of legacy, the idea that even the best carnival cannot be good enough if it does not bring some lasting public benefit as well.

Popular reactions to the World Cup in Brazil sharply dramatised these issues. There were large scale public protests at the misdirection of resources involved in its preparation, especially at the construction of massive stadiums (some of them with little likely future usefulness to justify their costs) in a country where many millions live in favelas or shanty-towns without basic amenities, and where disparities of income and wealth are among the highest in the world. (One must note, however, that the excitement of the Games themselves and some concessions made by the government seem to have successfully dispelled or diminished the scale of these protests.) What could be the justification of spending £11 billion this way (most of it by the Brazilian government), it was widely asked, when social and economic needs in the country are so great?

Similar questions can be asked about the London Games. Here the question of social justice goes well beyond the Games themselves, since the larger question is the huge disparity of public investment in London and the south-east, compared with the rest of the country. London and the south-east receive 84 per cent of all public investment in transport in England and Wales – £2,631 per person compared with £5 per person in the north east, according to a recent report (Cox and Davies 2014). Given its level of relative prosperity, there could hardly be an argument from social justice to confer on it the benefits of expenditures on the Olympic Games as well! One could argue that East London was relatively deprived, in the context of London and the south-east, but the development needs of many other parts of the United Kingdom are far greater. Why not, it might therefore be argued, have bid for the Olympics to be located in Manchester, Birmingham, Glasgow or Newcastle? The answer of course is that these cities would have had no chance of winning the competition. This global mode of production has its own priorities for investment, which favours the already advantaged, not the regions of nations and the world which are most deprived, whatever the compensatory rhetoric of 'legacies' might claim. FIFA's decision to award the 2022 World Cup to be played in Qatar, of all places in the middle of summer, indicates with startling clarity the priorities of these systems.

It may reasonably be argued that the benefits achieved by governments from their investments of many billions on events like the Olympics are much less than could have been achieved by expenditures of similar amounts on many other projects. This is similar to the argument that far greater improvements in connectivity could be achieved with the £43 billion assigned by the UK government to be invested in the High Speed 2 rail track (others claim its cost will be double) if all this money were spent on more widely spread improvements to the existing railway network. But on the other hand, it can also be said that the most likely alternative to governments' not spending money on events like the Olympics Games, or prestige projects like HS2, would not be more socially useful investments, but rather no investments at all, or at least far less of them. If Britain had lost its 2005 bid for the Games, who imagines that much of the £9 billion set aside for it would instead have instead been spent on the physical and human resources which could have increased active participation in sport?[3]

It is more widely true that while investments in mega-events are invariably highly selective between cities, often assigning even greater advantage to already favoured places (think of Barcelona and Sydney from this point of view), nevertheless they do provide a boost which, had these cities not become their locations, would probably have occurred nowhere. There are even Keynesian arguments for such expenditures, in an epoch of austerity, which is that most kinds of large investment are preferable to none.

The reality is that mega-events, as a new global mode of production, take place at the intersection of extremely complex economic and political forces, each struggling for influence and benefit within the systems of decision-making. The idea and practices of 'legacy' is but one element in this combination of pressures; 'Legacy effects' are never, in reality, the primary reason for staging events such as the Olympics Games, although they are sometimes misleadingly represented as if they were. But however limited or flawed in their outcomes, these claims for social benefit are, nevertheless, the most 'democratic' of those involved in the design and delivery of these spectacles. Therefore, it must be said that in the future, the more 'legacy' that is built into such events, the better.

References

Cox, E., and B. Davies (2013). *Still On the Wrong Track: An Updated Analysis of Transport Infrastructure Spending*. Newcastle-upon-Tyne: IPPR North.

Elias, N., and E. Dunning (1986). *The Quest for Excitement: Sport and Leisure in the Civilizing Process*. Oxford: Blackwell.

Jerram, B. and R. Wells (1995). 'Traversing the great divides: the North London Line and East London', in T. Butler and M. Rustin (eds.) (1995). *Rising in the East: The Regeneration of East London*. London: Lawrence and Wishart.

MacRury I. (2008). 'Rethinking the Legacy 2012: The Olympics as commodity and gift'. *20th Century Society*, Vol. 2 (3), pp. 297–312.

MacRury, I. (2009). 'Branding the Games: Commercialism and the Olympic city', in G. Poynter and I. MacRury (eds.) (2009). *Olympic Cities: 2012 and the Remaking of London*. Farnham: Ashgate.

Roche, M.C. (2000). *Mega-events and Modernity: Olympics and Expos in the Growth of Global Culture*. London: Routledge.

---

3   Consider for example the thousands of school playing fields that have been sold off since 1979, and continue to be. http://www.telegraph.co.uk/education/keep-the-flame-alive/10516870/One-school-playing-field-sold-off-every-three-weeks-since-Coalition-was-formed.html

Rustin, M.J. (2008). 'Introduction: Social science perspectives on the 2012 London Olympic Games'. *20th Century Society*, Vol. 2 (3), pp. 279–383.

Rustin, M.J. (2009). 'Sport, spectacle and society: Understanding the Olympics', in G.Poynter and I. MacRury (eds.) (2009). *Olympic Cities: 2012 and the Remaking of London* Farnham: Ashgate.

Urry, J. (2008). 'The London Olympics and global competition on the move'. *20th Century Society*, Vol. 2 (3), pp. 291–296.

Index

Printed and bound by CPI Group (UK) Ltd, Croydon, CR0 4YY

22/10/2024

01777639-0003